T0301853

Real Analysis

Textbooks in Mathematics

Series editors:

Al Boggess, Kenneth H. Rosen

Nonlinear Optimization

Models and Applications

William P. Fox

Linear Algebra

James R. Kirkwood, Bessie H. Kirkwood

Real Analysis

With Proof Strategies

Daniel W. Cunningham

Train Your Brain

Challenging Yet Elementary Mathematics

Bogumil Kaminski, Pawel Pralat

Contemporary Abstract Algebra, Tenth Edition

Joseph A. Gallian

Geometry and Its Applications

Walter J. Meyer

Linear Algebra

What you Need to Know

Hugo J. Woerdeman

Introduction to Real Analysis, 3rd Edition

Manfred Stoll

Discovering Dynamical Systems Through Experiment and Inquiry

Thomas LoFaro, Jeff Ford

Functional Linear Algebra

Hannah Robbins

Introduction to Financial Mathematics

With Computer Applications

Donald R. Chambers, Qin Lu

https://www.routledge.com/Textbooks-in-Mathematics/book-series/CANDHTEXBOOMTH

Real Analysis
With Proof Strategies

Daniel W. Cunningham

 **CRC Press**
Taylor & Francis Group
Boca Raton London New York

CRC Press is an imprint of the
Taylor & Francis Group, an **informa** business

A CHAPMAN & HALL BOOK

First edition published 2021
by CRC Press
6000 Broken Sound Parkway NW, Suite 300, Boca Raton, FL 33487-2742

and by CRC Press
2 Park Square, Milton Park, Abingdon, Oxon, OX14 4RN

© 2021 Daniel W. Cunningham

CRC Press is an imprint of Taylor & Francis Group, LLC

The right of Daniel W. Cunningham to be identified as author of this work has been asserted by him in accordance with sections 77 and 78 of the Copyright, Designs and Patents Act 1988.

Reasonable efforts have been made to publish reliable data and information, but the author and publisher cannot assume responsibility for the validity of all materials or the consequences of their use. The authors and publishers have attempted to trace the copyright holders of all material reproduced in this publication and apologize to copyright holders if permission to publish in this form has not been obtained. If any copyright material has not been acknowledged please write and let us know so we may rectify in any future reprint.

Except as permitted under U.S. Copyright Law, no part of this book may be reprinted, reproduced, transmitted, or utilized in any form by any electronic, mechanical, or other means, now known or hereafter invented, including photocopying, microfilming, and recording, or in any information storage or retrieval system, without written permission from the publishers.

For permission to photocopy or use material electronically from this work, access www.copyright.com or contact the Copyright Clearance Center, Inc. (CCC), 222 Rosewood Drive, Danvers, MA 01923, 978-750-8400. For works that are not available on CCC please contact mpkbookspermissions@tandf.co.uk

Trademark notice: Product or corporate names may be trademarks or registered trademarks and are used only for identification and explanation without intent to infringe.

Library of Congress Cataloging-in-Publication Data

Names: Cunningham, Daniel W., author.
Title: Real analysis : with proof strategies / Daniel W. Cunningham.
Description: First edition. | Boca Raton : Chapman & Hall, CRC Press, 2021.
| Series: Textbooks in mathematics | Includes bibliographical references and index.
Identifiers: LCCN 2020034986 (print) | LCCN 2020034987 (ebook) | ISBN 9780367549657 (hardback) | ISBN 9781003091363 (ebook)
Subjects: LCSH: Mathematical analysis--Textbooks. | Functions of real variables--Textbooks.
Classification: LCC QA300 .C86 2021 (print) | LCC QA300 (ebook) | DDC 515/.8--dc23
LC record available at https://lccn.loc.gov/2020034986
LC ebook record available at https://lccn.loc.gov/2020034987

ISBN: 9780367549657 (hbk)
ISBN: 9781003091363(ebk)

Typeset in Computer Modern font
by KnowledgeWorks Global Ltd.

Contents

Preface

Real analysis is the important branch of mathematics that investigates the properties of the real numbers and establishes the theory behind calculus, differential equations, probability, and other related subjects. The main concepts studied in real analysis are sets of real numbers, functions, limits, sequences, continuity, differentiation, integration, and sequences of functions. The study of these topics allows one to gain a much deeper understanding of the behavior and properties of real-valued functions, sequences, and sets of real numbers. Such depth allows one to really understand why the theorems of the calculus are true.

One of the main goals of a course in real analysis is to cover the proofs that were omitted in calculus. Calculus books usually prove a few of the easiest theorems. However, if you look carefully, you will see that most calculus textbooks do not prove, in the main body of the text, many of the most important theorems in the calculus. The Intermediate Value Theorem, the Extreme Value Theorem, and the integrability of continuous functions on a closed bounded interval are results all crucial to calculus and to higher analysis. Moreover, these theorems can be stated in terms that a calculus student can understand, but an actual proof cannot be achieved without addressing a fundamental question: What are the important properties of the set of real numbers? This text provides an answer to this question.

Typically, undergraduates view real analysis as one of the most difficult courses that a mathematics major is required to take. The main reason for this perception is twofold: One must comprehend new abstract concepts and learn to deal with these concepts on a level of rigor and proof not previously encountered. In particular, for many of these students, the mental gymnastics required to prove theorems about limits is initially formidable. They struggle to find the appropriate values (e.g., δ or N) that are necessary to compose a logically correct proof concerning the limit of a function or a sequence. This text offers a resolution to this difficulty. The book not only presents the central theorems of real analysis, but also shows the reader how to compose and produce the proofs of these theorems. This approach should be to the benefit of any undergraduate reader.

In real analysis, a facility for working with the supremum of a bounded set, the limit of a sequence, and the ε-δ definitions of continuity are essential for a student to be successful. This book is designed to show students how to overcome their initial hurdles and gain such a facility. I present proof strategies that explicitly show students how to deal with the fundamental definitions that one encounters in real analysis; each of which is followed by numerous examples of proofs that use these strategies. In Chapters 2–4, many of the introductory proofs are preceded by a "proof analysis" that carefully explains how to apply these strategies. The proof analysis is then followed by the actual proof. When a real analysis instructor presents a proof in class, these strategies can also help to expedite a student's understanding of the proof and its logical structure.

Roadmap to the Book

The book presents a mathematical theory that validates the calculus of a single variable. Moreover, the book is designed for students who are learning the rudiments of this theory for the first time. Students typically stumble when first asked to compose proofs concerning the core concepts covered in real analysis. To assist such students, we present specific strategies for each core concept that are designed to guide a student in the discovery of a correct proof. In fact, these strategies will be appreciated by students at all levels of preparation.

Chapter 1 reviews some of the important topics that a student usually learns in a discrete mathematics course and/or a transition to advanced mathematics course. This chapter is intended to act as a reference for the basic mathematical concepts that will appear in the remaining chapters of the book.

Chapter 2 presents the properties of the real number system. The first two sections deal with the algebraic and order properties of the real numbers and the crucial completeness property. In this chapter, we identify strategies that are designed to give the reader a guide on how to prove theorems that concern the supremum and infimum of a nonempty bounded set of real numbers. The chapter also shows how the completeness axiom implies the Archimedean Property,[1] the density of the rational numbers, and the nested intervals theorem. After delivering a succinct introduction to countable sets, we prove that the set of real numbers is uncountable. The argument is based on Cantor's first proof of this result.

Chapter 3 focuses on sequences of real numbers. The limit of a sequence is one of the most important concepts in mathematical analysis. Our initial emphasis is to identify the strategies that are used to prove limit theorems. The attention then turns to subsequences, monotone sequences, Cauchy sequences, and the limit superior and limit inferior of a bounded sequence.

Continuity is the central topic covered in Chapter 4, which begins on the principal technique that is used to prove that a function is continuous at a point. This is followed by the fundamental theorems on continuity. Then the sequential criterion for the continuity of function is covered. This is followed by a section on the limit of a function and its sequential equivalent. The chapter ends on the topic of uniform continuity.

In Chapter 5, we present a theoretical treatment of the derivative and derive its principal properties. We also identify and establish the following three significant results that concern the derivative: the Mean Value Theorem, Cauchy's Mean Value Theorem, and Taylor's Theorem. We also prove four versions of L'Hôpital's Rule, using the sequential criterion for limits.

---

[1]This property is named in honor of Archimedes, a great mathematician who is also regarded as one of the leading scientists in classical antiquity.

The Riemann integral is the main subject of Chapter 6. We follow the development of Jean Gaston Darboux, which simplifies Riemann's original idea. In this chapter, we show that continuous functions, monotone functions, and functions of bounded variation are Riemann integrable. After proving the two versions of the Fundamental Theorem of Calculus, we establish three techniques of antidifferentiation, the latter of which validates integration by trigonometric substitution and Weierstrass substitution. The chapter ends with a discussion on improper integrals.

Using results from Chapter 3 on the convergence of sequences and on the limit superior of a sequence, Chapter 7 confirms the important theorems that concern an infinite series of real numbers. The chapter ends by establishing results on the effect of regrouping and rearranging the terms of a series. Finally, in Chapter 8, we investigate sequences and series of functions, including power series and Taylor series.

In Appendix A, we provide a technical proof of a theorem on the integrability of a composite function. In Chapter 6, we just state and apply this useful theorem. Appendix B covers the topology of the real numbers, including compactness and the Heine–Borel Theorem. The appendix ends with a section on the Cantor set and some of its remarkable properties. The material in this appendix is not used in the text itself, but is included to present the reader with an optional topic that often appears in more advanced courses in analysis. It has been my experience that students have considerable difficulty with topological concepts. Such concepts tend to make a first course in real analysis more challenging than it needs to be. Since topology is not required to prove the theorems in this text, this topic is presented in an appendix. Finally, Appendix C provides a brief review of logic and the basic proof strategies that are typically covered in an introduction to proof course.

Exercises are given at the end of each section in a chapter and in Appendix B. Exercise notes often appear at the end of an exercise set. The notes offer suggestions for the more challenging problems and comments that relate to specific exercises. An exercise marked with an asterisk * is one that is cited, or referenced, elsewhere in the book. A referenced exercise without an adjoined page number appears in the exercise set below the reference.

How to Use the Book

The text assumes that the reader has completed a three-term calculus sequence and recalls how to differentiate and integrate the typical functions that appear in such a course sequence. It is also assumed that the reader is familiar with the standard techniques of proof; that is, the reader should know how to read and write a mathematical proof. These techniques are covered in a "transition course," which is typically a prerequisite for a course in real analysis. Nevertheless, my primary goal was to write a book for a reader who may not be sufficiently well-versed in the proof techniques that are often applied in real analysis. Consequently, each proof presented in the book favors detail over brevity. However, this detail is intended mainly for the reader and it can be abridged in a lecture setting.

Depending on the student's background, Chapter 1 can be briefly discussed in class or given as assigned reading. However, on the first day of class, one should discuss Sections 1.1.2, 1.1.3, 1.1.4, 1.1.5, and in particular the Sum and Product Principles of Inequality in Section 1.1.8. A one-semester course that ends with the Fundamental Theorem of Calculus can then present Chapters 2 through 5 together with Sections 6.1, 6.2.1, and 6.2.3 followed by Theorem 6.3.1, Corollary 6.3.2, and Theorems 6.4.2 and 6.4.4. On the other hand, one could thoroughly cover the first five chapters, and then pursue the remaining parts of the book, including Appendix B, in a second semester. In any case, Sections 2.5, 5.2.1, 6.3.3, 6.4.3, and 7.3 can be omitted without loss of continuity. Section 3.8 should be discussed if covering Sections 7.2.5, 8.3, and 8.4.

To the Student

As an undergraduate, I found my first real analysis course with its focus on proof to be very difficult. I thought that I might fail the course. But, in the end, I passed with a good grade. The reason for this is perfectly described by the following quote.

> *Patience and perseverance have a magical effect before which difficulties disappear and obstacles vanish.* – John Quincy Adams

This experience as an undergraduate inspired me to write this book, which is designed to increase your confidence by providing you with a guide for finding and writing proofs in real analysis. This guide involves a so-called "proof diagram." Such a guide is illustrated in Section 1.1.5. A proof diagram demonstrates the structure of the proof and provides a tool for writing a correct mathematical proof. Even with a guide, the work required to find a proof can be quite challenging. Professional mathematicians also have difficulty finding proofs; however, as I learned in my first real analysis course, persistence often pays off and thus, they do not easily give up.

To be successful in this course, review the material in Chapter 1 and Appendix C, even if it is not explicitly covered in class. Tenaciously read and understand every section, lemma, theorem, corollary, and proof discussed from the text and diligently work on the assigned exercises. The only way to acquire a deep understanding of real analysis is through the exercises. Take advantage of the exercise notes that are provided for many of the exercises. Use the proofs provided in the text as models for your own proofs. Follow this advice and you, too, can pass with a good grade.

Acknowledgments

I am grateful to Springer Science and Business Media for granting me copyright permission to use in this book some of the language, examples, and figures that I composed and created in Chapter 9 "Core Concepts in Real Analysis" of [1].

CHAPTER 1

Proofs, Sets, Functions, and Induction

In this introductory chapter, we review and identify the preliminaries that are essential for real analysis. In particular, we review inequalities, sets, functions, and proof by mathematical induction. These preliminary topics should be familiar, as should be the basic proof techniques that are used in mathematics. A review of logic and proof is presented in Appendix C on page 253. The most important proof strategies that are applied in this text can be found in the appendix starting on page 256.

1.1 PROOFS

1.1.1 Important Sets in Mathematics

The set concept is frequently used in mathematics. A set is a well-defined collection of objects. The items in such a collection are called the elements or members of the set. The symbol "$\in$" is used to indicate membership in a set. Thus, if A is a set, we write $x \in A$ to declare that x is an element of A. Moreover, we write $x \notin A$ to assert that x is not an element of A. In mathematics, a set is typically a collection of mathematical objects, for example, numbers, functions, or other sets. Certain sets are frequently used in mathematics. The most commonly used ones are the sets of natural numbers, integers, and rational and real numbers. These sets will be denoted by the following symbols:

1. $\mathbb{N} = \{1, 2, 3, \dots\}$ is the set of natural numbers.
2. $\mathbb{Z} = \{\dots, -3, -2, -1, 0, 1, 2, 3, \dots\}$ is the set of integers.
3. $\mathbb{Q}$ is the set of rational numbers. So, $\frac{3}{2} \in \mathbb{Q}$.
4. $\mathbb{R}$ is the set of real numbers and so, $\pi \in \mathbb{R}$.

In this text, we do **not** consider 0 to be a natural number.

A set can sometimes be identified by enclosing a list of its elements by curly brackets; for example, the set $A = \{1, 2, 3, 4, 5, 6, 7, 8, 9\}$ consists of a few natural numbers. More often, one forms a set by enclosing a particular expression within curly brackets, where the expression identifies the elements of the set. To illustrate this method of identifying a set, we can form a set B of even natural numbers, using the above set A, as follows:

$$B = \{n \in A : n \text{ is even}\} \qquad (\blacktriangle)$$

which is read as "the set of $n \in A$ such that n is even." Clearly, $B = \{2, 4, 6, 8\}$. So using the set A and the property "n is even," we formed the set B in ($\blacktriangle$). More generally, given a set C and a property $P(x)$, we can form the set $X = \{x \in C : P(x)\}$. Thus, X is the set of all objects in C that make $P(x)$ true. In the definition of X, the set C will be referred to as the *restriction* and X will be called a *restricted truth set*. Moreover, when the restriction is understood or unspecified, we will only write $X = \{x : P(x)\}$ and refer to X as a *truth set*.

Example. Consider the set of integers $\mathbb{Z}$. We evaluate the following sets:

1. $\{x \in \mathbb{Z} : x \text{ is a prime number}\} = \{2, 3, 5, 7, 11, \ldots\}$.
2. $\{x \in \mathbb{Z} : x \text{ is even}\} = \{\ldots, -8, -6, -4, -2, 0, 2, 4, 6, 8, \ldots\}$.
3. $\{z \in \mathbb{Z} : z^2 \leq 1\} = \{-1, 0, 1\}$.

For each of the sets $\mathbb{Z}$, $\mathbb{Q}$, and $\mathbb{R}$, we may add '+' or '−' as a superscript. The $+$ (or $-$) superscript indicates that only the positive (or negative) numbers will be allowed. For example, $\mathbb{Q}^+ = \{x \in \mathbb{Q} : x > 0\}$ and $\mathbb{R}^- = \{x \in \mathbb{R} : x < 0\}$.

Interval Notation

A *point* is a term that will be used to denote a real number.

Definition 1.1.1. An **interval** is a set I of real numbers that has at least two points and for any two points $x, y \in I$, every real number between x and y is also in I.

Of course, the set consisting of all the real numbers that lie between two real numbers is an interval. We now identify all of the possible forms of an interval.

1. The open interval (a, b) is defined to be $(a, b) = \{x \in \mathbb{R} : a < x < b\}$.
2. The closed interval $[a, b]$ is defined to be $[a, b] = \{x \in \mathbb{R} : a \leq x \leq b\}$.
3. The left-closed interval $[a, b)$ is defined to be $[a, b) = \{x \in \mathbb{R} : a \leq x < b\}$.
4. The right-closed interval $(a, b]$ is defined to be $(a, b] = \{x \in \mathbb{R} : a < x \leq b\}$.
5. The interval (a, ∞) is defined to be $(a, \infty) = \{x \in \mathbb{R} : a < x\}$.
6. The interval $[a, \infty)$ is defined to be $[a, \infty) = \{x \in \mathbb{R} : a \leq x\}$.
7. The interval $(-\infty, b)$ is defined to be $(-\infty, b) = \{x \in \mathbb{R} : x < b\}$.
8. The interval $(-\infty, b]$ is defined to be $(-\infty, b] = \{x \in \mathbb{R} : x \leq b\}$.
9. The interval $(-\infty, \infty)$ is $\mathbb{R}$, the set of all real numbers.

The numbers a and b, in items 1–8, are called the *endpoints* of the interval and a is called the *left endpoint* and b is said to be the *right endpoint*. A point in an interval that is not an endpoint is an *interior point*. An interval may or may not include its endpoints. The symbol ∞ denotes "infinity" and is not a number or an endpoint. The notation ∞ is a useful symbol that allows us to represent intervals that are "without an end." The notation $-\infty$ is used to denote an interval "without a beginning."

Some sets of real numbers, but not all, can be expressed in terms of an interval.

Example. We evaluate the sets (1) $\{x \in \mathbb{R} : x^2 - 1 < 3\}$ and (2) $\{x \in \mathbb{R}^- : x > \frac{1}{x}\}$ in terms of an interval as follows:

(1) Solving the inequality $x^2 - 1 < 3$ for x^2, we obtain $x^2 < 4$. The solution to this latter inequality is $-2 < x < 2$. Thus, $\{x \in \mathbb{R} : x^2 - 1 < 3\} = (-2, 2)$.

(2) We need to find all the real numbers $x < 0$ that satisfy $x > \frac{1}{x}$. We conclude that $x^2 < 1$. Thus, $-1 < x < 0$. So, $\{x \in \mathbb{R}^- : x > \frac{1}{x}\} = (-1, 0)$.

Definition. A positive rational number $\frac{m}{n}$ is in **reduced form** if $m \in \mathbb{N}$ and $n \in \mathbb{N}$ have no common factors greater than 1.

Example. The rational number $\frac{4}{3}$ is in reduced form and $\frac{7}{14}$ is not in reduced form because 7 and 14 have a common factor greater than 1. However, $\frac{7}{14} = \frac{1}{2}$ which is in reduced form.

As illustrated in the above example, each positive rational number can be expressed in terms of a ratio in reduced form.

Lemma 1.1.2. Let $a, b \in \mathbb{Z}$. If p is a prime and p divides ab, then either p divides a or p divides b.

Theorem 1.1.3. Let $p \in \mathbb{N}$ be a prime number. Then $\sqrt{p}$ is an irrational number.

Proof. Let $p \in \mathbb{N}$ be a prime. Assume, for a contradiction, that $\sqrt{p}$ is rational. Thus, $\sqrt{p} = \frac{m}{n}$ for some $m, n \in \mathbb{N}$ where $n \neq 0$ and $\frac{m}{n}$ is in reduced form. Since $\sqrt{p} = \frac{m}{n}$, we have that $p = \frac{m^2}{n^2}$ and $(\star)$ $m^2 = pn^2$. Hence, p evenly divides m^2. Since p is a prime, p evenly divides m by Lemma 1.1.2. So, $m = pk$ for some $k \in \mathbb{N}$. After substituting $m = pk$ into $(\star)$, we obtain $p^2 k^2 = pn^2$. Therefore, $n^2 = pk^2$. Thus, p evenly divides n^2 and so, p evenly divides n. Hence, m and n have p as a common factor. It follows that $\frac{m}{n}$ is not in reduced form. Contradiction. $\square$

1.1.2 How to Prove an Equation

Equations play an important role in mathematics. In this text we will establish many theorems that require us to correctly prove an equation. Since this is so fundamental, our first proof strategy presents two correct methods for proving equations.

Proof Strategy 1.1.4. To prove a new equation $\varphi = \psi$, there are two approaches:

(a) Start with one side of the equation and derive the other side.

(b) Perform operations on any given equations to derive the new equation.

We now apply strategy 1.1.4(a) to prove a well-known algebraic identity.

Theorem. Let x and y be arbitrary real numbers. Then $(x + y)(x - y) = x^2 - y^2$.

Proof. We start with the left hand side $(x + y)(x - y)$ and derive the right hand side as follows:

$$\begin{aligned} (x + y)(x - y) &= x(x - y) + y(x - y) && \text{by the distributive property} \\ &= x^2 - xy + yx - y^2 && \text{by the distributive property} \\ &= x^2 - y^2 && \text{by algebra.} \end{aligned}$$

Therefore, $(x + y)(x - y) = x^2 - y^2$. □

We now apply strategy 1.1.4(b) to prove an equation from some given equations.

Theorem. Let x, y, i, j be real numbers. Suppose that $x = 2i + 5$ and $y = 3j$. Then $xy = 6ij + 15j$.

Proof. Assume that $x = 2i+5$ and $y = 2j$. By multiplying corresponding sides of these two equations, we obtain $xy = (2i + 5)(3j)$. Thus, by algebra, $xy = 6ij + 15j$. □

Remark 1.1.5. To prove that an equation $\varphi = \psi$ is true, it is not a correct method of proof to *assume* the equation $\varphi = \psi$ and then derive an identity.

The method described in Remark 1.1.5 is invalid and, if applied, can produce false equations. For example, this incorrect method can be used to deduce the equation $-2 = 2$. To illustrate this, let us assume the equation $-2 = 2$. Now square both sides, obtaining $(-2)^2 = 2^2$ which results in the true equation $4 = 4$. The method cited in Remark 1.1.5 would allow us to conclude that $-2 = 2$. This is nonsense. **In mathematics, one never applies a method that can produce false results!**

1.1.3 How to Prove an Inequality

Many of the proofs in real analysis involve working with inequalities. In this section, we review the properties of inequality that are frequently applied in analysis. As you may recall, to prove a new inequality from some given inequalities is typically more difficult than proving an equation. The major reason for this added difficulty: one has to correctly use the Laws of Inequality.

Laws of Inequality 1.1.6. For all $a, b, c, d \in \mathbb{R}$, the following hold:

1. Exactly one of the following holds: $a < b$ or $a = b$ or $a > b$.   (Trichotomy)
2. If $a < b$ and $b < c$, then $a < c$.   (Transitivity Law)
3. If $a < b$, then $a + c < b + c$.   (Adding on both sides)
4. If $a < b$ and $c > 0$, then $ac < bc$.   (Multiplying by a positive)
5. If $a < b$ and $c < 0$, then $ac > bc$.   (Multiplying by a negative)
6. if $a < b$ and $c < d$, then $a + c < b + d$.   (Additivity)

We write $a > b$ when $b < a$, and $a \leq b$ states that $a < b$ or $a = b$. Similarly, $a \geq b$ means that $a > b$ or $a = b$. The Trichotomy Law allows us to assert that if $a \not< b$, then $a \geq b$. We note that one can actually prove laws 5 and 6 from laws 1 to 4. Moreover, one can also prove that $0 < 1$ and $-1 < 0$.

Theorem 1.1.7. Let a, b, c be real numbers where $a < b$. Then $a - c < b - c$.

Proof. Let a, b, c be real numbers such that $a < b$. From the inequality law 3, we obtain $a + (-c) < b + (-c)$. Thus, by algebra, we conclude that $a - c < b - c$. □

The following principles of inequality are based on the Laws of Inequality 1.1.6 and are frequently applied in real analysis.

Sum and Product Principles of Inequality 1.1.8. Let a, p, x, y be real numbers. Then the following hold:

(1) Given the sum $a + x$, if $x < y$, then you can conclude that $a + x < a + y$.
   (*Replacing a summand with a larger value yields a larger sum.*)
(2) Given the sum $a + x$, if $x > y$, then you can conclude that $a + x > a + y$.
   (*Replacing a summand with a smaller value yields a smaller sum.*)
(3) Given the product px where $p > 0$, if $x < y$, then you can infer that $px < py$.
   (*Replacing a factor with a larger value yields a larger product.*)
(4) Given the product px where $p > 0$, if $x > y$, then you can infer that $px > py$.
   (*Replacing a factor with a smaller value yields a smaller product.*)

Principle (1) holds for $\leq$ as well (i.e., upon replacing both occurrences of $<$ in (1) with $\leq$). The above (2) also holds for $\geq$. Moreover, (3) holds for $\leq$ when $p \geq 0$; and (4) holds for $\geq$ when $p \geq 0$.

We give two proofs of Theorems 1.1.9 and 1.1.10 below. Each first proof uses the Laws of Inequality 1.1.6, while each corresponding second proof uses the Sum and Product Principles of Inequality 1.1.8. Of these two proofs, which is easier to follow?

Theorem 1.1.9. Let a, b, c, d be real numbers and suppose that $a < b$ and $c < d$. Then $a + c < b + d$.

First Proof. Let a, b, c, d be real numbers satisfying (▲) $a < b$ and (♦) $c < d$. We prove that $a + c < b + d$. From (▲) and law 3 of the Laws of Inequality 1.1.6, we obtain $a + c < b + c$. From (♦) and law 3 again, we conclude that $b + c < b + d$. So, $a + c < b + c < b + d$. Therefore, $a + c < b + d$. □

Second Proof. Let a, b, c, d be real numbers satisfying $a < b$ and $c < d$. We show that $a + c < b + d$ as follows:

$$a + c < b + c \quad \text{as } a < b \text{ (see 1.1.8(1))}$$
$$< b + d \quad \text{as } c < d \text{ (see 1.1.8(1))}.$$

Therefore, $a + c < b + d$. □

Theorem 1.1.10. Suppose a and b are real numbers. If $0 < a < b$, then $a^2 < b^2$.

First Proof. Assume $0 < a < b$. We show that $a^2 < b^2$. Multiplying both sides of the inequality $a < b$ by the positive a yields the inequality (▲) $a^2 < ab$, and multiplying both sides of the inequality $a < b$ by the positive b gives the inequality (♦) $ab < b^2$. Thus, (▲) and (♦) imply $a^2 < ab < b^2$. Thus, $a^2 < b^2$. □

Second Proof. Assume $0 < a < b$. We show that $a^2 < b^2$ as follows:

$$a^2 = aa \qquad \text{by algebra}$$
$$< ab \qquad \text{as } a < b \text{ (see 1.1.8(3))}$$
$$< bb = b^2 \qquad \text{as } a < b \text{ (see 1.1.8(3))}.$$

Therefore, $a^2 < b^2$. □

In elementary mathematics we learned that for any real number $x \geq 0$, there is a unique real number $y \geq 0$, such that $y^2 = x$. We write $y = \sqrt{x}$ and say that y is the *square root* of x. Consequently, $(\sqrt{x})^2 = x$. Our next theorem shows that the square root operation preserves the inequality relation $<$ for positive numbers.

Theorem 1.1.11. Suppose a and b are real numbers. If $0 < a < b$, then $\sqrt{a} < \sqrt{b}$.

Proof. Suppose $0 < a < b$. We will prove that $\sqrt{a} < \sqrt{b}$. Suppose, for a contradiction, that $\sqrt{b} \leq \sqrt{a}$. If $\sqrt{b} = \sqrt{a}$, then $b = (\sqrt{b})^2 = (\sqrt{a})^2 = a$. Contradiction. If $\sqrt{b} < \sqrt{a}$, then $b = (\sqrt{b})^2 < (\sqrt{a})^2 = a$ by Theorem 1.1.10, and so $b < a$. Contradiction. □

One can also prove the following extension of Theorem 1.1.10 and Theorem 1.1.11 (see Exercises 3 and 7 on pages 26 and 27).

Theorem 1.1.12. If $0 < a < b$, then $a^n < b^n$ and $a^{\frac{1}{n}} < b^{\frac{1}{n}}$, for all $n \in \mathbb{N}$.

We end this section by establishing two fundamental properties of inequality.

Theorem 1.1.13. Let $a, b, c, d > 0$. If $a < b$ and $c < d$, then $ac < bd$.

Proof. Let a, b, c, d be positive real numbers satisfying (▲) $a < b$ and (♦) $c < d$. From (▲) we conclude that $ac < bc$ because $c > 0$. From (♦) we obtain $bc < bd$ as $b > 0$. So, $ac < bc < bd$. Therefore, $ac < bd$. □

Corollary 1.1.14. Suppose $a, b, x, y > 0$. If $a \leq b$ and $x \leq y$, then $ax \leq by$.

Proof. Let $a, b, x, y > 0$, $a \leq b$, and $x \leq y$. We prove that $ax \leq by$. There are several cases to consider. If $a = b$ and $x = y$, then $ax = by$ and so, $ax \leq by$. If $a = b$ and $x < y$, then $ax < ay = by$ and so, $ax \leq by$. If $a < b$ and $x = y$, then $ax < bx = by$ and so, $ax \leq by$. If $a < b$ and $x < y$, then $ax < by$ by Theorem 1.1.13. So, $ax \leq by$. □

1.1.4 Important Properties of Absolute Value

We shall now review the absolute value function and some of its properties. This function will be more formally discussed is Section 2.2.1. Given a real number x, the absolute value of x, denoted by $|x|$, is defined by

$$|x| = \begin{cases} x, & \text{if } x \geq 0; \\ -x, & \text{if } x < 0. \end{cases}$$

For all $a, b, x, c \in \mathbb{R}$, where $c > 0$, we have (see Theorems 2.2.6 and 2.2.7)

1. $|x| < c$ if and only if $-c < x < c$

2. $|x| > c$ if and only if $x < -c$ or $x > c$

3. $|-x| = |x|$

4. $x \le |x|$ and $-x \le |x|$

5. $|ab| = |a||b|$

6. $|a + b| \le |a| + |b|$

7. $\big| |a| - |b| \big| \le |a - b|$

8. $|a| - |b| \le |a - b|$

9. $|b| - |a| \le |a - b|$.

Item 6 is called the *triangle inequality*, and we shall refer to items 7–9 as the *backward triangle inequality*.

1.1.5   Proof Diagrams

In Section 1.4.1 we introduce a diagram that encapsulates proof by mathematical induction (see page 23). This diagram is obtained by identifying the logical structure of proof by induction. One can use the diagram as a guide to produce a correct proof by mathematical induction. All mathematical proofs can be analyzed in terms of the logical structure of the statement to be proven. This structure allows one to "diagram" a mathematical proof. The proof diagram demonstrates the structure of the proof and provides a tool for writing a correct mathematical proof.

We now illustrate how a proof diagram can be used to write a proof that concerns inequalities and the absolute value function. As will be seen, the proof applies the Sum and Product Principles of Inequality 1.1.8.

Theorem. For every $\varepsilon > 0$, there exists a $\delta > 0$ such that if $|x - 1| < \delta$, then $|(3x + 2) - 5| < \varepsilon$.

Before proving the above theorem, we present a "proof analysis" which discusses how to obtain the necessary values for a correct proof. This analysis is not part of the proof, but it does provide all the necessary ingredients that are needed to compose a valid proof.

Proof Analysis. The statement of the theorem has the logical form

$$(\forall \varepsilon > 0)(\exists \delta > 0)(|x - 1| < \delta \to |(3x + 2) - 5| < \varepsilon).$$

Given this logical form, by applying the strategies outlined on page 256, we obtain the following "proof diagram:"

> Let $\varepsilon > 0$.
> > Let $\delta =$ (the positive value you found).
> > > Assume $|x - 1| < \delta$.
> > > > Prove $|(3x + 2) - 5| < \varepsilon$.

The indentation indicates the proof's logical dependencies. So we have $\varepsilon > 0$, and we need to find a $\delta > 0$ so that if $|x - 1| < \delta$, then $|(3x + 2) - 5| < \varepsilon$. Using algebra on the expression $|(3x + 2) - 5|$, we extract out $|x - 1|$ as follows:

$$
\begin{aligned}
|(3x + 2) - 5| &= |3x - 3| \quad \text{by algebra} \\
&= 3\,|x - 1| \quad \text{by algebra and property of absolute value.}
\end{aligned}
$$

Thus, if we have that $3\,|x-1| < \varepsilon$, we can conclude that $|(3x+2)-5| < \varepsilon$. Solving the inequality $3\,|x-1| < \varepsilon$ for $|x-1|$, we obtain $|x-1| < \frac{\varepsilon}{3}$. So, we will let $\delta = \frac{\varepsilon}{3}$. We can now compose a correct proof, using the preceding proof diagram as a guide.

Proof. Let $\varepsilon > 0$. Let $\delta = \dfrac{\varepsilon}{3}$. Assume $(\star)$ $|x-1| < \delta$. We prove that $|(3x+2)-5| < \varepsilon$ as follows:

$$\begin{aligned}
|(3x+2)-5| &= |3(x-1)| \quad &&\text{by algebra}\\
&= 3\,|x-1| \quad &&\text{by property of absolute value}\\
&< 3\delta \quad &&\text{by } (\star) \text{ and property of inequality}\\
&= 3\frac{\varepsilon}{3} = \varepsilon \quad &&\text{since } \delta = \frac{\varepsilon}{3}.
\end{aligned}$$

Therefore, $|(3x+2)-5| < \varepsilon$. □

The above analysis and proof illustrates a theme that will reoccur in the text. We will be presenting proof diagrams that explicitly deal with the core definitions and proofs that are an essential part of real analysis. Such proofs will be preceded by a "proof analysis" and then followed by the actual proof. The reader will thus see how to develop and execute a plan of attack for finding and composing a correct proof.

Exercises 1.1

1. Let x and y be real numbers. Prove that $(x-y)(x^2+xy+y^2) = x^3 - y^3$.

2. Let x and y be real numbers. Prove that $(x+y)(x^2-xy+y^2) = x^3 + y^3$.

3. Let x and y be real numbers. Prove that $(x+y)^2 = x^2 + 2xy + y^2$.

4. Using Exercise 3, prove that $(x+y)^3 = x^3 + 3x^2y + 3xy^2 + y^3$ for all real numbers x and y.

5. Let φ be the positive real number satisfying $\varphi^2 - \varphi - 1 = 0$. Prove that $\varphi = \frac{1}{\varphi-1}$.

6. Let φ be as in Exercise 5. Let $a \neq b$ be real numbers satisfying $\frac{b}{a} = \varphi$. Prove that $\frac{a}{b-a} = \varphi$.

7. Let $c < 0$. Solve $-2cx - 10 > -2c - 10$ for x.

8. Let $c < 1$. Solve $\frac{x-3}{c-1} > c+1$ for x.

9. Let x be a real number such that $x > 1$. Prove that $x^2 > x$.

10. Let x be a real number where $x < 0$. Prove that $x^2 > 0$.

11. Let x be a real number where $x > 0$. Prove that $x^2 > 0$.

12. Let x be a nonzero real number. Using Exercises 10 and 11, prove that $x^2 > 0$.

13. Let a and b be distinct real numbers. Using Exercise 12 prove that $a^2 + b^2 > 2ab$.

14. Let x be a real number so that $x^2 > x$. Must we conclude that $x > 1$?

15. Let x be a real number satisfying $0 < x < 1$. Prove that $x^2 < x$.

16. Let x be a real number where $x^2 < x$. Must we conclude that $0 < x < 1$?

17. Let a and b are real numbers where $a < b$. Prove that $-a > -b$.

18. Let a, b be positive real numbers and let c, d be negative real numbers. Suppose $a < b$ and $c < d$. Prove that $ad > bc$.

19. Find a counterexample showing that the following conjecture is false: *Let a, b, c, d be natural numbers satisfying $\frac{a}{b} \leq \frac{c}{d}$. Then $a \leq c$ and $b \leq d$.*

20. Find a counterexample showing that the following conjecture is false: *Let $m \geq 0$ and $n \geq 0$ be integers. Then $m + n \leq m \cdot n$.*

21. Find a counterexample showing that the following conjecture is false: *Let $x \geq 0$ and $y \geq 0$ be real numbers. Then $\sqrt{x + y} = \sqrt{x} + \sqrt{y}$.*

22. Let a, b, c, d be real numbers. Suppose that $a + b = c + d$ and $a \leq c$. Prove that $d \leq b$. [Hint: $x \leq y$ if and only if $x - y \leq 0$.]

23. Let $a > 0$ be a real number. Prove that $\frac{1}{a} > 0$.

24. Suppose that $0 < a < b$. Prove that $\frac{1}{b} < \frac{1}{a}$.

25. Let x and y be real numbers where $x > 0$. Using Exercise 23, prove that if $xy > 0$, then $y > 0$.

26. Let $\delta > 0$. If $|x - 1| < \delta$, then $|3x + 5| < 3\delta + 8$.

27. Prove that if $|x - 1| < 2$, then $4 < |x + 5|$.

28. Prove that for all $\varepsilon > 0$ there exists a $\delta > 0$ such that if $|x + 1| < \delta$, then $|(2x + 5) - 3| < \varepsilon$.

29. Let x be a real number. Prove that $0 \leq x + |x| \leq 2|x|$.

30. Let $\delta > 0$. Prove that if $|x - 5| < \frac{\delta}{3}$, then $|3x - 15| < \delta$.

31. Let $\delta > 0$. Prove that if $|x - 5| < \delta$, then $|x + 3| < \delta + 8$.

32. Prove that if $|x + 5| < 1$, then $1 < |x + 3|$.

33. Let $\delta > 0$. Prove that if $|x - 3| < \delta$, then $|x^2 - 9| < \delta(\delta + 6)$.

34. Prove that for every real number $x > 3$, there exists a real number $y < 0$ such that $x = \frac{3y}{2 + y}$.

35. Prove that for all real numbers x, if $x > 1$, then $0 < \frac{1}{x} < 1$.

36. Evaluate each set in terms of an interval:

    (a) $\{x \in \mathbb{R}^+ : x > \frac{1}{x}\}$ (c) $\{x \in \mathbb{R}^+ : x > \frac{1}{x} \text{ and } x > 2\}$

    (b) $\{x \in \mathbb{R}^- : x^2 > \frac{1}{x}\}$ (d) $\{x \in \mathbb{R}^- : x > \frac{1}{x} \text{ and } x \not> -\frac{1}{2}\}$.

1.2 SETS

Many of the most important ideas in modern mathematics are expressed in term of sets. Thus, it is good to have a basic understanding of sets. In this section, we will review a few elementary facts about sets. A *set* is just a collection of objects. Again, these objects are referred to as the *elements* of the set. We will usually use upper case letters to denote sets and lower case letters to denote elements of a set.

1.2.1 Basic Definitions of Set Theory

Definition 1.2.1. The following set notation is used throughout mathematics:

1. For sets A and B, we write $A = B$ when both sets have exactly the same elements.

2. For sets A and B, we write $A \subseteq B$ when A is a subset of B; that is, every element of A is also an element of B.

3. For sets A and B, we write $A \subset B$ to state that A is a **proper** subset of B; that is, $A \subseteq B$ and $A \neq B$.

4. We write $\varnothing$ for the empty the set, that is, the set with no elements.

5. If A is a finite set, then $^{\#}A$ denotes the number of elements in A.

We are familiar with the following subset relationships $\mathbb{N} \subseteq \mathbb{Z} \subseteq \mathbb{Q} \subseteq \mathbb{R}$.

1.2.2 Set Operations

The language of set theory is used in the definitions of nearly all of mathematics. There are three important and fundamental operations on sets that we shall now discuss: the intersection, the union, and the difference of two sets. We illustrate these three set operations in Figure 1.1 using Venn diagrams, where shading is used to identify the result of each set operation. Venn diagrams are geometric shapes that can be used to depict sets and their relationships.

Definition 1.2.2. Given sets A and B, we can construct new sets using the following **set operations**:

1. $A \cup B = \{x : x \in A \text{ or } x \in B\}$ is the **union** of A and B.

2. $A \cap B = \{x : x \in A \text{ and } x \in B\}$ is the **intersection** of A and B.

3. $A \setminus B = \{x : x \in A \text{ and } x \notin B\}$ is the **set difference** of A and B.

The operation $A \setminus B$ is often stated as "A minus B." We will be mainly applying Definitions 1.2.1 and 1.2.2 to sets of real numbers.

Example. Let $A = \{1, 2, 3, 4, 5, 6\}$ and $B = \{2, 4, 6, 8, 10, 12\}$. Then

1. $A \cup B = \{1, 2, 3, 4, 5, 6, 8, 10, 12\}$.
2. $A \cap B = \{2, 4, 6\}$.
3. $A \setminus B = \{1, 3, 5\}$.
4. $B \setminus A = \{8, 10, 12\}$.

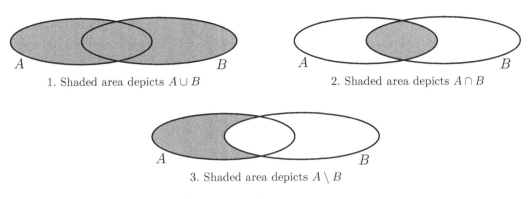

1. Shaded area depicts $A \cup B$

2. Shaded area depicts $A \cap B$

3. Shaded area depicts $A \setminus B$

Figure 1.1: Set operations.

1.2.3   Indexed Families of Sets

Given a property $P(x)$, recall that we can form the set $\{x : P(x)\}$ when the restriction is understood. However, there is another way to construct a set. For example, consider the set S of all even natural numbers, that is, the set of all integers of the form $2n$ for some natural number n. We can define S in two ways:

1. $S = \{x : (\exists n \in \mathbb{N})\, (x = 2n)\} = \{2, 4, 6, 8, 12, \cdots\}$.
2. $S = \{2n : n \in \mathbb{N}\} = \{2, 4, 6, 8, 12, \cdots\}$.

In item 1, we have expressed S as a truth set. Item 2 offers an alternative method for constructing the exact same set S. This alternative method is a special case of the following technique for constructing sets from the set $\mathbb{N}$ of natural numbers. Suppose for each $i \in \mathbb{N}$, that o_i is some specific object. Then we can form the set $S = \{o_i : i \in \mathbb{N}\}$ of all such objects. In this case, the set $\mathbb{N}$ is said to be the *index set* and S is called an *indexed set*. As this concept is often used in mathematics, we now formulate this idea in terms of a general definition.

Definition 1.2.3. let I be *any* set and for each $i \in I$, let o_i be some specific *object*. Then we can form the set $S = \{o_i : i \in I\}$ where I is referred to as the *index set* and S is called an *indexed set*.

For example, for each $i \in \mathbb{Q}$ we have the number $\cos(i)$. Thus, we can construct the indexed set $\{\cos(i) : i \in \mathbb{Q}\}$.

Problem. Explain what the following statements mean.

1. $y \in \{\cos(i) : i \in \mathbb{Q}\}$.
2. $\{x_i : i \in I\} \subseteq A$.
3. $\{x_i : i \in I\} \nsubseteq A$.

Solution. The first statement $y \in \{\cos(i) : i \in \mathbb{Q}\}$ means that $y = \cos(i)$ for some $i \in \mathbb{Q}$. The second statement $\{x_i : i \in I\} \subseteq A$ means that $x_i \in A$ for every $i \in I$. Finally, the third statement $\{x_i : i \in I\} \nsubseteq A$ means that $x_i \notin A$ for some $i \in I$.

Definition 1.2.4. A set $\mathcal{F}$, whose elements are sets, is a **family of sets**.

Definition 1.2.5. Let I be a set and for each $i \in I$, let C_i be a specific *set*. Then we can form the set $\mathcal{F} = \{C_i : i \in I\}$, where I is the **index set** and $\mathcal{F}$ is an **indexed family of sets**.

Example. Suppose for each natural number n, we define the set $A_n = \{0, 1, 2, \ldots, n\}$. Then $\mathcal{F} = \{A_n : n \in \mathbb{N}\} = \{A_1, A_2, A_3, \ldots\}$ is an indexed family of sets.

Example 1.2.6. For each real number $x > 0$, let $B_x = (-x, x + 1)$. We now can define an indexed family of sets by $\mathcal{F} = \{B_x : x \in \mathbb{R}^+\}$.

Example 1.2.7. Let $I = \{i \in \mathbb{R} : i > 1\}$. For each real number $i \in I$, let $B_i = [-i, \frac{1}{i}]$. Thus, $\mathcal{F} = \{B_i : i \in I\}$ is an indexed family of sets.

1.2.4 Generalized Unions and Intersections

For any two sets A and B, we can form the union $A \cup B$ and the intersection $A \cap B$ of these sets. In mathematics, one often forms the union and intersection of many more than just two sets. We will now generalize the operations of union and intersection so that they apply to more than just two sets. We first extend the notions of union and intersection to a finite number of sets, and then to any collection of sets.

We know that $x \in A \cup B$ means that x is in at least one of the two sets A and B. This notion of union can be easily extended to more than two sets. For finitely many sets, say $A_1, A_2, \ldots A_n$, we say that x is in the union

$$A_1 \cup A_2 \cup \cdots \cup A_n$$

when x is in *at least one of the sets* $A_1, A_2, \ldots A_n$; that is, $x \in A_i$ for *some* $1 \leq i \leq n$. Using $I = \{1, 2, \ldots, n\}$ as an index set, we write

$$\bigcup_{i \in I} A_i = A_1 \cup A_2 \cup \cdots \cup A_n$$

and so, $x \in \bigcup_{i \in I} A_i$ means that $x \in A_i$ for some $i \in I$.

Clearly, $x \in A \cap B$ means that x is in both of the two sets A and B. For finitely many sets, say $A_1, A_2, \ldots A_n$, we shall say that x is in the intersection

$$A_1 \cap A_2 \cap \cdots \cap A_n$$

when x is in *every one of the sets* $A_1, A_2, \ldots A_n$; that is, $x \in A_i$ for *every* $1 \leq i \leq n$. Using $I = \{1, 2, \ldots, n\}$ as an index set, we write

$$\bigcap_{i \in I} A_i = A_1 \cap A_2 \cap \cdots \cap A_n$$

and so, $x \in \bigcap_{i \in I} A_i$ means that $x \in A_i$ for every $i \in I$.

Similarly, we can form the union and intersection of any indexed family of sets $\{C_i : i \in I\}$, where I can be a finite or infinite set.

Definition 1.2.8. Let $\{C_i : i \in I\}$ be an indexed family of sets. The **union** $\bigcup\limits_{i \in I} C_i$ is the set of elements x such that $x \in C_i$ for at least one $i \in I$; that is,

$$\bigcup_{i \in I} C_i = \{x : x \in C_i \text{ for some } i \in I\}.$$

Definition 1.2.9. Let $\{C_i : i \in I\}$ be an indexed family of sets. The **intersection** $\bigcap\limits_{i \in I} C_i$ is the set of elements x such that $x \in C_i$ for every $i \in I$; that is,

$$\bigcap_{i \in I} C_i = \{x : x \in C_i \text{ for every } i \in I\}.$$

The next remark can be used to prove set identities that involve indexed unions and/or intersections. The phrase "iff" abbreviates the expression "if and only if."

Remark. Let $\{C_i : i \in I\}$ be an indexed family of sets. The following hold:

(1) $x \in \bigcup\limits_{i \in I} C_i$ iff $x \in C_i$ for some $i \in I$.   (3) $x \in \bigcap\limits_{i \in I} C_i$ iff $x \in C_i$ for every $i \in I$.

(2) $x \notin \bigcup\limits_{i \in I} C_i$ iff $x \notin C_i$ for every $i \in I$.   (4) $x \notin \bigcap\limits_{i \in I} C_i$ iff $x \notin C_i$ for some $i \in I$.

De Morgan's Laws for Indexed Families of Sets

Theorem 1.2.10. Let $\{B_i : i \in I\}$ be a family of sets and let A be a set. Then

(1) $A \setminus \bigcup\limits_{i \in I} B_i = \bigcap\limits_{i \in I} (A \setminus B_i)$.

(2) $A \setminus \bigcap\limits_{i \in I} B_i = \bigcup\limits_{i \in I} (A \setminus B_i)$.

Proof. We prove (1) and leave (2) as an exercise. For any x, we prove that

$$x \in A \setminus \bigcup_{i \in I} B_i \text{ iff } x \in \bigcap_{i \in I}(A \setminus B_i),$$

as follows:

$$x \in A \setminus \bigcup_{i \in I} B_i \text{ iff } x \in A \text{ and } x \notin \bigcup_{i \in I} B_i \qquad \text{by the definition of } \setminus$$

$$\text{iff } x \in A \text{ and } x \notin B_i \text{ for every } i \in I \quad \text{by the definition of } \bigcup$$

$$\text{iff } x \in A \setminus B_i \text{ for every } i \in I \qquad \text{by the definition of } \setminus$$

$$\text{iff } x \in \bigcap_{i \in I}(A \setminus B_i) \qquad \text{by the definition of } \bigcap.$$

Therefore, $A \setminus \bigcup\limits_{i \in I} B_i = \bigcap\limits_{i \in I} (A \setminus B_i)$. □

1.2.5   Unindexed Families of Sets

Indexed families of sets occur frequently in mathematics. Mathematicians also deal with families of sets (see Definition 1.2.4) that are not identified as an indexed set. When $\mathcal{F}$ is a family of sets, the **union** $\bigcup \mathcal{F}$ is the set of elements x such that $x \in C$ for some $C \in \mathcal{F}$; that is,

$$\bigcup \mathcal{F} = \{x : x \in C \text{ for some } C \in \mathcal{F}\}.$$

The **intersection** $\bigcap \mathcal{F}$ is the set of elements x such that $x \in C$ for every $C \in \mathcal{F}$; that is,

$$\bigcap \mathcal{F} = \{x : x \in C \text{ for every } C \in \mathcal{F}\}.$$

For example, let $\mathcal{F}$ be the family of sets defined by $\mathcal{F} = \{\{1,2,9\},\{2,9\},\{4,9\}\}$. Then $\bigcup \mathcal{F} = \{1,2,4,9\}$ and $\bigcap \mathcal{F} = \{9\}$. An "unindexed" version of De Morgan's Theorem 1.2.10 also holds.

Theorem 1.2.11. Suppose that A is a set and that $\mathcal{F}$ is a family of sets. Then

(1) $A \setminus \bigcup \mathcal{F} = \bigcap \{A \setminus B : B \in \mathcal{F}\}$,
(2) $A \setminus \bigcap \mathcal{F} = \bigcup \{A \setminus B : B \in \mathcal{F}\}$.

Exercises 1.2

1. Recalling our discussion on interval notation on page 2, evaluate the following set operations in terms of an interval:

   (a) $(-2,0) \cap (-\infty, 2)$.
   (b) $(-2,4) \cup (-\infty, 2)$.
   (c) $(-\infty, 0] \setminus (-\infty, 2]$.
   (d) $\mathbb{R} \setminus (2, \infty)$.
   (e) $(\mathbb{R} \setminus (-\infty, 2]) \cup (1, \infty)$.

2. Let $I = \{2,3,4,5\}$, and for each $i \in I$, let $C_i = \{i, i+1, i-1, 2i\}$.

   (a) For each $i \in I$, list the elements of C_i.
   (b) Find $\bigcap_{i \in I} C_i$ and $\bigcup_{i \in I} C_i$.

3. For each $n \in \mathbb{N}$, let O_n be the open interval $O_n = (1, 1 + \frac{1}{n})$. Then $\{O_n : n \in \mathbb{N}\}$ is an indexed family of sets. Evaluate the sets: $\bigcap_{n \in \mathbb{N}} O_n$, and $\bigcup_{n \in \mathbb{N}} O_n$.

4. Let $I = \{i \in \mathbb{R} : 1 \leq i\} = [1, \infty)$ and let $A_i = \{x \in \mathbb{R} : -\frac{1}{i} \leq x \leq 2 - \frac{1}{i}\}$ for each $i \in I$. Express $\bigcup_{i \in I} A_i$ in interval notation, if possible. Express $\bigcap_{i \in I} A_i$ in interval notation, if possible.

5. Prove Theorem 1.2.10(2).

6. Prove the following theorems:

(a) **Theorem.** Let $\{A_i : i \in I\}$ and $\{B_i : i \in I\}$ be indexed families of sets with indexed set I. If $A_i \subseteq B_i$ for all $i \in I$, then $\bigcup_{i \in I} A_i \subseteq \bigcup_{i \in I} B_i$ and $\bigcap_{i \in I} A_i \subseteq \bigcap_{i \in I} B_i$.

(b) **Theorem.** Let $\{A_i : i \in I\}$ and $\{B_j : j \in J\}$ be indexed families of sets. If $i_0 \in I$ is such that $A_{i_0} \subseteq B_j$ for all $j \in J$, then $\bigcap_{i \in I} A_i \subseteq \bigcap_{j \in J} B_j$.

(c) **Theorem.** Let A be a set and $\{B_i : i \in I\}$ be an indexed family of sets. Then $A \cap \bigcup_{i \in I} B_i = \bigcup_{i \in I} (A \cap B_i)$.

(d) **Theorem.** Let A be a set and $\{B_i : i \in I\}$ be an indexed family of sets. Then $A \cup \bigcap_{i \in I} B_i = \bigcap_{i \in I} (A \cup B_i)$.

(e) **Theorem.** Let A be a set and $\{B_i : i \in I\}$ be an indexed family of sets. Then $A \setminus \bigcap_{i \in I} B_i = \bigcup_{i \in I} (A \setminus B_i)$.

7. Prove Theorem 1.2.11.

8. Let $\{B_x : x \in \mathbb{R}^+\}$ be as in Example 1.2.6. Evaluate $\bigcap_{x \in \mathbb{R}^+} B_x$ and $\bigcup_{x \in \mathbb{R}^+} B_x$.

9. Let $\{B_i : i \in I\}$ be as in Example 1.2.7. Evaluate $\bigcap_{i \in I} B_i$ and $\bigcup_{i \in I} B_i$.

1.3 FUNCTIONS

Another fundamental concept in mathematics is the notion of a function. One can formally define this concept using set-theoretic principles; however, as the reader is already familiar with functions, we will essentially just review the key features of a function. Recall that a function is a way of associating each element of a nonempty set X with exactly one element of another set Y. In calculus, functions are typically defined in terms of a formula, but in real analysis one needs to view functions in a more general setting.

Definition 1.3.1. We write $f \colon A \to B$ to mean that f is a **function** from the set A to the set B, that is, for every element $x \in A$ there is exactly one element $f(x)$ in B. The value $f(x)$ is called "f of x," or "the image of x under f." The set A is the **domain** of the function f and the set B is the **co-domain** of the function f. In addition, we shall say that $x \in A$ is an *input* for the function f and that $f(x)$ is the resulting *output*. We will also say that x gets *mapped* to $f(x)$.

In the above definition, keep in mind that the notation f represents a function and the notation $f(x)$ identifies the value of the function at x. In particular, $f \neq f(x)$.

Remark. If $f \colon A \to B$, then each $x \in A$ is assigned to exactly one element $f(x)$ in B. So f is **single-valued**, that is, for all $x \in A$ and $z \in A$, if $x = z$, then $f(x) = f(z)$.

Definition 1.3.2. The **range** of a function $f \colon A \to B$, denoted by $\mathrm{ran}(f)$, is the set

$$\mathrm{ran}(f) = \{f(a) : a \in A\} = \{b \in B : b = f(a) \text{ for some } a \in A\}.$$

The range of a function f is the set of all the "output" values produced by f.

Question. Let $h\colon X \to Y$ be a function. What does it mean to say that $b \in \operatorname{ran}(h)$?
Answer: $b \in \operatorname{ran}(h)$ means that $b = f(x)$ for some $x \in A$.

Example. Let $f\colon \mathbb{R} \to \mathbb{R}$ be the function in Figure 1.2 defined by the formula $f(x) = x^2 - 1$. Then $\operatorname{ran}(f) = \{f(x) : x \in \mathbb{R}\} = \{x^2 - 1 : x \in \mathbb{R}\} = [-1, \infty)$.

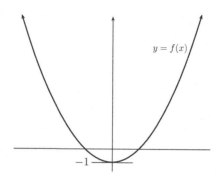

Figure 1.2: Graph of $f(x) = x^2 - 1$.

1.3.1 Real-Valued Functions

Real-valued functions are the focus in a calculus course and they will be the focus in this text as well. A **real-valued function** is one that has the form $f\colon D \to \mathbb{R}$, that is, the output values of the function f are real numbers. In this book, the domain of a real-valued function will typically be a set of real numbers. A **polynomial function** of **degree** n is a real-valued function of the form

$$f(x) = a_n x^n + a_{n-1} x^{n-1} + \cdots a_1 x + a_0,$$

where $a_n, \ldots, a_1, a_0$ are real number constants, n is a natural number, and $a_n \neq 0$. A **rational function** is one that is defined as the ratio of polynomials. A **constant function** has the form $f(x) = a$ for all $x \in D$ where a is a constant. The **zero function** is the constant function where $a = 0$.

Definition 1.3.3. A $f\colon \mathbb{R} \to \mathbb{R}$ is an **even** function if $f(-x) = f(x)$ for all $x \in \mathbb{R}$. If $f(-x) = -f(x)$ for all $x \in \mathbb{R}$, then f is an **odd** function.

1.3.2 Injections and Surjections

There are two fundamental properties that a function may possess; namely, a function may be injective and/or surjective.

Definition. A function $f\colon X \to Y$ is an **injection** (or **one-to-one**), if distinct elements in X get mapped to distinct elements in Y; that is,

$$\text{for all } a, b \in X, \text{ if } a \neq b, \text{ then } f(a) \neq f(b),$$

or equivalently,

$$\text{for all } a, b \in X, \text{ if } f(a) = f(b), \text{ then } a = b.$$

Definition. A function $f\colon X \to Y$ is a **surjection** (or **onto** Y), if for each $y \in Y$, there is an $x \in X$ such that $f(x) = y$.

Definition. A function $f\colon X \to Y$ is a **bijection**, if f is an injection and surjection.

Concerning the above three definitions, we make three observations:

- $f\colon X \to Y$ is an *injection* if and only if for each $y \in Y$ there is *at most one* $x \in X$ such that $f(x) = y$. An injection is also said to be *injective*.

- $f\colon X \to Y$ is a *surjection* if and only if for each $y \in Y$ there is *at least one* $x \in X$ such that $f(x) = y$. A surjection is often referred as being *surjective*.

- $f\colon X \to Y$ is a *bijection* if and only if for each $y \in Y$ there is *exactly one* $x \in X$ such that $f(x) = y$. A bijection is also described as being *bijective*.

1.3.3 Composition of Functions

If $f\colon X \to Y$ and $g\colon Y \to Z$, then for any $x \in X$, we have that $f(x) \in Y$. So $g(f(x))$ is an element in Z (see Figure 1.3). This "composition" $g(f(x))$ allows us to assign elements in X to elements in Z. Thus, we can define a new function.

Definition. Given functions $f\colon X \to Y$ and $g\colon Y \to Z$, one forms the **composition** $(g \circ f)\colon X \to Z$ by defining $(g \circ f)(x) = g(f(x))$ for all $x \in X$. The function $g \circ f$ is a **composite function**.

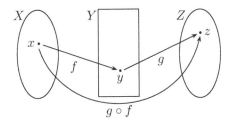

Figure 1.3: $f(x) = y$, $g(y) = z$, and $g(f(x)) = z$.

We now present results on the composition of injections and surjections.

Theorem 1.3.4. If $f\colon X \to Y$ and $g\colon Y \to Z$ are injections, then $(g \circ f)\colon X \to Z$ is an injection.

Proof. Let $f\colon X \to Y$ and $g\colon Y \to Z$ be injections. Let $a, b \in X$. Assume that $(g \circ f)(a) = (g \circ f)(b)$. Thus, $(\star)$ $g(f(a)) = g(f(b))$ by the definition of composition. Since g is an injection, $(\star)$ implies that $f(a) = f(b)$. As f is an injection, we conclude that $a = b$. Therefore, $(g \circ f)$ is an injection. □

Theorem 1.3.5. If $f\colon X \to Y$ and $g\colon Y \to Z$ are surjections, then $(g \circ f)\colon X \to Z$ is a surjection.

Proof. Let $f\colon X \to Y$ and $g\colon Y \to Z$ be surjections. Let $z \in Z$. Since g is a surjection, there exists a $y \in Y$ such that $g(y) = z$. As $y \in Y$ and f is a surjection, there is an $x \in X$ such that $f(x) = y$. Thus, $(g \circ f)(x) = g(f(x)) = g(y) = z$. Hence, $(g \circ f)$ is a surjection. □

1.3.4   Inverse Functions

A function $f\colon X \to Y$ assigns elements in X to elements in Y. This assignment can be partially reversed if f is an injection.

Theorem 1.3.6. Let $f\colon X \to Y$ be an injection and let $R = \mathrm{ran}(f)$. Then there is a function $f^{-1}\colon R \to X$ defined as follows: For each $y \in R$, $f^{-1}(y)$ is the unique element x in X such that $f(x) = y$. Thus, for all $y \in R$ and $x \in X$

$$f^{-1}(y) = x \text{ iff } f(x) = y. \tag{1.1}$$

Proof. Let $f\colon X \to Y$ be an injection and let $R = \mathrm{ran}(f)$. Let $y \in R$. Thus, there is an $x \in X$ such that $f(x) = y$. Suppose that $x' \in X$ also satisfies $f(x') = y$. Thus, $f(x) = f(x')$. Because f is injective, it follows that $x = x'$. So for every $y \in R$ there is exactly one $x \in X$ such that $f(x) = y$. Hence, the formula $f(x) = y$ used in (1.1) defines the function $f^{-1}\colon R \to X$. □

Definition 1.3.7. Given an injection $f\colon X \to Y$, let $R = \mathrm{ran}(f)$. Then the function $f^{-1}\colon R \to X$, satisfying (1.1) for all $y \in R$, is the **inverse function** of f.

We now show that if a function and its inverse are functionally composed, then the result is an identity function i, that is, $i(x) = x$ for all x in the domain of i.

Theorem 1.3.8. Let $f\colon X \to Y$ be injective. Let $R = \mathrm{ran}(f)$ and let $f^{-1}\colon R \to X$ be the inverse of f. Then $(f^{-1} \circ f)\colon X \to X$ and $(f \circ f^{-1})\colon R \to R$. Moreover,

(a) $f\colon X \to R$ is a bijection,
(b) $(f^{-1} \circ f)(x) = x$, for all $x \in X$,
(c) $(f \circ f^{-1})(y) = y$, for all $y \in R$.

Proof. Item (a) should be clear. Let $x \in X$. Since $f(x) \in Y$, let $y \in Y$ be such that $f(x) = y$. Theorem 1.3.6 implies that (∗) $f^{-1}(y) = x$. After substituting $y = f(x)$ into equation (∗), we see that $f^{-1}(f(x)) = x$. So (b) holds. To prove (c), let $y \in R$. Since $f^{-1}(y) \in X$, let $x \in X$ be such that $f^{-1}(y) = x$. Thus, (†) $f(x) = y$ by Theorem 1.3.6. Upon substituting $x = f^{-1}(y)$ into equation (†), we obtain $f(f^{-1}(y)) = y$. □

We end this section by showing that the inverse of a bijection is also a bijection.

Theorem 1.3.9. Suppose that $f\colon X \to Y$ is a bijection. Let $f^{-1}\colon Y \to X$ be the inverse of f. Then f^{-1} is a bijection.

Proof. Let $f\colon X \to Y$ be a bijection. So $Y = \mathrm{ran}(f)$. We first prove that $f^{-1}\colon Y \to X$ is injective. Let $y, y' \in Y$. Assume $f^{-1}(y) = f^{-1}(y')$. Let $x \in X$ be this common value. Thus, $f^{-1}(y) = x$ and $f^{-1}(y') = x$. So $f(x) = y$ and $f(x) = y'$, by (1.1). Since f is a function, we conclude that $y = y'$. Hence, f^{-1} is an injection.

To prove that $f^{-1}\colon Y \to X$ is surjective, let $x \in X$. So let $y \in Y$ be such that $f(x) = y$. By (1.1), $f^{-1}(y) = x$. Therefore, f^{-1} is a surjection. □

Thus, if $f\colon X \to Y$ is injective, then $f\colon X \to R$ and $f^{-1}\colon R \to X$ are bijections, where $R = \mathrm{ran}(f)$.

1.3.5 Functions Acting on Sets

Often in real analysis, one is more interested in what a function does to an entire subset of its domain, rather than how the function affects each individual element in its domain. This brings us to the following definition.

Definition 1.3.10 (Image of a Set). Let $f\colon X \to Y$ be a function and let $S \subseteq X$. The set $f[S]$, called the **image** of S, is defined by

$$f[S] = \{f(x) : x \in S\} = \{y \in Y : y = f(x) \text{ for some } x \in S\}.$$

(Note that $f[X] = \operatorname{ran}(f)$.)

Figure 1.4 illustrates Definition 1.3.10. The square S represents a subset of the domain of the function f. The image $f[S]$ is represented by a rectangle.

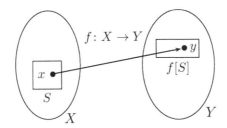

Figure 1.4: Starting with $S \subseteq X$, we obtain the new set $f[S] \subseteq Y$.

The functional image of a set can lead to a better understanding of the function itself and can reveal some properties concerning its domain and range.

Remark. In some mathematics textbooks, $f[S]$ is written as $f(S)$; however, this latter notation is ambiguous and its meaning must be inferred from the context. The conventional functional notation $f(x)$ means that x is an element of the domain of f and it does not mean that x is a subset of the domain. Here, in this book, we will avoid such vague notation by denoting the image of set S by $f[S]$.

Example 1.3.11. Let $f\colon \mathbb{R} \to \mathbb{R}$ be defined by $f(x) = |x|$. Let $S = \{-4, -3, 2, 3\}$. Then the image of S is $f[S] = \{f(x) : x \in S\} = \{|x| : x \in S\} = \{2, 3, 4\}$.

Given a subset S of the domain of a function f, Definition 1.3.10 identifies a subset $f[S]$ of the co-domain of f. This process can be inverted. We can start with a subset T of the co-domain and then use it to define a subset of the domain.

Definition 1.3.12 (Inverse Image of a Set). Let $f\colon X \to Y$ be a function and let $T \subseteq Y$. The set $f^{-1}[T]$ is the subset of X defined by $f^{-1}[T] = \{x \in X : f(x) \in T\}$. The set $f^{-1}[T]$ is the **inverse image** of T.

A depiction of Definition 1.3.12 is given in Figure 1.5. The circle T depicts a subset of the co-domain of f. The inverse image $f^{-1}[T]$ is represented by an ellipse.

Example 1.3.13. Let $f\colon \mathbb{R} \to \mathbb{R}$ be defined by $f(x) = |x|$ and let $T = \{-8, 2, 3\}$. The inverse image of T is $f^{-1}[T] = \{x \in \mathbb{R} : f(x) \in T\} = \{-3, -2, 2, 3\}$.

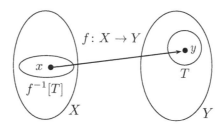

Figure 1.5: Starting with $T \subseteq Y$, we obtain the new set $f^{-1}[T] \subseteq X$.

The notation f^{-1} used in Definition 1.3.12, should not be confused with that of an inverse function. Theorem 1.3.6 implies that the inverse function exists if and only if the original function is an injection. Definition 1.3.12 applies to all functions, even those that are not injective. Given a subset of the co-domain of *any* function f, the inverse image of this subset defines a subset of the function's domain.

On the other hand, suppose that $f\colon X \to Y$ is an injection and $R = \mathrm{ran}(f) \subseteq Y$. Then the inverse function $f^{-1}\colon R \to X$ exists. Now let $U \subseteq R$. In this case, there are two ways that the notation $f^{-1}[U]$ can be interpreted:

1. $f^{-1}[U] = \{f^{-1}(y) : y \in U\}$, the image of U under the inverse function f^{-1}; and
2. $f^{-1}[U] = \{x \in X : f(x) \in U\}$, the inverse image of U under the function f.

However, using (1.1), one can show that the two sets in 1 and 2 (on the right side of the equality) are equal. So, if the inverse function exists, both interpretations yield the same set, and thus, one can use either interpretation without ambiguity.

The following remark states three observations that can be very useful when working with the image, or the inverse image, of a set.

Remark 1.3.14. Let $f\colon X \to Y$, $S \subseteq X$, $T \subseteq Y$, $a \in X$ and $b \in Y$.

1. If $a \in S$, then $f(a) \in f[S]$.
2. $b \in f[S]$ if and only if $b = f(x)$ for some $x \in S$.
3. $a \in f^{-1}[T]$ if and only if $f(a) \in T$.

Remark (Image Warning). If $f(a) \in f[S]$, then we can conclude that $f(a) = f(x)$ for some $x \in S$, by item 2 of Remark 1.3.14; however, we **cannot necessarily conclude** that $a \in S$. In Example 1.3.11, we have that $f(4) \in f[S]$ and yet $4 \notin S$.

Given a function $f\colon X \to Y$, we now identify four relationships that hold for the image and inverse image of subsets of X and Y, respectively.

Theorem 1.3.15. Let $f\colon X \to Y$ be a function. Let C and D be subsets of X, and let U and V be subsets of Y. Then

(a) $f[C \cap D] \subseteq f[C] \cap f[D]$ (c) $f^{-1}[U \cap V] = f^{-1}[U] \cap f^{-1}[V]$

(b) $f[C \cup D] = f[C] \cup f[D]$ (d) $f^{-1}[U \cup V] = f^{-1}[U] \cup f^{-1}[V]$.

Proof. Let $f\colon X \to Y$ be a function. We only prove (a) and (d). Let C and D be subsets of X and let U and V be subsets of Y.

(a). We prove $f[C \cap D] \subseteq f[C] \cap f[D]$. Let $y \in f[C \cap D]$. Since $y \in f[C \cap D]$, there is an $x \in C \cap D$ such that $y = f(x)$. Because $x \in C \cap D$, we see that $x \in C$ and $x \in D$. Therefore, $y = f(x) \in f[C]$ and $y = f(x) \in f[D]$. Thus, $y \in f[C] \cap f[D]$.

(d). We prove $f^{-1}[U \cup V] = f^{-1}[U] \cup f^{-1}[V]$. Let x be arbitrary. We shall prove that $x \in f^{-1}[U \cup V]$ if and only if $x \in f^{-1}[U] \cup f^{-1}[V]$, as follows:

$$
\begin{aligned}
x \in f^{-1}[U \cup V] \text{ iff } & f(x) \in U \cup V && \text{by definition of inverse image} \\
\text{iff } & f(x) \in U \text{ or } f(x) \in V && \text{by definition of } \cup \\
\text{iff } & x \in f^{-1}[U] \text{ or } x \in f^{-1}[V] && \text{by definition of inverse image} \\
\text{iff } & x \in f^{-1}[U] \cup f^{-1}[V] && \text{by definition of } \cup.
\end{aligned}
$$

Therefore, $f^{-1}[U \cup V] = f^{-1}[U] \cup f^{-1}[V]$. □

When a function is injective, then the subset relationship in Theorem 1.3.15(a) can be replaced with equality.

Theorem 1.3.16. Let $f\colon X \to Y$ be a function. Let C and D be subsets of X. If f is an injection, then $f[C \cap D] = f[C] \cap f[D]$.

Proof. Let $f\colon X \to Y$ be injective and C, D be subsets of X. By Theorem 1.3.15(a), $f[C \cap D] \subseteq f[C] \cap f[D]$. To prove that $f[C] \cap f[D] \subseteq f[C \cap D]$, let $y \in f[C] \cap f[D]$. Thus, $y \in f[C]$ and $y \in f[D]$. Because $y \in f[C]$, there is an $x_1 \in C$ such that $f(x_1) = y$. Also, since $y \in f[D]$, there is an $x_2 \in D$ such that $f(x_2) = y$. Hence, $y = f(x_1) = f(x_2)$. Since f is injective, we infer that $x_1 = x_2$. Thus, $x_1 \in D$. So, $x_1 \in C \cap D$ and therefore, $y = f(x_1) \in f[C \cap D]$. Hence, $f[C \cap D] = f[C] \cap f[D]$. □

Remark 1.3.17. If $f\colon D \to \mathbb{R}$ and $[a, b] \subseteq D$, then we denote the image $f[[a, b]]$ by $f([a, b])$, as an aid to readability.

Exercises 1.3 ────────────────────────────────

1. Using Definitions 1.3.10 and 1.3.12, explain why items 1–3 of Remark 1.3.14 hold.

2. Let $f\colon X \to Y$ be a function. Let S be a subset of X, and let T be a subset of Y. Prove that $f[S] \subseteq T$ if and only if for all $x \in S$ we have $f(x) \in T$.

3. Prove item (b) of Theorem 1.3.15.

4. Prove item (c) of Theorem 1.3.15.

5. Given $a, b \in \mathbb{R}$ with $a > 0$, define the function $f\colon \mathbb{R} \to \mathbb{R}$ by $f(x) = ax + b$. Let $U = [2, 3]$. Evaluate $f[U]$ and $f^{-1}[U]$ in terms of intervals.

6. Let $f\colon X \to Y$ be a function and let $A \subseteq X$ and $B \subseteq X$. Prove that if $A \subseteq B$, then $f[A] \subseteq f[B]$.

7. Let $f\colon \mathbb{R} \to \mathbb{R}$ be the function defined in Example 1.3.11. Find $A \subseteq \mathbb{R}$ and $B \subseteq \mathbb{R}$ such that $f[A] \subseteq f[B]$ and $A \not\subseteq B$.

8. Suppose $f\colon X \to Y$ is an injection and let $A \subseteq X$ and $B \subseteq X$. Prove that if $f[A] \subseteq f[B]$, then $A \subseteq B$.

*9. Let $f\colon X \to Y$ be a function and let $C \subseteq Y$ and $D \subseteq Y$. Prove that if $C \subseteq D$, then $f^{-1}[C] \subseteq f^{-1}[D]$.

10. Let $f\colon \mathbb{R} \to \mathbb{R}$ be the function defined in Example 1.3.13. Find $C \subseteq \mathbb{R}$ and $D \subseteq \mathbb{R}$ such that $f^{-1}[C] \subseteq f^{-1}[D]$ and $C \not\subseteq D$.

11. Let $f\colon X \to Y$ be onto Y and let $C \subseteq Y$ and $D \subseteq Y$. Prove if $f^{-1}[C] \subseteq f^{-1}[D]$, then $C \subseteq D$.

12. Define the function $f\colon \mathbb{R} \to \mathbb{R}$ by $f(x) = x^2$ and let $U = [-1, 4]$. Show that

    (a) $f[f^{-1}[U]] \neq U$,
    (b) $f^{-1}[f[U]] \neq U$,
    (c) $f[f^{-1}[U]] \neq f^{-1}[f[U]]$.

*13. Let $f\colon X \to Y$ be a function and let $A \subseteq X$ and $C \subseteq Y$. Prove that $f[A] \subseteq C$ if and only if $A \subseteq f^{-1}[C]$.

14. Let $f\colon X \to Y$ be a function. Let A be a subset of X. Prove that $A \subseteq f^{-1}[f[A]]$.

15. Suppose $f\colon X \to Y$ is an injection. Let $A \subseteq X$ and $x \in X$. Prove if $f(x) \in f[A]$, then $x \in A$.

16. Suppose that $f\colon X \to Y$ is injective. Let $A \subseteq X$. Prove that $A = f^{-1}[f[A]]$.

17. Let $f\colon X \to Y$. Suppose $A = f^{-1}[f[A]]$ for all finite subsets A of X. Prove f is injective.

18. Let $f\colon X \to Y$ be a function. Let C be a subset of Y. Prove that $f[f^{-1}[C]] \subseteq C$.

19. Assume that $f\colon X \to Y$ is a surjection. Let $C \subseteq Y$. Prove that $f[f^{-1}[C]] = C$.

20. Given $a, b \in \mathbb{R}$ with $a > 0$, define the function $f\colon \mathbb{R} \to \mathbb{R}$ by $f(x) = ax + b$. Using Exercises 16 and 19, prove that $f[f^{-1}[U]] = f^{-1}[f[U]]$ for every $U \subseteq \mathbb{R}$.

21. Let $f\colon X \to Y$ be a function. Let $\{V_i : i \in I\}$ be an indexed family of sets where $V_i \subseteq Y$ for all $i \in I$. Prove that $f^{-1}\left[\bigcup_{i \in I} V_i\right] = \bigcup_{i \in I} f^{-1}[V_i]$.

Exercise Notes: Review Remark 1.3.14. Exercise 7 verifies that the converse of Exercise 6 is not true for all functions. Exercise 8, however, shows that this converse is true for all injective functions. Similarly, Exercise 10 shows that the converse of Exercise 9 is not true for all functions. Exercise 11 then shows that this converse is true for all surjective functions.

1.4 MATHEMATICAL INDUCTION

Mathematical induction is a potent method of proof that is often applied to establish that certain statements are true of all the natural numbers. Before presenting this method of proof, we first recall the well-ordering principle.

1.4.1 The Well-Ordering Principle

Every nonempty subset of $\mathbb{N}$ has a least element. This important property is stated as a principle that is routinely applied in mathematical proofs.

Well-Ordering Principle 1.4.1. Every nonempty subset of $\mathbb{N}$ has a least element.

The well-ordering principle implies the next theorem which is typically covered in a discrete mathematics course or an introduction to proof course.

Definition 1.4.2. For a natural number $n > 1$, we say that $n = p_1^{a_1} p_2^{a_2} \cdots p_k^{a_k}$ is a prime factorization of n, when $p_1, \ldots, p_k$ are distinct prime numbers and $a_1, \ldots, a_k$ are natural numbers. If $p_i < p_j$ when $1 \leq i < j \leq k$, then the factorization shall be referred to as an **ascending prime factorization**.

Theorem 1.4.3 (Fundamental Theorem of Arithmetic). Let $n > 1$ be a natural number. There exist primes $p_1 < p_2 < \cdots < p_k$ and natural numbers $a_1, a_2, \ldots, a_k$ such that $n = p_1^{a_1} p_2^{a_2} \cdots p_k^{a_k}$. Moreover, if $n = q_1^{b_1} q_2^{b_2} \cdots q_\ell^{b_\ell}$ is any ascending prime factorization, then $\ell = k$, $p_1 = q_1$, $\ldots$, $p_k = q_\ell$ and $a_1 = b_1$, $\ldots$, $a_k = b_\ell$.

The well-ordering principle also implies that proof by mathematical induction is a valid technique of proof (see Exercise 18).

1.4.2 Proof by Mathematical Induction

Mathematical induction is a powerful method for proving theorems about the natural numbers. Suppose we have a statement about every integer greater than or equal to the integer b. How does one prove this statement by mathematical induction? First prove that the statement definitely holds for b. Then prove that whenever the statement holds for an integer $n \geq b$, then it must hold for the next integer $n + 1$ as well. We now present a proof strategy that summarizes proof by mathematical induction.

Proof Strategy 1.4.4. Let b be a fixed integer. To prove a statement of the form $(\forall n \geq b)\, P(n)$ by mathematical induction, use the diagram:

> *Base step:*   Prove $P(b)$.
> *Inductive step:* Let $n \geq b$ be arbitrary.
>   Assume $P(n + 1)$.
>   Prove $P(n)$.

In a proof by Mathematical Induction, the proof of $P(b)$ is called the *base step* and the proof of $(\forall n \geq b)(P(n) \to P(n+1))$ is called the *inductive step*. The assumption

$P(n)$ is the **induction hypothesis** and the statement $P(n+1)$ is the **induction conclusion**. A proof which uses the Principle of Mathematical Induction is called an **induction proof** or **proof by induction**.

In the *base step* of an induction proof, one must show that the statement $P(b)$ is true. To do so, simply replace n by b everywhere in $P(n)$ and verify that $P(b)$ holds.

The *inductive step* is more challenging as it requires one to show that $P(n+1)$ follows from the assumption $P(n)$. This usually involves expressing the statement $P(n+1)$ in terms that relate to the assumption $P(n)$. Then one can apply the assumption $P(n)$, which is referred to as **using the induction hypothesis**. After establishing that $P(n+1)$ holds, the proof is complete.

We will prove our next two theorems by mathematical induction. In the proof of the following lemma and theorem, we will implicitly apply the Sum and Product Principles of Inequality 1.1.8.

Lemma 1.4.5. If $n \geq 4$, then $n^2 > 2n + 1$.

Proof. Assume that $(\star)$ $n \geq 4$. Since $n > 0$, we conclude that $n^2 \geq 4n = 2n + 2n$. From $(\star)$, we also see that $2n \geq 8 > 1$. Therefore, $n^2 > 2n + 1$. $\qquad\square$

Theorem 1.4.6. For every natural number $n \geq 5$, $2^n > n^2$.

Proof. We prove, by mathematical induction, that $2^n > n^2$ for all $n \geq 5$.

Base step: For $n = 5$, we see that $2^n = 32$ and $n^2 = 25$. Thus, $2^5 > 5^2$.

Inductive step: Let $n \geq 5$ and assume the induction hypothesis that (IH) $2^n > n^2$. We show that $2^{n+1} > (n+1)^2$. Note that $2^{n+1} = 2 \cdot 2^n = 2^n + 2^n$. Hence

$$\begin{aligned}
2^{n+1} &= 2^n + 2^n && \text{by algebra} \\
&> n^2 + n^2 && \text{by the induction hypothesis (IH)} \\
&> n^2 + 2n + 1 && n^2 > 2n + 1 \text{ by Lemma 1.4.5} \\
&= (n+1)^2 && \text{by factoring.}
\end{aligned}$$

Hence, $2^{n+1} > (n+1)^2$ and the proof is complete. $\qquad\square$

Let $m \leq n$ be integers and let $h \colon \{k \in \mathbb{Z} : m \leq k \leq n\} \to \mathbb{R}$ be a function. Recall the summation notation, or Σ notation, $\sum_{k=m}^{n} h(i) = h(m) + h(m+1) + \cdots + h(n)$. Our next theorem will be applied later in the text.

Theorem 1.4.7 (Geometric Summation). Let $r \neq 1$ be a fixed real number. For every integer $n \geq 0$, we have $\sum_{k=0}^{n} r^k = \frac{r^{n+1}-1}{r-1}$.

Proof. Let $r \neq 1$. We prove that $\sum_{k=0}^{n} r^k = \frac{r^{n+1}-1}{r-1}$ for all $n \geq 0$, by induction.

Base step: For $n = 0$, $\sum_{k=0}^{0} r^k = r^0 = 1$ and $\frac{r^{0+1}-1}{r-1} = 1$. Thus, $\sum_{k=0}^{0} r^k = \frac{r^{0+1}-1}{r-1}$.

Inductive step: Let $n \geq 0$ and assume the induction hypothesis that

$$\sum_{k=0}^{n} r^k = \frac{r^{n+1} - 1}{r - 1}. \qquad \text{(IH)}$$

We show that $\sum_{k=0}^{n+1} r^k = \frac{r^{(n+1)+1} - 1}{r-1}$, that is, we show that $\sum_{k=0}^{n+1} r^k = \frac{r^{n+2}-1}{r-1}$ as follows:

$$\sum_{k=0}^{n+1} r^k = \left(\sum_{k=0}^{n} r^k \right) + r^{n+1} \quad \text{by property of } \Sigma \text{ notation}$$

$$= \frac{r^{n+1} - 1}{r - 1} + r^{n+1} \quad \text{by induction hyp. (IH)}$$

$$= \frac{r^{n+2} - 1}{r - 1} \quad \text{by algebra.}$$

Hence, $\sum_{k=0}^{n+1} r^k = \frac{r^{n+2}-1}{r-1}$ and the proof is complete. $\qquad \square$

The shift rule, below, allows one to change the lower limit of a sum without changing the value of the sum. This rule will be applied several times in the book.

The Shift Rule 1.4.8. Consider the sum $\sum_{k=m}^{n} h(k)$ with lower limit m. To rewrite this sum as one with the lower limit s, compute $d = s - m$ and then

$$\sum_{k=m}^{n} h(k) = \sum_{k=m+d}^{n+d} h(k - d) = \sum_{k=s}^{n+d} h(k - d).$$

Example. Let us shift the sum $\sum_{k=3}^{n} \frac{1}{k^5}$ to an equal sum with lower limit 1. We obtain $d = 1 - 3 = -2$. We now shift the upper and lower limit values **down** by 2 and modify the summand by shifting k **up** by 2, to obtain $\sum_{k=3}^{n} \frac{1}{k^5} = \sum_{k=1}^{n-2} \frac{1}{(k+2)^5}$.

In mathematics, the factorial of a natural number n is the product of all the natural numbers less than or equal to n. This is written as $n!$ and is called "n factorial."

Definition 1.4.9. For each natural number n, the value $n!$ is defined to be

$$n! = n(n - 1)(n - 2) \cdots 2 \cdot 1.$$

By convention, we define $0! = 1$ (which is used to simplify mathematical formulas).

Example. Note the following:

1. $n! = n(n - 1)!$ for $n \geq 1$.
2. $(n + 1)! = (n + 1)n!$ for $n \geq 0$.
3. $\dfrac{(n + 1)!}{(n - 1)!} = \dfrac{(n + 1)(n)(n - 1)!}{(n - 1)!} = n(n + 1)$, for $n \geq 1$.

We now introduce the *binomial coefficient* $\binom{n}{k}$, an important tool.

Definition 1.4.10. Let $n \geq k \geq 0$ be integers. Then $\binom{n}{k} = \frac{n!}{k!(n-k)!}$.

Remark. Since $0! = 1$, we obtain $\binom{n}{0} = 1$ and $\binom{n}{n} = 1$, for all integers $n \geq 0$. Moreover, when $n \geq k \geq 1$, we see that

$$\binom{n}{k} = \frac{n!}{k!(n-k)!} = \frac{n(n-1)\cdots(n-(k-1))}{k!}.$$

The binomial coefficient $\binom{n}{k}$ is used frequently in many areas of mathematics. In combinatorics, the notation $\binom{n}{k}$ is usually read as "n choose k" because there are $\binom{n}{k}$ ways to choose an subset of k elements from a set of n elements.

The following famous theorem identifies the resulting algebraic expansion of the expression $(x+y)^n$ when n is a natural number, where we let $a^0 = 1$ for all $a \in \mathbb{R}$,

Theorem 1.4.11 (Binomial Theorem). Let x and y be real numbers. Then for every integer $n \geq 1$, $(x+y)^n = \sum_{k=0}^{n} \binom{n}{k} x^{n-k} y^k = x^n + \binom{n}{1} x^{n-1} y + \cdots + \binom{n}{n-1} xy^{n-1} + y^n$.

From the binomial theorem, we can derive a useful inequality.

Corollary 1.4.12. Let $y \geq 0$. Then $(1+y)^n \geq 1 + \frac{n(n-1)}{2} y^2$, for every integer $n \geq 2$.

Proof. By Theorem 1.4.11,

$$(1+y)^n = \sum_{k=0}^{n} \binom{n}{k} 1^{n-k} y^k \geq 1 + \binom{n}{2} y^2 = 1 + \frac{n(n-1)}{2} y^2$$

when $y \geq 0$ and $n \geq 2$. $\qquad\square$

We end this section with a general description of an inductively defined sequence. The first term of such a sequence is identified and thereafter, each successive term is defined in terms of the previous term.

Inductively Defined Sequences. Let $M[x]$ be a "method" or function to obtain a number m from another number x, denoted by $m = M[x]$. Let b be a given number and let i be a fixed integer. We define an infinite sequence $a_i, a_{i+1}, a_{i+2}, \ldots, a_n, \ldots$ of numbers, starting at i, **by induction** as follows:

(1) $a_i = b$

(2) $a_{n+1} = M[a_n]$ for all $n \geq i$.

Exercises 1.4 ──────────────────────────────

1. Let $1 \leq a$. Prove that for every natural number $n \geq 1$, $a \leq a^n$.

*2. Prove that for every natural number $n \geq 1$, $2^n > n$.

*3. Let $a, b \in \mathbb{R}$ be positive where $a < b$. Prove that $a^n < b^n$, for every $n \in \mathbb{N}$.

4. Prove that $2^n < n!$ for all $n \geq 4$.

*5. (Bernoulli's inequality) Let $x > -1$ be a real number. Prove that $(1 + x)^n \geq 1 + nx$ for all natural numbers $n \geq 1$.

6. Let $a_1, a_2, \ldots$ be an infinite sequence of real numbers where $a_n \geq 0$ for all $n \geq 1$. Prove that $(1 + a_1)(1 + a_2) \cdots (1 + a_n) \geq 1 + a_1 + a_2 + \cdots + a_n$ for all $n \geq 1$.

*7. Let a, b be positive real numbers. Let $n \geq 1$ be a natural number.

    (a) Using Exercise 3, prove that $a < b$ if and only if $a^n < b^n$.

    (b) Prove that $a < b$ if and only if $a^{\frac{1}{n}} < b^{\frac{1}{n}}$.

8. Let $c < d$ be real numbers. Let $a_1, a_2, \ldots$ is a sequence of numbers such that $c \leq a_n \leq d$ for all $n \geq 1$. Prove that $c \leq \frac{a_1 + a_2 + \cdots + a_n}{n} \leq d$ for all $n \geq 1$.

*9. (Telescoping Sum) Let $a_0, a_1, a_2, \ldots$ be an infinite sequence of real numbers. Prove for all natural numbers $n \geq 1$, that

$$\sum_{i=1}^{n} (a_i - a_{i-1}) = a_n - a_0.$$

10. Prove that for every natural number $n \geq 1$, $1 + 3 + 5 + \cdots + (2n - 1) = n^2$.

11. Prove that for every natural number $n \geq 0$, $1 + 2 + 2^2 + \cdots + 2^n = 2^{n+1} - 1$.

12. Prove that for every natural number $n \geq 1$, $2 + 6 + 18 + \cdots + 2 \cdot 3^{n-1} = 3^n - 1$.

13. Prove that for every natural number $n \geq 2$, $\left(1 - \frac{1}{4}\right)\left(1 - \frac{1}{9}\right) \cdots \left(1 - \frac{1}{n^2}\right) = \frac{n+1}{2n}$.

14. Prove that for every natural number $n \geq 1$, $\sum_{k=1}^{n} k^2 = \frac{n(n+1)(2n+1)}{6}$.

15. Prove that for every natural number $n \geq 1$, $\sum_{k=1}^{n} \frac{1}{4k^2 - 1} = \frac{n}{2n+1}$.

    [Hint: $4(n + 1)^2 - 1 = (2n + 1)(2n + 3)$.]

16. Prove that for every natural number $n \geq 1$, $\sum_{k=1}^{n} k \cdot k! = (n + 1)! - 1$.

17. Prove that for every natural number $n \geq 1$,

$$\sum_{k=1}^{n} k^3 = \left[\frac{n(n + 1)}{2}\right]^2.$$

*18. Let $P(n)$ be a statement that is defined for all integers $n \geq 1$. Suppose that

    (a) $P(1)$ is true, and

    (b) for all integers $n \geq 1$, if $P(n)$ holds, then $P(n + 1)$ also holds.

    Using the Well-Ordering Principle 1.4.1, prove that $P(n)$ holds for all $n \in \mathbb{N}$.

19. Let $1 \leq k \leq n$ be integers. Prove that

$$\binom{n + 1}{k} = \binom{n}{k} + \binom{n}{k - 1}.$$

20. Let $1 \le k \le n$ be integers. Justify the following sequence of equalities:

$$x \sum_{k=0}^{n} \binom{n}{k} x^{n-k} y^k + y \sum_{k=0}^{n} \binom{n}{k} x^{n-k} y^k$$

$$= x \left[x^n + \sum_{k=1}^{n} \binom{n}{k} x^{n-k} y^k \right] + y \left[\sum_{k=0}^{n-1} \binom{n}{k} x^{n-k} y^k + y^n \right]$$

$$= x^{n+1} + \left(x \sum_{k=1}^{n} \binom{n}{k} x^{n-k} y^k \right) + \left(y \sum_{k=0}^{n-1} \binom{n}{k} x^{n-k} y^k \right) + y^{n+1}$$

$$= x^{n+1} + \left(\sum_{k=1}^{n} \binom{n}{k} x^{n+1-k} y^k \right) + \left(\sum_{k=0}^{n-1} \binom{n}{k} x^{n-k} y^{k+1} \right) + y^{n+1}$$

$$= x^{n+1} + \left(\sum_{k=1}^{n} \binom{n}{k} x^{n+1-k} y^k \right) + \left(\sum_{k=1}^{n} \binom{n}{k-1} x^{(n+1)-k} y^k \right) + y^{n+1}$$

$$= x^{n+1} + \left(\sum_{k=1}^{n} \left[\binom{n}{k} + \binom{n}{k-1} \right] x^{(n+1)-k} y^k \right) + y^{n+1}$$

$$= x^{n+1} + \left(\sum_{k=1}^{n} \binom{n+1}{k} x^{(n+1)-k} y^k \right) + y^{n+1}$$

$$= \sum_{k=0}^{n+1} \binom{n+1}{k} x^{(n+1)-k} y^k$$

21. Using Exercise 20 in the inductive step, prove Theorem 1.4.11 by mathematical induction.

Exercise Notes: For Exercise 7, no induction is needed. For Exercise 8, multiply the induction hypothesis inequality by n and use the assumption $c \le a_{n+1} \le d$. For Exercise 18, using proof by contradiction, let $N \ge 1$ be the least natural number such that $P(N)$ is false. Explain why $N = n + 1$ for some $n \ge 1$ and observe that $n < N$.

CHAPTER 2

The Real Numbers

2.1 INTRODUCTION

The differential and integral calculus was developed in the 17th century by Gottfried Leibniz and Isaac Newton in terms of quantities, called infinitesimals, which were infinitely small nonzero values. For example, the derivative was originally defined to be a ratio of two infinitesimals. In his calculations, Newton used a number o which, being infinitely small, could be multiplied by any real number and still be negligible. But it was necessary to divide by o, and so it had to be nonzero. Leibniz's number dx was positive and less than any assignable quantity, yet it was nonzero. Thus, the early development of the calculus was based on a number system that was not well understood (at the time) by mathematicians, including Newton and Leibniz.

In the 18th century, the development of the calculus using infinitesimals was attacked by the philosopher Bishop Berkeley. He declared that the calculus involved a logical fallacy: infinitesimals were treated as though they were equal to 0 and not equal to 0. Bishop Berkeley pointed out that the usual way to calculate derivatives, requires one to first assume a non-zero infinitesimal and then after a division by this infinitesimal, one gets the result by setting it to 0. So two assumptions in direct contradiction were being used. Bishop Berkeley comments:

> And what are these same evanescent increments? They are neither finite quantities, nor quantities infinitely small, nor yet nothing. May we not call them ghosts of departed quantities?

Berkeley's criticisms were well-founded and, as a result, mathematicians were motivated to abandon infinitesimals and to pursue a logically correct development of the calculus. It should be noted that Euler discovered many important theorems that are taught in calculus courses. Euler established his results by using infinitesimals in a liberal and free-swinging manner. As long as the infinitesimal method led to correct results, Euler (and others) argued that the use of infinitesimals must be fundamentally sound. Others, however, were much more skeptical and insisted that the infinitesimal concept, which produces contradictions, should not be used to validate the calculus.

Mathematicians eventually discovered a more sophisticated development of the calculus, one that avoids the use of infinitesimals. In the 19th century, the foundations of calculus were eventually formulated axiomatically in terms of limits rather than infinitesimals and thus, the debate subsided. Cauchy in the early 19th century and Weierstrass, Dedekind, and Riemann in the late 19th century presented this rigorous development of the calculus. It is their ideas and proofs that we will cover here, in this real analysis text. We will also examine the real numbers, sets of real numbers, and real-valued functions that are defined on sets of real numbers. Moreover, we will provide a theoretical foundation that validates the fundamental principles of the calculus.

We now begin this validation by identifying the intrinsic properties that characterize the real number system. In particular, we will focus on the fact that this system satisfies the ordered field axioms and the completeness axiom. Using these axioms and the fundamental topics discussed in Chapter 1, we can correctly derive all of the theorems of the calculus.

2.2 $\mathbb{R}$ IS AN ORDERED FIELD

Among the first topics covered in a typical real analysis course are the ordered field axioms. These axioms form a basis for the algebraic operations and order properties upon which the calculus is based. At this point, the real number system $(\mathbb{R}, +, \cdot, <)$ should be quite familiar. Moreover, the algebraic operations of this number system are typically used correctly by most students of mathematics. However, in formal mathematics, it is important to identify the essential algebraic properties of the real number system that can be used to derive all of its other algebraic properties. These essential properties are called the *field axioms*. Of course, we will not derive *all* of the familiar algebraic properties from the field axioms, but we will demonstrate how this can be done with a few examples.

Field Axioms. The two binary operations + (addition) and · (multiplication) on the set of real numbers $\mathbb{R}$ satisfy the following axioms (properties):

A1. For all $x, y \in \mathbb{R}$, $x + y = y + x$.

A2. For all $x, y, z \in \mathbb{R}$, $x + (y + z) = (x + y) + z$.

A3. There is a unique number 0 such that for all $x \in \mathbb{R}$, $x + 0 = x$.

A4. For all $x \in \mathbb{R}$, there exists a unique $y \in \mathbb{R}$ such that $x + y = 0$. $(y = -x)$

M1. For all $x, y \in \mathbb{R}$, $x \cdot y = y \cdot x$.

M2. For all $x, y, z \in \mathbb{R}$, $x \cdot (y \cdot z) = (x \cdot y) \cdot z$.

M3. There is a unique number 1 such that $1 \neq 0$ and for all $x \in \mathbb{R}$, $x \cdot 1 = x$.

M4. For all nonzero $x \in \mathbb{R}$, there is a unique $y \in \mathbb{R}$ such that $x \cdot y = 1$. $(y = x^{-1})$

D1. For all $x, y, z \in \mathbb{R}$, $x \cdot (y + z) = (x \cdot y) + (x \cdot z)$.

Axioms A1 and M1 are the familiar *commutative laws*; Axioms A2 and M2 are the *associate laws*; and Axioms A3 and M3 assert that additive and multiplicative identity elements exist. The existence of additive and multiplicative inverse elements is declared, respectively, by Axioms A4 and M4. Finally, Axiom D1 is called the *distributive law*.

Again, the field axioms imply all of the well-known algebraic relationships that hold for the real numbers. For example, note that A3 implies that $0 + 0 = 0$; and M3 implies that $1 \cdot 1 = 1$. Moreover, by A4, -0 is the *unique* real number such that $0 + (-0) = 0$. Since $0 + 0 = 0$, we conclude that $-0 = 0$.

The proof of our next theorem demonstrates how these axioms can be used to derive other familiar properties. As this proof is presented only to illustrate the deductive strength of the field axioms, it can be bypassed without loss of continuity.

Theorem 2.2.1. For all $x, y, z \in \mathbb{R}$, the following items hold:

(a) If $x + z = y + z$, then $x = y$.   (d) $x \cdot y = 0$ if and only if $x = 0$ or $y = 0$.

(b) $x \cdot 0 = 0$.   (e) $(-x)y = -xy$.

(c) $(-1) \cdot x = -x$.   (f) $(-x)(-y) = xy$.

Proof. Let $x, y, z \in \mathbb{R}$. We now prove items (a)–(d), and leave (e) and (f) as exercises.

(a) Assume that $(\star)$ $x + z = y + z$. By A4, let $-z$ be such that $(\blacktriangle)$ $z + (-z) = 0$. Then

$$
\begin{aligned}
(x + z) + (-z) &= (y + z) + (-z) && \text{by } (\star) \\
x + (z + (-z)) &= y + (z + (-z)) && \text{by A2} \\
x + 0 &= y + 0 && \text{by } (\blacktriangle) \\
x &= y && \text{by A4.}
\end{aligned}
$$

(b) By A3, we have that $0 + 0 = 0$. Thus,

$$
\begin{aligned}
x \cdot (0 + 0) &= x \cdot 0 && \text{by } (\star) \\
x \cdot 0 + x \cdot 0 &= x \cdot 0 && \text{by D1} \\
x \cdot 0 + x \cdot 0 &= x \cdot 0 + 0 && \text{by A3} \\
x \cdot 0 &= 0 && \text{by (a).}
\end{aligned}
$$

(c) We show that $x + (-1) \cdot x = 0$. Axiom A4 will then imply that $(-1) \cdot x = -x$.

$$
\begin{aligned}
x + (-1) \cdot x &= x + x \cdot (-1) && \text{by M1} \\
&= x \cdot 1 + x \cdot (-1) && \text{by M3} \\
&= x \cdot (1 + (-1)) && \text{by D1} \\
&= x \cdot 0 && \text{by A4} \\
&= 0 && \text{by (b).}
\end{aligned}
$$

(d) If $x = 0$, then $x \cdot y = 0$ by (b). If $y = 0$, then $x \cdot y = 0$ by M1 and (b). Assume that $x \cdot y = 0$. Then Exercise 1 shows that either $x = 0$ or $y = 0$.   □

In addition to the field axioms, the set of real numbers also satisfies four order axioms from which one can derive, together with the field axioms, all of the ordering properties that hold in the real number system.

Order Axioms. The ordering relation $<$ (less than) on the set of real numbers $\mathbb{R}$ satisfies the following axioms:

O1. For all $x, y \in \mathbb{R}$, exactly one of the following holds: $x < y$, $y < x$, or $x = y$.

O2. For all $x, y, z \in \mathbb{R}$, if $x < y$ and $y < z$, then $x < z$.

O3. For all $x, y, z \in \mathbb{R}$, if $x < y$, then $x + z < y + z$.

O4. For all $x, y, z \in \mathbb{R}$, if $x < y$ and $z > 0$, then $x \cdot z < y \cdot z$.

Axiom O1 is called the *trichotomy law*, and Axiom O2 is said to be the *transitive law*. Because the real number system $(\mathbb{R}, +, \cdot, <)$ satisfies the field and order axioms, it is said to be an *ordered field*.

In the next theorem, we show that two key properties of inequality are derivable from the field and order axioms. Again, a proof is given only to support the contention that these axioms imply all the familiar properties of the real numbers. It too can be bypassed.

Theorem 2.2.2. For all $x, y, z \in \mathbb{R}$, the following items hold:

(1) $x < y$ if and only if $-y < -x$.   (2) If $x < y$ and $z < 0$, then $x \cdot z > y \cdot z$.

Proof. Let $x, y, z \in \mathbb{R}$. We now prove items (1) and (2).

(1) Assume that (▼) $x < y$. We show that $-y < -x$, as follows:

$$
\begin{aligned}
x + ((-x) + (-y)) &< y + ((-x) + (-y)) && \text{by (▼) and O3} \\
x + ((-x) + (-y)) &< y + ((-y) + (-x)) && \text{by A1} \\
(x + (-x)) + (-y) &< (y + (-y)) + (-x) && \text{by A2} \\
0 + (-y) &< 0 + (-x) && \text{by A3} \\
-y &< -x && \text{by A1 and A3.}
\end{aligned}
$$

The converse follows similarly.

(2) Assume $x < y$ and $z < 0$. Then $0 < -z$ by (1) and the identity $-0 = 0$. So,

$$
\begin{aligned}
x &< y && \text{by assumption} \\
x \cdot (-z) &< y \cdot (-z) && \text{by O4} \\
x \cdot ((-1)z) &< y \cdot ((-1)z) && \text{by Theorem 2.2.1(c)} \\
(-1)(x \cdot z) &< (-1)(y \cdot z) && \text{by M2 and M2} \\
-(x \cdot z) &< -(y \cdot z) && \text{by Theorem 2.2.1(c)} \\
y \cdot z &< x \cdot z && \text{by (1).} && \square
\end{aligned}
$$

Recall the standard notation connected with the ordering relation $<$ (less than): $x > y$ means that $y < x$; $x \leq y$ means that either $x < y$ or $x = y$; and $x \geq y$ means that either $x > y$ or $x = y$.

This ends our discussion of the field and order axioms. In the remaining body of the text, we will not directly cite any of the above theorems and we will assume that the reader is familiar with the properties of equality and inequality discussed in this section and in Section 1.1.3. However, before moving on to the next topic, we need to identify one unfamiliar property of inequality that is often applied in real analysis.

Theorem 2.2.3. Let $x, y \in \mathbb{R}$. Suppose $x \le y + \varepsilon$ for all $\varepsilon > 0$. Then $x \le y$.

Proof. Let $x, y \in \mathbb{R}$ and assume that $x \le y + \varepsilon$ for all $\varepsilon > 0$. We will prove that $x \le y$. Suppose, for a contradiction, that $x > y$. Thus, $x - y > 0$ and $\varepsilon = \frac{x-y}{2} > 0$. So, $x \le y + \varepsilon$ by our assumption. Since $\varepsilon = \frac{x-y}{2} < x - y$, we obtain

$$x \le y + \varepsilon = y + \left(\frac{x-y}{2}\right) < y + (x - y) = x. \tag{2.1}$$

Therefore, from (2.1) we conclude that $x < x$, a contradiction. □

Corollary 2.2.4. If $0 \le x$ and $x \le \varepsilon$ for all $\varepsilon > 0$, then $x = 0$.

2.2.1 The Absolute Value Function

A very important function on the real numbers is the *absolute value function*. Many of the proofs in real analysis apply this function and its properties. In this section, we will identify and confirm these properties. The definition of the absolute value was given in Section 1.1.4, but we now repeat it here for clarity.

Definition 2.2.5 (Absolute Value). Given a real number x, the **absolute value** of x, denoted by $|x|$, is defined by

$$|x| = \begin{cases} x, & \text{if } x \ge 0; \\ -x, & \text{if } x < 0. \end{cases}$$

For any real number x, the number $|x|$ can be viewed as the distance, on the real line, from x to the point 0. Our next theorem identifies the basic properties of the absolute value function that will be regularly applied throughout the text.

Theorem 2.2.6 (Basic Properties of Absolute Value). For all $a, x \in \mathbb{R}$, where $a \ge 0$, the following hold:

(a) $0 \le |x|$, $x \le |x|$, $-x \le |x|$, $|-x| = |x|$.

(b) $|x| = 0$ if and only if $x = 0$.

(c) $|x| \le a$ if and only if $-a \le x \le a$.

(d) $|xy| = |x||y|$.

(e) $|x|^2 = x^2$.

(f) $|x + y| \le |x| + |y|$.

Proof. Items (a)-(e) follow directly from Definition 2.2.5. We shall now prove (f) as follows: Since $|x + y|^2 = (x + y)^2$ by (e), we conclude from (a), (d), and (e) that

$$|x + y|^2 = x^2 + 2xy + y^2 \le x^2 + |2xy| + y^2 = |x|^2 + 2|x||y| + |y|^2 = (|x| + |y|)^2.$$

Thus, $|x + y|^2 \le (|x| + |y|)^2$. Theorem 1.1.11 implies that $|x + y| \le |x| + |y|$. □

Item (f) of Theorem 2.2.6 is called the **triangle inequality**. If we view real numbers as points on the real line, then $|a - b|$ is the distance between the points a and b. Let c be another point. From the triangle inequality, we can deduce that

$$|a - b| = |a - c + c - b| \le |a - c| + |c - b|. \tag{2.2}$$

Thus, $|a - b| \leq |a - c| + |c - a|$ holds for any real numbers a, b, c. As we will see, the derivation in (2.2) is often applied in real analysis. Here are a few more properties that are regularly employed.

Theorem 2.2.7 (More Properties of Absolute Value). For all $x, y, k \in \mathbb{R}$, where $k > 0$, we have

(i) $|x| < k$ if and only if $-k < x < k$.   (iv) $|y| - |x| \leq |x - y|$.

(ii) $|x| > k$ if and only if $x < -k$ or $x > k$.   (v) $\big| |x| - |y| \big| \leq |x - y|$.

(iii) $|x| - |y| \leq |x - y|$.

Proof. Items (i)-(ii) follow easily from Definition 2.2.5. Items (iii) and (iv) imply (v), using Definition 2.2.5. We now prove (iii) and (iv). First we prove (iii). Observe that $|x| = |x - y + y|$. Hence, the triangle inequality implies that

$$|x| = |x - y + y| \leq |x - y| + |y|.$$

So, $|x| \leq |x - y| + |y|$ and thus, $|x| - |y| \leq |x - y|$. To prove item (iv), note that $|x - y| = |y - x|$. Thus, (iii) implies that $|y| - |x| \leq |x - y|$. □

Items (iii)-(v) of Theorem 2.2.7 will all be referred to as the **backward triangle inequality**. The triangle inequality and the three versions of the backward triangle inequality are important tools that will be used throughout the book.

Given a finite nonempty set of real numbers A, we let $\max A$, or $\max(A)$, denote the maximum number in A. Similarly, we define $\min A$, or $\min(A)$, to be the minimum number in A. For example, $\max\{-1, 2, \pi, 3\} = \pi$ and $\min\{-1, 2, \pi, 3\} = -1$.

The next four results on the absolute value function are applied later in the text.

Lemma 2.2.8. Let x, a, and b be real numbers. If $a \leq x \leq b$, then $|x| \leq \max\{|a|, |b|\}$.

Proof. Assume that $a \leq x \leq b$. First suppose that $x \geq 0$. So, as $x \leq b$, we conclude that $b \geq 0$. So, $|x| = x$ and $|b| = b$. Since $x \leq b$, we see that $|x| \leq |b|$ and therefore, $|x| \leq \max\{|a|, |b|\}$. Now suppose that $x < 0$. Then, because $a \leq x$, we see that $a < 0$. So, $|x| = -x$ and $|a| = -a$. Since $a \leq x$, we see that $-x \leq -a$ and so, $|x| \leq |a|$. Therefore, $|x| \leq \max\{|a|, |b|\}$. □

Lemma 2.2.9. Let $a \leq x \leq b$ and $a \leq y \leq b$. Then $|x - y| \leq |b - a|$.

Proof. Assume (▲) $a \leq x \leq b$; and $a \leq y \leq b$. Thus, (◆) $-b \leq -y \leq -a$. Adding (▲) and (◆), we obtain $-(b - a) \leq x - y \leq (b - a)$. Thus, $|x - y| \leq |b - a|$ by Theorem 2.2.6(c). □

Corollary 2.2.10. Let a, b, and L be real numbers.

(1) If $b \geq 0$ and $a \leq L \leq a + b$, then $|L - (a + b)| \leq |b|$.

(2) If $b \leq 0$ and $a + b \leq L \leq a$, then $|L - (a + b)| \leq |b|$.

Proof. To prove (1), let $b \geq 0$ and assume that $a \leq L \leq a + b$. Since $a \leq a + b \leq a + b$, Lemma 2.2.9 implies that $|L - (a + b)| \leq |a + b - a| = |b|$. The proof of item (2) is left for Exercise 14. □

Lemma 2.2.11. Let $\varepsilon > 0$ and $\ell \in \mathbb{R}$. Suppose that $a \leq b \leq c$. If $|a - \ell| < \varepsilon$ and $|c - \ell| < \varepsilon$, then $|b - \ell| < \varepsilon$.

Proof. Let $\varepsilon > 0$, $\ell \in \mathbb{R}$, and $a \leq b \leq c$. Assume that $|a - \ell| < \varepsilon$ and $|c - \ell| < \varepsilon$. Theorem 2.2.7(i) implies that $\ell - \varepsilon < a < \ell + \varepsilon$ and $\ell - \varepsilon < c < \ell + \varepsilon$. Since $a \leq b \leq c$, it follows that $\ell - \varepsilon < b < \ell + \varepsilon$. Thus, $|b - \ell| < \varepsilon$. □

Exercises 2.2

*1. Assume that $x \cdot y = 0$. To show that either $x = 0$ or $y = 0$, assume that $x \neq 0$. Thus, $x \cdot x^{-1} = 1$ by axiom M4. Justify the steps below, as in the proof of Theorem 2.2.1, to show that $y = 0$.

$$y = y \cdot 1 \qquad \rule{3cm}{0.4pt}$$
$$= 1 \cdot y \qquad \rule{3cm}{0.4pt}$$
$$= (x \cdot x^{-1}) \cdot y \qquad \rule{3cm}{0.4pt}$$
$$= (x^{-1} \cdot x) \cdot y \qquad \rule{3cm}{0.4pt}$$
$$= x^{-1} \cdot (x \cdot y) \qquad \rule{3cm}{0.4pt}$$
$$= x^{-1} \cdot 0 \qquad \rule{3cm}{0.4pt}$$
$$= 0 \qquad \rule{3cm}{0.4pt}$$

2. Prove (e) of Theorem 2.2.1.

3. Prove (f) of Theorem 2.2.1.

4. Prove Corollary 2.2.4.

5. Prove (c) of Theorem 2.2.6.

6. Prove (i) to (iii) of Theorem 2.2.7.

7. Justify the equalities and inequalities used in the proof of Theorem 2.2.6(f).

8. Let x and y be real numbers. Show that $|x| - |y| \leq |x + y|$.

9. Let $a > 0$. Prove that if $|x - a| < a$, then $x > 0$.

10. Suppose that $x > 0$ and $y > 0$. Prove that $x^2 + y^2 < (x + y)^2$.

11. Suppose that $x < 0$ and $y < 0$. Prove that $x^2 + y^2 < (x + y)^2$.

12. Suppose that $x > 0$ and $y > 0$. Prove that $\sqrt{x + y} < \sqrt{x} + \sqrt{y}$.

13. Let $a < x < b$ and $a < y < b$. Prove that $|x - y| < |b - a|$.

*14. Prove Corollary 2.2.10(2).

15. Let x be between a and b. Let $c \in \mathbb{R}$. Show that $|x - c| \leq \max\{|a - c|, |b - c|\}$.

16. Let x and y be real numbers. Show that the following hold:

(a) $\max\{x, y\} = \frac{1}{2}(x + y) + \frac{1}{2}|x - y|$.   (b) $\min\{x, y\} = \frac{1}{2}(x + y) - \frac{1}{2}|x - y|$.

17. Let $x \in \mathbb{R}$, $c \in \mathbb{R}$, and $\varepsilon > 0$. Suppose that $|x - c| < \varepsilon$.

   (a) Prove that $|x| < \varepsilon + |c|$.   (b) Prove that $|c| - \varepsilon < |x|$.

Exercise Notes: For Exercise 9, use Theorem 2.2.7(i). For Exercise 12, use proof by contradiction and Theorem 1.1.10. For Exercise 15, apply Lemma 2.2.8.

2.3   THE COMPLETENESS AXIOM

In this section we present the completeness axiom, which is an assertion that implies that there are no gaps or holes in the real number line. We shall see that this axiom identifies an essential property of the real number system that will allow us to prove the most important theorems in the calculus; for example, the Cauchy Convergence Criterion, the Intermediate Value Theorem, and the Fundamental Theorem of Calculus. Before stating the completeness axiom, we need a few preliminary definitions.

Definition 2.3.1 (Upper Bound and Lower Bound). Let $S \subseteq \mathbb{R}$ be nonempty.

- Suppose there is a $b \in \mathbb{R}$ such that every real number in S is less than or equal to b, that is, for all $x \in S$, we have $x \leq b$. Then we shall say that b is an **upper bound** for S and that S is **bounded above.**

- Suppose there is an $a \in \mathbb{R}$ such that a is less than or equal to every real number in S, that is, for all $x \in S$, we have $a \leq x$. Then we will say that a is a **lower bound** for S and that S is **bounded below.**

- If S has both a lower bound and an upper bound, then we say that S is **bounded.**

Let $S = \{\frac{2}{n} : n \in \mathbb{N}\}$ (see Figure 2.1). Since $\frac{2}{n} \leq 3$ for all $n \in \mathbb{N}$, we see that the set S is bounded above. Furthermore, because $0 \leq \frac{2}{n}$ for all $n \in \mathbb{N}$, the set S is bounded below.

Figure 2.1: The set $\{\frac{2}{n} : n \in \mathbb{N}\}$ plotted on the real line.

We present a few more examples.

1. Let $a, b \in \mathbb{R}$ be such that $a < b$. Then b is an upper bound and a is a lower bound for each of the sets (a, b), $[a, b)$, $[a, b]$.

2. $\mathbb{Z}^+$ and $\mathbb{Q}^+$ are not bounded above.

3. -2 is a lower bound for the sets $\mathbb{Z}^+$ and $\mathbb{Q}^+$.

4. The set $\{\frac{x}{x+1} : x \in \mathbb{Q}^+\}$ is bounded, because $0 < \frac{x}{x+1} < 1$ for all $x \in \mathbb{Q}^+$.

5. The set $\{x \in \mathbb{R} : x < 0\}$ is bounded above, but it is not bounded.

The following result is often applied to show that a set of real numbers is bounded.

Theorem 2.3.2. Let $S \subseteq \mathbb{R}$ be nonempty. Then S is bounded if and only if there is an $M > 0$ so that $|x| \leq M$ for all $x \in S$.

Proof. Let $S \subseteq \mathbb{R}$ be nonempty. Assume that S is bounded. Thus, there are nonzero real numbers a, b such that $a \leq x \leq b$ for all $x \in S$. Let $M = \max\{|a|, |b|\} > 0$. By Lemma 2.2.8, $|x| \leq M$ for all $x \in S$. To prove the converse, assume that $|x| \leq M$ for all $x \in S$, where $M > 0$. Thus, $-M \leq x \leq M$ for all $x \in S$. So S is bounded. □

For example, let $S = \{x \in \mathbb{R} : |3x - 1| < 5\}$. We will apply Theorem 2.3.2 to show that S is bounded. For each $x \in S$ we have that $|3x - 1| < 5$. Thus, by the backward triangle inequality, we conclude that $|3x| - 1 \leq |3x - 1| < 5$. Hence, $3|x| - 1 < 5$ and so, $|x| < 2$. Therefore, $|x| < 2$ for all $x \in S$ and S is bounded.

A set S that is not bounded is said to be **unbounded**. By Theorem 2.3.2, a set S is unbounded if and only if for all $M > 0$ there is an $x \in S$ such that $|x| > M$.

Let b be an upper bound for a nonempty set S. So, $x \leq b$ for all $x \in S$. It follows that any number greater than b is also an upper bound for S. Our next definition will allow us to focus on the smallest, or least, upper bound for S.

Definition 2.3.3 (Least Upper Bound and Greatest Lower Bound). Let $S \subseteq \mathbb{R}$ be nonempty.

- Suppose that β is an upper bound for S. If β is the **least upper bound** for S, then β is said to be the **supremum** of S and we write $\beta = \sup(S)$.

- Suppose that α is a lower bound for S. If α is the **greatest lower bound** for S, then α is called the **infimum** of S and we write $\alpha = \inf(S)$.

Example. Consider the set $S = \{\frac{2}{n} : n \in \mathbb{N}\}$ (see Figure 2.1). As we saw earlier, the set S is bounded. We note that 0 is the greatest lower bound for S and that 2 is the least upper bound for S. Thus, $\inf(S) = 0$ and $\sup(S) = 2$.

Let $S \subseteq \mathbb{R}$ be nonempty. The equation $\beta = \sup(S)$ means that the following two items hold: (a) β is an upper bound for S and (b) β is the smallest upper bound for S. Similarly, the equation $\alpha = \inf(S)$ means that (a) α is a lower bound for S and (b) α is the largest lower bound for S. The next remark repeats and clarifies these observations. This remark will be used to identify two important proof strategies.

Remark 2.3.4. Let $S \subseteq \mathbb{R}$ be nonempty and let $\alpha, \beta \in \mathbb{R}$. Then we have the following (see Figure 2.2):

1. $\beta = \sup(S)$, if and only if the following two conditions hold:

    (a) For all $x \in S$, $x \leq \beta$.
    (b) For all real numbers b, if b is an upper bound for S, then $\beta \leq b$.

2. $\alpha = \inf(S)$, if and only if the following two conditions hold:

    (a) For all $x \in S$, $\alpha \leq x$.
    (b) For all real numbers a, if a is a lower bound for S, then $a \leq \alpha$.

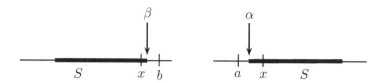

Figure 2.2: Illustrations for parts 1 and 2 of Remark 2.3.4, respectively.

Before we continue, we need to distinguish between the notions of a maximum element and the supremum. We have defined the maximum and minimum elements of a nonempty finite set of real numbers (see page 34). Now, let S be an infinite set of real numbers. Suppose that there exists an $m \in S$ such that $x \leq m$ for all $x \in S$. Thus, m is a maximum element of S. Also observe that $m = \sup(S)$. It turns out that S has a maximum element if and only if $\sup(S) \in S$. Thus, if $\sup(S) \notin S$, then S does not have a maximum element. A similar observation can be made concerning the minimum and infimum of S.

Definition 2.3.5 (Maximum Element and Minimum Element). Let $S \subseteq \mathbb{R}$.

- Let $\beta = \sup(S)$. If $\beta \in S$, then β is the **maximum element** of S, denoted by $\beta = \max(S)$.

- Let $\alpha = \inf(S)$. If $\alpha \in S$, then α is the **minimum element** of S, denoted by $\alpha = \min(S)$.

For example, let $S = [2,5]$. Because $\sup(S) = 5$ and $5 \in S$, we observe that $\max(S) = 5$. Since $\inf(S) = 2$ and $2 \in S$, we see that $\min(S) = 2$. For another example, let $T = (2,5)$. Because $\sup(T) = 5$ and $5 \notin T$, we have that $\max(T)$ is undefined; that is, T does not have a maximum element. Similarly, $\min(T)$ is undefined. Finally, let $S = \{\frac{2}{n} : n \in \mathbb{N}\}$ (see Figure 2.1). We see that $\max(S) = 2$ and $\min(S)$ is undefined.

We are now ready to present one of the most important properties of the set of real numbers, namely, the completeness axiom. This axiom ensures that every nonempty set of real numbers that is bounded above has a supremum. As we will see, the completeness axiom forms the basis for all the limit concepts that are applied throughout the text.

Completeness Axiom. Every nonempty set of real numbers that is bounded above has a least upper bound.

The completeness axiom just asserts that if a nonempty set $S \subseteq \mathbb{R}$ is bounded above, then there is a real number β that satisfies the equation $\beta = \sup(S)$. Thus, the completeness axiom implies that there are no gaps, or "missing points," in the set of real numbers; however, there are number systems that do have gaps. For example, let $\ell \in \mathbb{R}$, $\mathbb{R}^* = \mathbb{R} \setminus \{\ell\}$, and $S \subseteq \mathbb{R}^*$ be defined by $S = \{x \in \mathbb{R}^* : x < \ell\}$. So $\sup(S) = \ell$ and $\ell \notin \mathbb{R}^*$. Thus, $\mathbb{R}^*$ has a gap and $\mathbb{R}^*$ does not satisfy the completeness axiom. Moreover, the number system $\mathbb{Q}$ also has gaps (see Remark 2.4.7). Thus, $\mathbb{Q}$ fails to satisfy the completeness axiom as well.

Of course, the completeness axiom is very important; but, rather than just accept the axiom, can one actually prove it? The answer is yes. The proof requires one to first construct the set $\mathbb{R}$ of real numbers from the rational numbers. Such a construction is based on set theoretic tools that were independently developed by Georg Cantor and Richard Dedekind. Using these tools, Cantor and Dedekind were able to prove the completeness axiom. However, we will not present such a proof as it would take us too far astray. Thus, we will just accept the axiom and pursue its consequences.

2.3.1 Proofs on the Supremum of a Set

Remark 2.3.4(1) inspires a strategy for proving equations of the form $\sup(S) = \beta$.

Proof Strategy 2.3.6. Given a real number β and a nonempty $S \subseteq \mathbb{R}$, to prove that $\sup(S) = \beta$ we may use the two-step proof diagram:

Step (1): Prove $x \leq \beta$ for all $x \in S$.
Step (2): Assume b is an upper bound for S.
Prove $\beta \leq b$.

In other words, to prove that $\sup(S) = \beta$ you must first prove that β is an upper bound for S and then prove that β is the smallest upper for S. Similarly, Remark 2.3.4(1) yields a strategy that allows one to take advantage of an assumption having the form $\sup(S) = \beta$.

Assumption Strategy 2.3.7. Let $S \subseteq \mathbb{R}$ be nonempty and let $\beta \in \mathbb{R}$. Suppose that you are *assuming* that $\sup(S) = \beta$. Then (1) $x \leq \beta$ for all $x \in S$, and (2) whenever b is an upper bound for S, you can conclude that $\beta \leq b$.

Our proof of the next lemma employs both proof strategy 2.3.6 and assumption strategy 2.3.7. Given a set $S \subseteq \mathbb{R}$ and a real number $k \in \mathbb{R}$, we can form the new set of real numbers defined by $A = \{kx : x \in S\}$. The set A is often denoted by kS.

Lemma 2.3.8. Let $S \subseteq \mathbb{R}$ be nonempty and bounded above, and let $k \geq 0$. Define the set A by $A = \{kx : x \in S\}$. Then the set A is bounded above and $\sup(A) = k\sup(S)$.

Proof Analysis. Let $S \subseteq \mathbb{R}$ be bounded above. Hence, $\sup(S) = \beta$ exists by the completeness axiom. Let $k \geq 0$. We must prove that $\sup(A) = k\beta$. There are two cases to consider: $k = 0$ and $k > 0$. When $k = 0$ the proof follows easily (see proof below). Let $k > 0$. First note that every element in A has the form kx for some $x \in S$. So to prove that $k\beta$ is an upper bound for A, we must show that $kx \leq k\beta$ for all $x \in S$. Appealing to proof strategy 2.3.6, we construct the following proof diagram:

Assume $\beta = \sup(S)$.
Prove $kx \leq k\beta$ for all $x \in S$.
Assume c is an upper bound for A.
Prove $k\beta \leq c$.

Given that $\beta = \sup(S)$, we can use assumption strategy 2.3.7. Also note that if c is an upper bound for A, then $kx \leq c$ for all $x \in S$. We now have all of the ingredients required to write a correct proof of the lemma.

Proof of Lemma 2.3.8. Let $S \subseteq \mathbb{R}$ be nonempty and bounded above. Thus, by the completeness axiom, $\beta = \sup(S)$ exists. Thus, $x \le \beta$ for all $x \in S$. Let $k \ge 0$. We note that every element of $A = \{kx : x \in S\}$ has the form kx for some $x \in S$. There are two cases to consider.

CASE 1: Assume $k = 0$. Then $A = \{0\}$ and we clearly have $\sup(A) = k \sup(S)$.

CASE 2: Assume $k > 0$. We first prove that $k\beta$ is an upper bound for A. Since $x \le \beta$ for all $x \in S$ and $k > 0$, we have that $kx \le k\beta$ for all $x \in S$. Therefore, $k\beta$ is an upper bound for A. Let c be an upper bound for A. We shall prove that $k\beta \le c$. Since c is an upper bound for A, we have that $kx \le c$ for all $x \in S$. As $k > 0$, we see that $x \le \frac{c}{k}$ for all $x \in S$. Thus, $\frac{c}{k}$ is an upper bound for S. Since β is the smallest upper bound for S, we conclude that $\beta \le \frac{c}{k}$ and so, $k\beta \le c$. Therefore, $\sup(A) = k\beta$. □

2.3.2 Proofs on the Infimum of a Set

Remark 2.3.4(2) motivates a strategy for proving equations of the form $\inf(S) = \alpha$.

Proof Strategy 2.3.9. Given a real number α and a nonempty $S \subseteq \mathbb{R}$, to prove that $\inf(S) = \alpha$ we may use the two-step proof diagram:

Step (1): Prove $\alpha \le x$ for all $x \in S$.
Step (2): Assume a is a lower bound for S.
Prove $a \le \alpha$.

Let S be a nonempty set of real numbers that is bounded below. The completeness axiom makes no explicit assertion that the infimum of S exists; however, using the axiom one can prove that this infimum exists. In particular, the proof of the following theorem shows that the completeness axiom implies that there is a real number α that satisfies the equation $\alpha = \inf(S)$. Before reading the proof of this theorem, one should read Remark 2.3.4 and proof strategy 2.3.9.

Theorem 2.3.10. Every nonempty $S \subseteq \mathbb{R}$ that is bound below, has a greatest lower bound.

Proof. Let $S \subseteq \mathbb{R}$ be nonempty with lower bound a. Let $S^* = \{-x : x \in S\}$. Then every element of S^* has the form $-x$ for an $x \in S$. Since a is an lower bound for S, we have $a \le x$ for all $x \in S$. Thus, $-x \le -a$ for all $x \in S$. Hence, $-a$ is an upper bound for S^*. By the completeness axiom, $\sup(S^*) = \beta$ exists. Thus, $-x \le \beta$ for all $x \in S$, and β is the smallest such upper bound.

We now prove that $\inf(S) = -\beta$. First we prove that $-\beta$ is a lower bound for S. As $-x \le \beta$ for all $x \in S$, it follows that $-\beta \le x$ for all $x \in S$. Thus, $-\beta$ is a lower bound for S. To prove that $-\beta$ is the largest lower bound for S, let a be a lower bound for S. By the argument in the first paragraph, we see that $-a$ is an upper bound for the set S^*. Since β is the least such upper bound for S^*, it follows that $\beta \le -a$. Hence, $a \le -\beta$. Therefore, $\inf(S) = -\beta$. □

Assumption Strategy 2.3.11. Let $S \subseteq \mathbb{R}$ be nonempty and let $\alpha \in \mathbb{R}$. Suppose that you are *assuming* that $\inf(S) = \alpha$. Then (1) $\alpha \le x$ for all $x \in S$, and (2) whenever a is a lower bound for S, you can conclude that $a \le \alpha$.

The proof of the next lemma shows how to employ strategies 2.3.9 and 2.3.11. Moreover, this proof is analogous to Case 2 in the proof of Lemma 2.3.8; however, "the inequalities are reversed."

Lemma 2.3.12. Let $S \subseteq \mathbb{R}$ be nonempty and bounded below, and let $k > 0$. Let $A = \{kx : x \in S\}$. Then A is bounded below and $\inf(A) = k \inf(S)$.

Proof Analysis. Since every element of $A = \{kx : x \in S\}$ has the form kx for an $x \in S$, the logical structure of our proof will be as follows:

$$\text{Assume } \alpha = \inf(S).$$
$$\text{Prove } k\alpha \le kx \text{ for all } x \in S.$$
$$\text{Assume } c \text{ is an lower bound for } A.$$
$$\text{Prove } c \le k\alpha.$$

We now present a correct proof of the lemma.

Proof. Let $S \subseteq \mathbb{R}$ be nonempty and bounded below. Theorem 2.3.10 implies that S has a greatest lower bound α. So $\alpha = \inf(S)$. Thus, $\alpha \le x$ for all $x \in S$. Let $k > 0$. First we prove that $k\alpha$ is a lower bound for A. Since $\alpha \le x$ for all $x \in S$ and $k > 0$, we have that $k\alpha \le kx$ for all $x \in S$. Therefore, $k\alpha$ is a lower bound for A.

Let c be a lower bound for A. We will prove that $c \le k\alpha$. As c is a lower bound for A, we see that $c \le kx$ for all $x \in S$. Since $k > 0$, we have that $\frac{c}{k} \le x$ for all $x \in S$. Thus, $\frac{c}{k}$ is a lower bound for S. As α is the largest lower bound for S, we conclude that $\frac{c}{k} \le \alpha$. Hence, $c \le k\alpha$ and therefore, $\inf(A) = k\alpha$. □

We can now formally combine the results of Lemmas 2.3.8 and 2.3.12.

Theorem 2.3.13. Let $S \subseteq \mathbb{R}$ be nonempty and bounded, and let $k \in \mathbb{R}$. Define the set kS by $kS = \{kx : x \in S\}$. Then the set kS is bounded and

(a) if $k \ge 0$, then $\sup(kS) = k \sup(S)$ and $\inf(kS) = k \inf(S)$,
(b) if $k < 0$, then $\sup(kS) = k \inf(S)$ and $\inf(kS) = k \sup(S)$.

Proof. Lemmas 2.3.8 and 2.3.12 imply (a). The proof of (b) is left for Exercise 9. □

Our attention will now turn to a concept that plays a prominent role in real analysis, namely, the concept of a bounded function.

Definition 2.3.14. A function $f : D \to \mathbb{R}$ is **bounded** if the set

$$f[D] = \{f(x) : x \in D\}$$

is bounded. If $S \subseteq D$ and $f[S]$ is bounded, then f is said to be **bounded on** S.

If $f : D \to \mathbb{R}$ is bounded, then $\beta = \sup(f[D])$ and $\alpha = \inf(f[D])$ exist, and $\alpha \le f(x) \le \beta$ for all $x \in D$ (see Figure 2.3).

For example, let $D = (0, 2)$ and $f : D \to \mathbb{R}$ be defined by $f(x) = x^2$. Then f is bounded. Furthermore, $f[D] = (0, 4)$, $\sup(f[D]) = 4$, and $\inf(f[D]) = 0$. On the other hand, the function $g : D \to \mathbb{R}$ defined by $g(x) = \frac{1}{x}$ is not bounded.

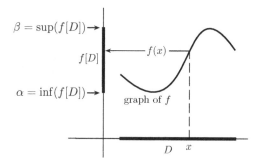

Figure 2.3: Illustration of a bounded function.

Remark 2.3.15. For a function $f\colon D \to \mathbb{R}$, Theorem 2.3.2 and the definition of $f[D]$ imply that there are two equivalent ways of saying that f is bounded; namely:

(1) there are real numbers a, b such that $a \le f(x) \le b$ for all $x \in D$,

(2) there is a real number $M > 0$ such that $|f(x)| \le M$ for all $x \in D$.

Our next theorem establishes a supremum and infimum relationship between two bounded functions and their sum. This relationship will be used to prove a sum rule for integration (see Lemma 6.2.4).

Theorem 2.3.16. Let $f\colon D \to \mathbb{R}$ and $g\colon D \to \mathbb{R}$ be bounded. Define $(f+g)\colon D \to \mathbb{R}$ by $(f + g)(x) = f(x) + g(x)$ for all $x \in D$. Then $(f + g)\colon D \to \mathbb{R}$ is bounded, and

(a) $\sup((f + g)[D]) \le \sup(f[D]) + \sup(g[D])$,

(b) $\inf(f[D]) + \inf(g[D]) \le \inf((f + g)[D])$.

Proof. Let $f\colon D \to \mathbb{R}$ and $g\colon D \to \mathbb{R}$ be bounded. Let $M_1 > 0$ and $M_2 > 0$ be such that $|f(x)| \le M_1$ and $|g(x)| \le M_2$ for all $x \in D$. By the triangle inequality we obtain

$$|(f + g)(x)| = |f(x) + g(x)| \le |f(x)| + |g(x)| \le M_1 + M_2$$

for all $x \in D$. Hence, $(f + g)$ is bounded. We prove (b) and leave (a) for Exercise 17. Since $f[D]$, $g[D]$, and $(f + g)[D]$ are bounded, let $\varepsilon = \inf(f[D])$, $\delta = \inf(g[D])$, and $\gamma = \inf((f+g)[D])$. Thus, $(*)$ $\varepsilon \le f(x)$ and $(**)$ $\delta \le g(x)$, for all $x \in D$. Inequalities $(*)$ and $(**)$ imply that $\varepsilon + \delta \le f(x) + g(x)$ for all $x \in D$. Thus, $\varepsilon + \delta \le (f + g)(x)$ for all $x \in D$. Hence, $\varepsilon + \delta$ is a lower bound for $(f+g)[D]$. Therefore, $\varepsilon + \delta \le \gamma$, that is, $\inf(f[D]) + \inf(g[D]) \le \inf((f + g)[D])$. $\square$

The inequalities in Theorem 2.3.16 can be strict (see Exercise 12). We end this section with two more theorems that involve the supremum and infimum. Viewing Figure 2.4 below, the first theorem is very easy to believe and understand. This result will allow us to prove a result concerning the supremum and infimum of Riemann sums (see Theorem 6.1.17).

Theorem 2.3.17. Let $A \subseteq \mathbb{R}$ and $B \subseteq \mathbb{R}$ be non-empty. Suppose for all $x \in A$ and all $y \in B$, we have that $x \le y$. Then A is bounded above, B is bounded below, and $\sup(A) \le \inf(B)$.

Proof. Suppose that
$$x \leq y \text{ for all } x \in A \text{ and all } y \in B. \tag{2.3}$$

Let $y \in B$ be arbitrary. From (2.3), we have that $x \leq y$ for all $x \in A$. Thus, y is an upper bound for A (see Figure 2.4). The completeness axiom implies that $\gamma = \sup(A)$ exists. Since y is an upper bound for A and γ is the least such upper bound, we conclude that $\gamma \leq y$. Because $y \in B$ was arbitrarily chosen, it follows that $\gamma \leq y$ for all $y \in B$. Thus, γ is a lower bound for B. Theorem 2.3.10 implies that $\delta = \inf(B)$ exists. As γ is a lower bound for B and δ is the greatest such lower bound, we see that $\gamma \leq \delta$. Therefore, $\sup(A) \leq \inf(B)$. □

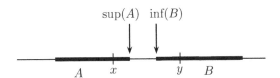

Figure 2.4: An illustration for the proof of Theorem 2.3.17.

The proof of Theorem 2.3.17 adapts to prove (see Exercise 16) our next result which will enable us to prove that a particular type of composite function is integrable (see Theorem 6.2.14).

Theorem 2.3.18. Let $A \subseteq \mathbb{R}$ be non-empty. Let $c \in \mathbb{R}$. Suppose for all $x \in A$ and all $y \in A$, we have that $x - y < c$. Then A is bounded and $\sup(A) - \inf(A) \leq c$.

2.3.3 Alternative Proof Strategies

The proofs of our earlier results on the supremum and infimum relied on Remark 2.3.4 in which we observed that the equation $\beta = \sup(S)$ means that (a) β is an upper bound for S and (b) β is the smallest upper bound for S. Given (a), another way to say (b) is that "every number $r < \beta$ is not an upper bound for S." A similar observation can be made concerning the infimum of a set. We formally record these observations in the following remark.

Remark 2.3.19. Let $S \subseteq \mathbb{R}$ be nonempty and let $\alpha, \beta \in \mathbb{R}$. Then (see Figure 2.5)

1. $\beta = \sup(S)$, if and only if the following two conditions hold:

   (a) For all $x \in S$, $x \leq \beta$.
   (b) For all real numbers r, if $r < \beta$, then there is an $x \in S$ so that $r < x$.

2. $\alpha = \inf(S)$, if and only if the following two conditions hold:

   (a) For all $x \in S$, $\alpha \leq x$.
   (b) For all real numbers q, if $\alpha < q$, then there is an $x \in S$ so that $x < q$.

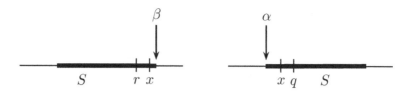

Figure 2.5: Illustrations for parts 1 and 2 of Remark 2.3.19, respectively.

Remark 2.3.19(1) motivates the following alternative strategies for dealing with an equation of the form $\sup(S) = \beta$.

Proof Strategy 2.3.20. Given a real number β and a nonempty $S \subseteq \mathbb{R}$, to prove that $\sup(S) = \beta$ we may use the two-step proof diagram:

> Step (1): Prove $x \leq \beta$ for all $x \in S$.
> Step (2): Let $r < \beta$.
> > Prove $r < x$ for some $x \in S$.

Assumption Strategy 2.3.21. Let $S \subseteq \mathbb{R}$ be nonempty and let $\beta \in \mathbb{R}$. Suppose that you are *assuming* that $\sup(S) = \beta$. Then (1) $x \leq \beta$ for all $x \in S$, and (2) whenever $r < \beta$ there is an $x \in S$ such that $r < x$.

We will present two proofs of our next theorem. In the first proof, we shall apply the above alternative strategies 2.3.20 and 2.3.21. In our second proof, we apply strategies 2.3.6 and 2.3.7.

Theorem 2.3.22. Let A and B be nonempty subsets of $\mathbb{R}$ that are both bounded above. Let $C = \{x + y : x \in A \text{ and } y \in B\}$. Then C is bounded above and

$$\sup(C) = \sup(A) + \sup(B).$$

Proof Analysis. Let A and B be bounded above. Thus, $\alpha = \sup(A)$ and $\beta = \sup(B)$ exist by the completeness axiom. So, we must prove that $\sup(C) = \alpha + \beta$. First note that every element in C has the form $x + y$ for some $x \in A$ and $y \in B$. So to prove that $\alpha + \beta$ is the least upper bound for C, we must fulfill the following proof diagram:

> Assume $\alpha = \sup(A)$ and $\beta = \sup(B)$.
> > Prove $x + y \leq \alpha + \beta$ for all $x \in A$ and $y \in B$.
> Let $r < \alpha + \beta$.
> > Prove $r < x + y$ for some $x \in A$ and $y \in B$.

We now present a correct proof of the theorem.

First Proof of Theorem 2.3.22. Let A and B be as stated. Thus, let $\alpha = \sup(A)$ and $\beta = \sup(B)$. Let $z \in C$. So $z = x + y$ for some $x \in A$ and some $y \in B$. As $x \leq \alpha$ and $y \leq \beta$, we see that $z = (x + y) \leq (\alpha + \beta)$. So $\alpha + \beta$ is an upper bound for C.

To prove that $\alpha + \beta$ is the least upper bound for C, let $r < \alpha + \beta$. Thus, $r - \beta < \alpha$. Since $\alpha = \sup(A)$, there is an $x \in A$ such that $r - \beta < x$. Thus, $r - x < \beta$. Since $\beta = \sup(B)$, there is an $y \in B$ such that $r - x < y$. Hence, $r < x + y$ and $x + y \in C$. Therefore, $\sup(C) = \alpha + \beta$. □

We will now give another proof of Theorem 2.3.22 using strategies 2.3.6 and 2.3.7. Using these strategies, the logical structure of our second proof will be as follows:

Assume $\alpha = \sup(A)$ and $\beta = \sup(B)$.
Prove $x + y \le \alpha + \beta$ for all $x \in A$ and $y \in B$.
Let c be an upper bound for C.
Prove $\alpha + \beta \le c$.

Second Proof of Theorem 2.3.22. Let A and B be as stated, and let $\alpha = \sup(A)$ and $\beta = \sup(B)$. Let $z \in C$. So $z = x + y$ for some $x \in A$ and some $y \in B$. Since $x \le \alpha$ and $y \le \beta$, we have that $z = (x + y) \le (\alpha + \beta)$. So $\alpha + \beta$ is an upper bound for C.

To prove that $\alpha + \beta$ is the least upper bound for C, let c be an upper bound for C. We shall prove that $\alpha + \beta \le c$. Since c is an upper bound for C, we have that

$$x + y \le c \text{ for all } x \in A \text{ and all } y \in B. \tag{2.4}$$

Let $y \in B$ be arbitrary. By (2.4), we have that $x + y \le c$ for all $x \in A$. Hence, $x \le c - y$ for all $x \in A$. Thus, $c - y$ is an upper bound for A. Since $\alpha = \sup(A)$, we infer that $\alpha \le c - y$. Since $y \in B$ was arbitrarily chosen, it follows that $y \le c - \alpha$ for all $y \in B$. Therefore, $c - \alpha$ is an upper bound for B. Since $\beta = \sup(B)$, we conclude that $\beta \le c - \alpha$. Thus, $\alpha + \beta \le c$. Therefore, $\sup(C) = \alpha + \beta$. □

Remark 2.3.19(2) also provides us with alternative strategies for working with an equation of the form $\inf(S) = \alpha$.

Proof Strategy 2.3.23. Given a real number α and a nonempty $S \subseteq \mathbb{R}$, to prove that $\inf(S) = \alpha$ we may use the two-step proof diagram:

Step (1): Prove $\alpha \le x$ for all $x \in S$.
Step (2): Let $\alpha < q$.
Prove $x < q$ for some $x \in S$.

Assumption Strategy 2.3.24. Let $S \subseteq \mathbb{R}$ be nonempty and let $\alpha \in \mathbb{R}$. Suppose that you are *assuming* that $\inf(S) = \alpha$. Then (1) $\alpha \le x$ for all $x \in S$, and (2) whenever $\alpha < q$ there is an $x \in S$ such that $x < q$.

Exercises 2.3

1. For each of the following subsets S of $\mathbb{R}$, answer the questions: Is the set S bounded above? Is the set S bounded below?
   (a) $S = [2, 5]$
   (b) $S = [2, 5)$
   (c) $S = (2, \infty)$
   (d) $S = \mathbb{N}$
   (e) $S = \{x \in \mathbb{R} : (x^2 + 1)^{-1} > \frac{1}{2}\}$
   (f) $S = \{\frac{1}{n} : n \in \mathbb{N}\}$
   (g) $S = \{q \in \mathbb{Q} : 0 \le q \le \sqrt{2}\}$
   (h) $S = \{x \in \mathbb{R} : |2x + 1| < 5\}$.

2. For each subset S of $\mathbb{R}$, identify the $\sup(S)$ and $\inf(S)$, if they exist.
   (a) $S = [2, 5]$
   (b) $S = [2, 5)$
   (c) $S = (2, \infty)$
   (d) $S = \mathbb{N}$
   (e) $S = \{x \in \mathbb{R} : (x^2 + 1)^{-1} > \frac{1}{2}\}$
   (f) $S = \{\frac{1}{n} : n \in \mathbb{N}\}$
   (g) $S = \{q \in \mathbb{Q} : 0 \le q \le \sqrt{2}\}$
   (h) $S = \{x \in \mathbb{R} : |2x + 1| < 5\}$.

3. For each subset S of $\mathbb{R}$, identify the $\max(S)$ and $\min(S)$ if they exist.

   (a) $S = \{2, 5, 6\}$ (b) $S = \mathbb{N}$
   (c) $S = (2, \infty)$ (d) $S = \{q \in \mathbb{Q} : 0 \leq q \leq \sqrt{2}\}$.

*4. Suppose $S \subseteq \mathbb{R}$ is nonempty and bounded. Let $A \subseteq S$ be nonempty. Prove that A is bounded. Then prove that $\sup(A) \leq \sup(S)$ and $\inf(S) \leq \inf(A)$.

5. Let $A \subseteq \mathbb{R}$ and $B \subseteq \mathbb{R}$ be non-empty bounded sets. Suppose that $\sup(A) = \inf(B) = \ell$. Prove that $x \leq \ell \leq y$ for all $x \in A$ and all $y \in B$.

6. Let $A \subseteq \mathbb{R}$ and $B \subseteq \mathbb{R}$ be non-empty sets. Suppose that B is bounded above and that for all $x \in A$ there is a $y \in B$ such that $x \leq y$. Prove that A is bounded above and $\sup(A) \leq \sup(B)$.

*7. Suppose $S \subseteq \mathbb{R}$ is nonempty and bounded. Let $\beta = \sup(S)$. Prove that for all $\varepsilon > 0$, there exists an $x \in S$ such that $\beta - \varepsilon < x$.

*8. Suppose $S \subseteq \mathbb{R}$ is nonempty and bounded. Let $\alpha = \inf(S)$. Prove that for all $\varepsilon > 0$, there exists an $x \in S$ such that $x < \alpha + \varepsilon$.

*9. Complete the proof of Lemma 2.3.13, that is, suppose that $S \subseteq \mathbb{R}$ is bounded and nonempty let $k < 0$. Prove that

   (a) $\sup(kS) = k \inf(S)$ (b) $\inf(kS) = k \sup(S)$.

10. Let $S^* = \{a \in \mathbb{R} : a \text{ is a lower bound for } S\}$ where $S \subseteq \mathbb{R}$ is nonempty and bounded below. Prove that S^* is bounded above and $\sup(S^*) = \inf(S)$.

*11. Let $f \colon D \to \mathbb{R}$ be bounded and $k \in \mathbb{R}$. Define $(kf) \colon D \to \mathbb{R}$ by $(kf)(x) = kf(x)$ for all $x \in D$. Prove that (kf) is bounded, $(kf)[D] = kf[D]$, and

   (a) if $k \geq 0$, then $\sup(kf[D]) = k \sup(f[D])$ and $\inf(kf[D]) = k \inf(f[D])$
   (b) if $k < 0$, then $\sup(kf[D]) = k \inf(f[D])$ and $\inf(kf[D]) = k \sup(f[D])$.

*12. Find functions $f \colon D \to \mathbb{R}$ and $g \colon D \to \mathbb{R}$ such that

   (a) $(f+g)[D] \neq f[D] + g[D]$ where $f[D] + g[D] = \{f(x) + g(y) : x \in D \,\&\, y \in D\}$,
   (b) $\sup((f+g)[D]) < \sup(f[D]) + \sup(g[D])$.

13. Let $S \subseteq \mathbb{R}$ be nonempty and bounded, and let $k \in \mathbb{R}$. Define $A = \{k + x : x \in S\}$. Prove that

   (a) $\sup(A) = k + \sup(S)$,   (b) $\inf(A) = k + \inf(S)$.

14. Let S be a bounded (nonempty) subset of $\mathbb{R}$, $k > 0$, and $c \in \mathbb{R}$. Let A be the set $A = \{kx + c : x \in S\}$. Prove that A is bounded and

   (a) $\sup(A) = k \sup(S) + c$,   (b) $\inf(A) = k \inf(S) + c$.

15. Let $S \subseteq \mathbb{R}$ and $T \subseteq \mathbb{R}$ be nonempty and bounded. Show that $S \cup T$ is bounded and then prove that

   (a) $\sup(S \cup T) = \max\{\sup(S), \sup(T)\}$,
   (b) $\inf(S \cup T) = \min\{\inf(S), \inf(T)\}$.

*16. Prove Theorem 2.3.18.

*17. Prove part (a) of Theorem 2.3.16.

18. Let $S \subseteq \mathbb{R}^+$ and $T \subseteq \mathbb{R}^+$ be nonempty and bounded above. Consider the set $P = \{xy : x \in S \text{ and } y \in T\}$. Prove that $\sup(P) = \sup(S) \cdot \sup(T)$.

19. Let $f \colon D \to \mathbb{R}$ be bounded. Let $E \subseteq D$ be nonempty. Prove that the set $f[E]$ is bounded. Then prove that $\sup(f[E]) \leq \sup(f[D])$ and $\inf(f[D]) \leq \inf(f[E])$.

20. Let $f \colon D \to \mathbb{R}$ and $g \colon D \to \mathbb{R}$ be bounded. Assume that $f(x) \leq g(x)$ for all $x \in D$. Prove that $\sup(f[D]) \leq \sup(g[D])$ and $\inf(f[D]) \leq \inf(g[D])$.

21. Let A and B be nonempty subsets of $\mathbb{R}$ which are both bounded below. Define the set $C = \{x + y : x \in A \text{ and } y \in B\}$. Prove that C is bounded below and that $\inf(C) = \inf(A) + \inf(B)$.

22. Let $S \subseteq \mathbb{R}^+$ and $T \subseteq \mathbb{R}^+$ be nonempty, and let $P = \{xy : x \in S \text{ and } y \in T\}$. Prove that $\inf(P) = \inf(S) \cdot \inf(T)$.

23. Let $f \colon D \to \mathbb{R}$ and $g \colon D \to \mathbb{R}$ be functions such that $f(x) \leq g(y)$ for all $x, y \in D$. Prove that $\sup(f[D]) \leq \inf(g[D])$.

Exercise Notes: For Exercise 1, $|2x + 1| < 5$ is equivalent to $-5 < 2x + 1 < 5$. For Exercises 7 and 8, see Remark 2.3.19. For Exercise 9, read the proof of Theorem 2.3.10. For Exercise 11, apply Theorem 2.3.13. For Exercises 13 and 14 review the analysis and proofs for Lemmas 2.3.8 and 2.3.12. For Exercise 18 review the analysis and proof of Theorem 2.3.22. For Exercise 21, review strategies 2.3.23 and 2.3.24. For Exercise 22, first show that $\alpha = \inf(S) \geq 0$ and $\beta = \inf(T) \geq 0$; and review strategies 2.3.9 and 2.3.11. Let c be a lower bound for P. If $c \leq 0$, then clearly $c \leq \alpha\beta$. Suppose $c > 0$ and let $y \in T$. Prove that $\frac{c}{y}$ is a lower bound for S.

2.4   THE ARCHIMEDEAN PROPERTY

In this section, we will explore some consequences of the completeness axiom. The first such consequence is called the *Archimedean property*, which asserts that every real number is strictly less than some natural number. Of course, this seems to be quite obvious, but this property actually depends on the completeness axiom.

Theorem 2.4.1 (Archimedean Property of $\mathbb{R}$). For each $x \in \mathbb{R}$, there is an $n \in \mathbb{N}$ such that $x < n$.

Proof. Let $x \in \mathbb{R}$. We prove that there exists a natural number $n \in \mathbb{N}$ such that $x < n$. Suppose, for a contradiction, that $n \leq x$ for all $n \in \mathbb{N}$. Thus, $\mathbb{N} \subseteq \mathbb{R}$ is bounded above. By the completeness axiom, $\beta = \sup(\mathbb{N})$ exists. Since $\beta - 1 < \beta$, there is an $m \in \mathbb{N}$ such that $\beta - 1 < m$ (see Remark 2.3.19(1-b)). Therefore, $\beta < m + 1 = n \in \mathbb{N}$; contradicting the fact that β is an upper bound for $\mathbb{N}$. This completes the proof.   □

In our next theorem, we show that the Archimedean property implies two useful results. The first result states that for any two positive real numbers there is an integer multiple of one that exceeds the other. The second result tells us that for any positive real number x, there is a natural number whose reciprocal is less than x.

Theorem 2.4.2. Each of the following statements hold:

(a) For all $x \in \mathbb{R}$ and $y \in \mathbb{R}$, if $x > 0$, then there is an $n \in \mathbb{N}$ such that $y < nx$.

(b) For all $x \in \mathbb{R}$, if $x > 0$, then there is an $n \in \mathbb{N}$ such that $0 < 1/n < x$.

Proof. We first prove (a). Let $x, y \in \mathbb{R}$ where $x > 0$. Consider the real number y/x. By Theorem 2.4.1, there is an $n \in \mathbb{N}$ such that $y/x < n$. Thus, $y < nx$. To prove (b), let $x > 0$. From (a) (where we take $y = 1$), we conclude that there is an $n \in \mathbb{N}$ such that $0 < 1 < nx$. Thus, $0 < 1/n < x$. □

The Archimedean property also implies the following two lemmas, the latter of which will be used in the next section. These two lemmas should be obvious to anyone who is familiar with the elementary properties of the real numbers; nevertheless, we will formally prove these two results.

Lemma 2.4.3. For all $x \in \mathbb{R}$, there exists an integer m such that $m - 1 \le x < m$.

Proof. Let $x \in \mathbb{R}$. If $x \in \mathbb{Z}$, then $m = x + 1 \in \mathbb{Z}$ and $m - 1 \le x < m$. Let $x \notin \mathbb{Z}$. So $x \ne 0$. First suppose that $x > 0$. If $x < 1$, then $m = 1$ is as required. If $x \ge 1$, then (by Theorem 2.4.1 and the Well-Ordering Principle 1.4.1) let m be the least natural number such that $x < m$. Since the natural number $m - 1$ is less than m, it follows that $m - 1 \le x$. Therefore, $m - 1 \le x < m$.

Now suppose that $x < 0$, then $-x > 0$ and the argument in the previous paragraph implies that there is an $n \in \mathbb{N}$ such that $n - 1 \le -x < n$. Thus, by applying properties of inequalities, we conclude that $-n < x \le 1 - n$. Since $x \notin \mathbb{Z}$, we see that $-n < x < 1 - n$. Let $m = 1 - n$. Thus, $m \in \mathbb{Z}$, $m - 1 = -n$ and $m - 1 \le x < m$. □

Lemma 2.4.4. For all $a, b \in \mathbb{R}$, if $1 < b - a$, then there exists an $m \in \mathbb{Z}$ such that $a < m < b$.

Proof. Let $a, b \in \mathbb{R}$. Assume that $1 < b - a$. Since $1 < b - a$, we have that (i) $1 + a < b$. By Lemma 2.4.3, there is an $m \in \mathbb{Z}$ such that (ii) $m - 1 \le a$ and (iii) $a < m$. Note that (i) and (ii) imply that $m < b$. Thus, (iii) implies that $a < m < b$. □

2.4.1   The Density of the Rational Numbers

The completeness axiom, by means of the Archimedean property, implies that the set of rational numbers is dense in $\mathbb{R}$; that is, between any two real numbers there exists a rational number.

Definition 2.4.5. Let $D \subseteq \mathbb{R}$. We say that D **is dense in** $\mathbb{R}$, if for all $x, y \in \mathbb{R}$, if $x < y$, then there exists a $d \in D$ such that $x < d < y$.

Theorem 2.4.6. The set $\mathbb{Q}$ of rational numbers is dense in $\mathbb{R}$.

Proof. Let $x, y \in \mathbb{R}$ be such that $x < y$. Therefore, $(y - x) > 0$. Theorem 2.4.2(a) implies that there is an $n \in \mathbb{N}$ such that $1 < n(y - x)$. Thus, $1 < ny - nx$. Lemma 2.4.4 states that there is an $m \in \mathbb{Z}$ such that $nx < m < ny$. Thus, $x < \frac{m}{n} < y$ and $q = \frac{m}{n} \in \mathbb{Q}$ is as required. □

Theorem 2.4.6 implies that every open interval contains infinitely many rational numbers. So no matter where one may be on the real number line, there are rational numbers nearby. Thus, one may be tempted to conclude that there are just as many rational numbers as there are real numbers. However, in the next section, we will show that the set $\mathbb{R}$ of real numbers is much larger than the set of rational numbers $\mathbb{Q}$.

Remark 2.4.7. Let $S = \{x \in \mathbb{Q} : x < \sqrt{2}\}$ and $T = \{x \in \mathbb{Q} : x > \sqrt{2}\}$. Theorem 2.4.6 implies that $\sup(S) = \sqrt{2}$ and $\inf(T) = \sqrt{2}$. Since $\sqrt{2} \notin \mathbb{Q}$, it follows that there is a "gap" in the set of rational numbers (see page 38). In fact, if we view the set of rational numbers as a number line, then this line is quite porous, as it contains many gaps–one for each irrational number. On the other hand, the completeness axiom ensures that the set of real numbers has no gaps, of any kind.

Using Theorem 2.4.6, we will next show that the set of irrational numbers is also dense in $\mathbb{R}$. First we present a simple property about the product of two real numbers, one of which is rational and the other is irrational (see Exercise 3).

Lemma 2.4.8. Let $x \in \mathbb{Q}$ be nonzero and $y \in \mathbb{R}$ be irrational. Then xy is irrational.

Theorem 2.4.9 (Density of the Irrationals in $\mathbb{R}$). For all $x, y \in \mathbb{R}$, if $x < y$, then there exists an irrational w such that $x < w < y$.

Proof. Let $x, y \in \mathbb{R}$ be such that $x < y$. Therefore, $\frac{x}{\sqrt{2}} < \frac{y}{\sqrt{2}}$. Theorem 2.4.6 implies there is a nonzero $q \in \mathbb{Q}$ such that $\frac{x}{\sqrt{2}} < q < \frac{y}{\sqrt{2}}$. Hence, $x < q\sqrt{2} < y$. Since $\sqrt{2}$ is irrational, Lemma 2.4.8 implies that $q\sqrt{2}$ is irrational. □

Thus, the set of rational numbers $\mathbb{Q}$ and the set of irrational numbers $\mathbb{R} \setminus \mathbb{Q}$ are both dense in $\mathbb{R}$. Therefore, between any two real numbers there are infinitely many rational numbers and infinitely many irrational numbers.

Exercises 2.4 ───

1. Let $a > 0$ and $b > 0$. Prove that there exists an $n \in \mathbb{N}$ such that $\frac{b}{n} < a$.

2. Let $a < b$ be real numbers. Prove that there exists an $n \in \mathbb{N}$ such that $a + \frac{1}{n} < b$.

*3. Prove Lemma 2.4.8.

4. Let $A = \{x \in \mathbb{Q} : x < 2\}$. Prove that $\sup(A) = 2$.

5. Let $A = \{x \in \mathbb{Q} : x^2 < 2\}$. Prove that $\sup(A) = \sqrt{2}$.

6. Let $q > 0$ be a rational number. Prove that for all $x, y \in \mathbb{R}$, if $x < y$, then there exists an irrational w such that $x < qw < y$.

7. Let $x < y < a < b$ be real numbers. Prove that there exists rational numbers q and r such that $x < q < y$ and $a < q + r < b$.

8. Let $a \in \mathbb{R}$. Suppose that $a \leq \frac{1}{n}$ for all $n \in \mathbb{N}$. Prove that $a \leq 0$.

9. Let $a, b \in \mathbb{R}$. Suppose that $a \leq b + \frac{1}{n}$ for all $n \in \mathbb{N}$. Prove that $a \leq b$.

10. Let $x > 0$ be a real number. Prove that there is an $n \in \mathbb{N}$ such that $\frac{1}{n} < x < n$.

11. Let a and b be real numbers such that $a < b$. Show that there are infinitely many rational numbers between a and b.

Exercise Notes: For Exercises 4 and 5, apply proof strategy 2.3.20.

2.5   NESTED INTERVALS THEOREM

We conclude this chapter with two additional applications of the completeness axiom, namely, the nested intervals theorem and a theorem of Georg Cantor that shows that the set of real numbers is uncountable. These two interesting results are sometimes applied in real analysis.

Definition 2.5.1. Let $\{I_n : n \in \mathbb{N}\}$ be an indexed set of closed intervals of $\mathbb{R}$ such that $I_n \supseteq I_{n+1}$ for all $n \in \mathbb{N}$; that is,

$$I_1 \supseteq I_2 \supseteq I_3 \supseteq I_4 \supseteq I_5 \supseteq \cdots$$

Then $\{I_n : n \in \mathbb{N}\}$ is called a family (or set) of **nested closed intervals**.

Of course, the notation $A \supseteq B$ means $B \subseteq A$. The completeness axiom will be used in the proof of our next theorem, which states that for any nested family of closed intervals, there is a point that belongs to all of the intervals in the family.

Theorem 2.5.2 (Nested Intervals Theorem). Let $\{I_n : n \in \mathbb{N}\}$ be a family of nested closed intervals of $\mathbb{R}$. Then $\bigcap\limits_{n \in \mathbb{N}} I_n \neq \varnothing$, that is, there exists an $x \in \mathbb{R}$ such that $x \in I_n$ for all $n \in \mathbb{N}$.

Proof. Let $\{I_n : n \in \mathbb{N}\}$ be a family of nested closed intervals. For each $n \in \mathbb{N}$, let $I_n = [a_n, b_n]$ where $a_n < b_n$ are real numbers. Thus, for all $i, j \in \mathbb{N}$,

$$\text{if } i \leq j, \text{ then } a_i \leq a_j < b_j \leq b_i,$$

as illustrated in Figure 2.6. Let $A = \{a_1, a_2, a_3, \dots\}$. The set A is bounded above. In

Figure 2.6: Family of nested closed intervals.

fact, each b_n is an upper bound for A (see Exercise 1). Let $\beta = \sup(A)$. Since β is an upper bound for A, we have that $a_n \leq \beta$ for all $n \in \mathbb{N}$. Moreover, as each b_n is an upper bound for A and β is the smallest such upper bound, we conclude that $\beta \leq b_n$ for all $n \in \mathbb{N}$. Hence, $a_n \leq \beta \leq b_n$ for all $n \in \mathbb{N}$; that is, $\beta \in [a_n, b_n]$ for all $n \in \mathbb{N}$. Therefore, $\bigcap\limits_{n \in \mathbb{N}} I_n \neq \varnothing$. □

Definition 2.5.1 concerns a nested family of closed intervals. This definition can clearly be generalized to arbitrary sets of real numbers.

Definition 2.5.3. Let $\{A_n : n \in \mathbb{N}\}$ be an indexed set of subsets of $\mathbb{R}$ such that $A_n \supseteq A_{n+1}$ for all $n \in \mathbb{N}$; that is,

$$A_1 \supseteq A_2 \supseteq A_3 \supseteq A_4 \supseteq A_5 \supseteq \cdots$$

Then $\{A_n : n \in \mathbb{N}\}$ is a family (or set) of **nested sets**. If $A_n \supset A_{n+1}$ for all $n \in \mathbb{N}$, the family is said to be **proper**.

The proof of the above Theorem 2.5.2 clearly applies the completeness axiom. Recall that the set of rational numbers $\mathbb{Q}$ does not satisfy the completeness axiom (see page 38). Let us define the notion of a closed interval in the number system $\mathbb{Q}$. Given $a \in \mathbb{Q}$ and $b \in \mathbb{Q}$, where $a < b$, let $\mathcal{I} = \{x \in \mathbb{Q} : a \leq x \leq b\}$. Let us call $\mathcal{I}$ a *closed $\mathbb{Q}$-interval*. Using Theorem 2.4.6 and Remark 2.4.7, one can define a family $\{\mathcal{I}_n : n \in \mathbb{N}\}$ of nested closed $\mathbb{Q}$-intervals such that $\bigcap_{n \in \mathbb{N}} \mathcal{I}_n = \varnothing$. It follows that any valid proof of Theorem 2.5.2 requires the completeness axiom.

2.5.1 $\mathbb{R}$ is Uncountable

We end this chapter with a curious application of Theorem 2.5.2 due to Georg Cantor, a brilliant mathematician who single-handedly created the branch of mathematics now called set theory. Cantor first observed that two sets A and B have the same size if and only if there is a one-to-one correspondence between A and B, that is, there is a way of evenly matching the elements in A with the elements in B. In other words, Cantor noted that A and B have the same size if and only if there exists a bijection $f \colon A \to B$. In this case, Cantor said that A and B have the same *cardinality*.

During his set theoretic investigations, Cantor introduced the following definition.

Definition 2.5.4. A set A is said to be **countable** if $A = \varnothing$ or there exists a surjection $f \colon \mathbb{N} \to A$. If A is not countable, then A is said to be **uncountable**.

Let A be a nonempty countable set and let $f \colon \mathbb{N} \to A$ be a surjection. For each $n \in \mathbb{N}$, let $a_n = f(n)$. We can now itemize the elements of A in the following list:

$$a_1, a_2, a_3, a_4, \ldots.$$

As $f \colon \mathbb{N} \to A$ is onto A, it follows that *every* element in A appears in the above list. The proofs of the next two results are outlined in Exercises 11 and 12, respectively.

Lemma 2.5.5. If $f \colon \mathbb{N} \to A$ is a surjection, then there is an injection $g \colon A \to \mathbb{N}$.

Theorem 2.5.6. If A is a countable infinite set, then there is a bijection $f \colon \mathbb{N} \to A$.

Theorem 2.5.6 implies that a countable infinite set has just as many elements as the set $\mathbb{N}$ of natural numbers. It thus follows that a set is countable if and only if it either finite or has the same cardinality as the set of natural numbers.

After introducing the notion of a countable set, Cantor asked a simple question:

Is every infinite set countable? In the pursuit of this question, Cantor proved that the set of rational numbers is countable (see Exercise 10). Cantor then conjectured that the set of real numbers is also countable. It came a complete surprise to Cantor when he discovered the next remarkable theorem. First, we establish a simple lemma.

Lemma 2.5.7. Let $[a, b]$ be a closed interval and let $x \in \mathbb{R}$. Then there is a closed subinterval $[a', b'] \subseteq [a, b]$ such that $x \notin [a', b']$.

Proof. Let $[a, b]$ be a closed interval and let $x \in \mathbb{R}$. If $x \notin [a, b]$, then let $[a', b'] = [a, b]$. If $x \in [a, b]$, let $c \in (a, b)$ be such that $c \neq x$. If $c < x$, then let $[a', b'] = [a, c]$. If $x < c$, then let $[a', b'] = [c, b]$. In each case, there is an $[a', b'] \subseteq [a, b]$ where $x \notin [a', b']$. □

Theorem 2.5.8 (Cantor). The set of real numbers $\mathbb{R}$ is uncountable.

Proof. Suppose, for a contradiction, that $\mathbb{R}$ is countable and let $f \colon \mathbb{N} \to \mathbb{R}$ be a surjection. For each $n \in \mathbb{N}$, let $x_n = f(n)$. Thus, as f is onto $\mathbb{R}$, every real number appears in the list

$$x_1, x_2, x_3, x_4, \ldots \tag{2.5}$$

Let $[a_1, b_1]$ be a closed interval such that $x_1 \notin [a_1, b_1]$. By Lemma 2.5.7, there is a closed interval $[a_2, b_2] \subseteq [a_1, b_1]$ such that $x_2 \notin [a_2, b_2]$. Continuing in this manner, we obtain a nested family $\{[a_n, b_n] : n \in \mathbb{N}\}$ of closed intervals such that (▼) $x_n \notin [a_n, b_n]$ for every $n \in \mathbb{N}$. Theorem 2.5.2 implies that there is a real number x such that (▲) $x \in [a_n, b_n]$ for all $n \in \mathbb{N}$. Thus, by (▼) and (▲), x does not appear in the list (2.5). Hence, f is not onto $\mathbb{R}$. This contradiction implies that $\mathbb{R}$ must be uncountable. □

We now know that $\mathbb{Q}$ is countable and $\mathbb{R}$ is uncountable. Both of these sets are infinite, but Theorem 2.5.8 implies that $\mathbb{R}$ is "more infinite" than the set of rational numbers $\mathbb{Q}$. After proving that the set of real numbers is uncountable, Cantor was able to prove that there is an increasing sequence of larger and larger infinite sets. In other words, Cantor showed that there are "infinitely many different infinities," a result with clear philosophical and mathematical significance.

Exercises 2.5

*1. Let $\{I_n : n \in \mathbb{N}\}$ be a family of nested closed intervals. For each $n \in \mathbb{N}$, let $I_n = [a_n, b_n]$ where $a_n < b_n$ are real numbers. Thus, for all $i, j \in \mathbb{N}$,

$$\text{if } i \leq j, \text{ then } a_i \leq a_j < b_j \leq b_i.$$

Let $n \in \mathbb{N}$. Show that $a_k < b_n$ for all $k \in \mathbb{N}$.

2. Let $\{I_n : n \in \mathbb{N}\}$ be a nested family of closed intervals of $\mathbb{R}$, where $I_n = [a_n, b_n]$. Let $A = \{a_1, a_2, a_3, \ldots\}$. In the proof of Theorem 2.5.2 it is shown that the supremum $\beta = \sup(A)$ exists. Now, let $B = \{b_1, b_2, b_3, \ldots\}$.

   (a) Show that $\gamma = \inf(B)$ exists.

   (b) Show that $\beta \leq \gamma$.

   (c) Show that if $\beta < \gamma$, then $\bigcap_{n \in \mathbb{N}} I_n = [\beta, \gamma]$; and if $\beta = \gamma$, then $\bigcap_{n \in \mathbb{N}} I_n = \{\beta\}$.

3. For each $n \in \mathbb{N}$, let O_n be the open interval $O_n = (a_n, b_n)$ where $a_n < b_n$ are real numbers. Suppose that $a_n < a_{n+1}$ and $b_{n+1} < b_n$ for all $n \in \mathbb{N}$, that is,

$$a_1 < a_2 < a_3 < \cdots < a_n < \cdots \quad \cdots < b_n < \cdots < b_3 < b_2 < b_1.$$

Prove that $\bigcap_{n \in \mathbb{N}} O_n \neq \varnothing$.

4. Find a nested proper family of open intervals whose intersection is empty.

5. Find a nested proper family of closed intervals whose intersection has exactly one point.

6. Find a nested proper family of closed intervals whose intersection is the closed interval $[1, 2]$.

7. Let $\{I_n : n \in \mathbb{N}\}$ be the nested family of intervals of $\mathbb{R}$ defined by $I_n = [n, \infty)$ for all $n \in \mathbb{N}$. Show that $\bigcap_{n \in \mathbb{N}} I_n = \varnothing$.

8. Let $f : \mathbb{N} \to \mathbb{Q}$ be a surjection and let $\{I_n : n \in \mathbb{N}\}$ be a family of nested closed intervals such that $f(n) \notin I_n$ for all $n \in \mathbb{N}$. Let $\beta \leq \gamma$ be as in Exercise 2. Show that $\beta = \gamma$ and that β is an irrational number.

9. Let $f : \mathbb{N} \to \mathbb{Q}$ be a surjection. By modifying the proof of Theorem 2.5.8, show that there is a family $\{\mathcal{I}_n : n \in \mathbb{N}\}$ of nested closed $\mathbb{Q}$-intervals (page 51) such that $f(n) \notin \mathcal{I}_n$ for all $n \in \mathbb{N}$. Conclude that $\bigcap_{n \in \mathbb{N}} \mathcal{I}_n = \varnothing$.

*10. Consider the function $f : \mathbb{N} \to \mathbb{Q}$ defined by

$$f(n) = \begin{cases} (-1)^c \frac{a}{b}, & \text{if } n = 2^a 3^b 5^c \text{ and } a, b, c \in \mathbb{N}; \\ 0, & \text{otherwise.} \end{cases}$$

Show that f is onto $\mathbb{Q}$. So $\mathbb{Q}$ is countable. (By Theorem 1.4.3, f is well-defined.)

*11. Let $f : \mathbb{N} \to A$ be a surjection. Define $g : A \to \mathbb{N}$ by

$$g(a) = n \text{ if and only if } n \text{ is the least natural number such that } f(n) = a,$$

for each $a \in A$. Prove that g is one-to-one.

*12. Let A be an infinite countable set. By Definition 2.5.4 and Lemma 2.5.5, there is an injection $g : A \to \mathbb{N}$. Since A is infinite, the range of g must be an infinite subset of $\mathbb{N}$. Let $R = \operatorname{ran}(g)$. Since R is an infinite set of natural numbers, we can list the elements of R in strictly increasing order, say $R = \{n_1, n_2, n_3, \dots\}$ where $n_i < n_{i+1}$ for all $i \in \mathbb{N}$.[1] Define $f : \mathbb{N} \to A$ by $f(i) = g^{-1}(n_i)$, for all $i \in \mathbb{N}$. Prove that $f : \mathbb{N} \to A$ is a bijection.

Exercise Notes: For Exercise 1, there are two cases: either $n \leq k$ or $k \leq n$. Exercise 4 shows that the conclusion of Theorem 2.5.2 does not necessarily hold if the nested intervals are not closed.

---

[1] The sequence $n_1, n_2, n_3, \dots$ can be defined by induction using the Well-Ordering Principle 1.4.1.

CHAPTER 3

Sequences

A sequence is an enumerated infinite list of real numbers. In this chapter, we will investigate sequences and prove a number of important theorems about the limit of a sequence. The proofs will often require the completeness axiom. Sequences are fundamental in real analysis and, while you may already be familiar with sequences, it is useful to have a formal definition. We shall define a sequence to be a function from the set of natural numbers into the set of real numbers $\mathbb{R}$.

Definition 3.0.1. A **sequence** is a function $s\colon \mathbb{N} \to \mathbb{R}$. We shall denote the value $s(n)$ by s_n, where s_n is called the n**th term** of the sequence. We will write s as $\langle s_n \rangle$, as $\langle s_1, s_2, s_3, \dots \rangle$ or as $\langle s_n \rangle_{n=1}^{\infty}$ when we want to emphasize that the index variable n begins with 1. Moreover, $\langle s_n \rangle$ is said to be a sequence **of distinct points** if $s_n \neq s_m$, whenever $n \neq m$.

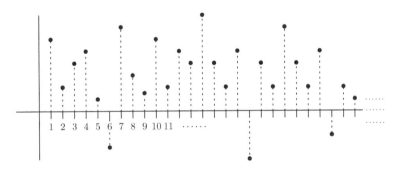

Figure 3.1: Functional representation of a sequence: $s(4) = s_4 > 0$ and $s(6) = s_6 < 0$.

Consider the sequence $\langle s_n \rangle$ where $s_n = \frac{1}{n}$. Then we can write $\langle s_n \rangle$ as $\langle \frac{1}{n} \rangle$ or as $\langle 1, \frac{1}{2}, \frac{1}{3}, \frac{1}{4}, \dots \rangle$. A *constant sequence* is denoted by $\langle a \rangle$ or $\langle a, a, a, a, \dots \rangle$, where a is a fixed real number.

One can also have sequences of the form $\langle s_n \rangle_{n=k}^{\infty} = \langle s_k, s_{k+1}, \dots \rangle$ where $k > 1$ or $k = 0$. However, one can easily re-express such a sequence as one starting at 1. Define $\langle t_n \rangle_{n=1}^{\infty}$ by $t_n = s_{n+k-1}$ for all $n \geq 1$. Then $\langle t_n \rangle_{n=1}^{\infty} = \langle s_k, s_{k+1}, \dots \rangle$. For example, $\langle \frac{1}{n-1} \rangle_{n=2}^{\infty} = \langle \frac{1}{n} \rangle_{n=1}^{\infty}$.

3.1   CONVERGENCE

The limit concept is one of the oldest and among the most important concepts in mathematical analysis. The formal definition of the limit of a sequence has a long and deep history. Around 250 BC, Archimedes informally used the convergence of sequences, via the method of exhaustion, to compute approximations of π and to show that πr^2 equals the area enclosed by a circle of radius r. The mathematically precise notion of the limit of a sequence that we use today was not formulated until the 19th century by Cauchy and Weierstrass.

The limit of a sequence $\langle s_n \rangle$ identifies the long-term behavior of the sequence; that is, what happens to s_n for larger and larger values of n. If the terms s_n get closer and closer to a real number ℓ, then we say that $\langle s_n \rangle$ *converges* to ℓ and that ℓ is the *limit* of the sequence. However, the notion of a sequence getting closer and closer to a real number seems to be quite vague and very likely cannot be be used to rigorously prove theorems about convergence. It took over a hundred years after Newton and Leibniz before the following question was correctly addressed:

> *Can the notion of sequential convergence be given a precise mathematical definition?*

In 1821, Cauchy responded to this question with the following definition:

> *When the values successively approach ... a fixed value, eventually differing from it by as little as one could wish, that fixed value is called the limit of all the others.*

Cauchy's definition of convergence is correct but it lacks sufficient precision. This shortcoming was addressed around 1870, when Karl Weierstrass proposed our next definition, which is now the standard definition that is used in real analysis and in calculus.

Definition 3.1.1. A sequence $\langle s_n \rangle$ is said to **converge** to the real number ℓ provided that for all $\varepsilon > 0$, there exists a natural number N such that for all $n \in \mathbb{N}$, if $n > N$, then $|s_n - \ell| < \varepsilon$.

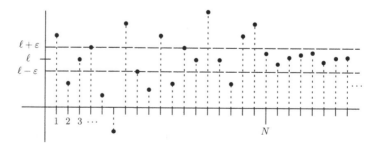

Figure 3.2: For all $n > N$, we have $|s_n - \ell| < \varepsilon$.

Figure 3.2 illustrates Definition 3.1.1. If a sequence $\langle s_n \rangle$ converges to ℓ, then ℓ

is called the **limit** of the sequence $\langle s_n \rangle$ and we write $\lim\limits_{n \to \infty} s_n = \ell$. If a sequence $\langle s_n \rangle$ does not converge, then we shall say that $\langle s_n \rangle$ **diverges**. The logical form of Definition 3.1.1 can be expressed as

$$(\forall \varepsilon > 0)(\exists N \in \mathbb{N})(\forall n \in \mathbb{N})(n > N \to |s_n - \ell| < \varepsilon) \tag{3.1}$$

and it is this logical form that motivates the following proof strategy.

Proof Strategy 3.1.2. To prove that $\lim\limits_{n \to \infty} s_n = \ell$, use the proof diagram:

> Let $\varepsilon > 0$ be an arbitrary real number.
> Let $N = $ (the natural number you found).
> Let $n > N$ be an arbitrary natural number.
> Prove $|s_n - \ell| < \varepsilon$.

To apply proof strategy 3.1.2 on a sequence defined by an explicit formula, we first let $\varepsilon > 0$. Then we find a natural number N such that when $n > N$, we can prove that $|s_n - \ell| < \varepsilon$. To find the desired N, we undertake the following:

> Using algebra and properties of inequality on the expression $|s_n - \ell|$, we "extract out" a larger value that resembles $\frac{1}{n}$.

We shall then use this larger value to find N so that when $n > N$, we will have that $|s_n - \ell| < \varepsilon$. We will illustrate this idea in our proof analysis of the next four theorems. Before we discuss these theorems, we identify three properties of inequality that are very useful when proving theorems about convergence.

Quotient Principles of Inequality 3.1.3. Let a, b, c, d be positive real numbers.

(1) Given the ratio $\frac{a}{b}$, if $a < c$, then you can conclude that $\frac{a}{b} < \frac{c}{b}$.
   (*Replacing a numerator with a larger value yields a larger ratio.*)

(2) Given the ratio $\frac{a}{b}$, if $d < b$, then you can conclude that $\frac{a}{b} < \frac{a}{d}$.
   (*Replacing a denominator with a smaller value yields a larger ratio.*)

(3) Given the ratio $\frac{a}{b}$, if $a \leq c$ and $d \leq b$, then you can infer that $\frac{a}{b} \leq \frac{c}{d}$.
   (*Replacing a numerator with a larger value and a denominator with a smaller value yields a larger ratio.*)

Example. The above property 3.1.3(2) implies the following assertions:

1. $\frac{1}{n} < \frac{1}{N}$ when $n > N > 0$.
2. $\frac{1}{\sqrt{n}} < \frac{1}{\sqrt{N}}$ when $n > N > 0$, by Theorem 1.1.11 on page 6.
3. $\frac{1}{2^n} < \frac{1}{n}$ when $2^n > n > 0$.

Theorem 3.1.4. $\lim\limits_{n \to \infty} \frac{1}{n} = 0$.

Proof Analysis. We apply proof strategy 3.1.2 to the sequence $\langle \frac{1}{n} \rangle$. Let $\varepsilon > 0$. We need an $N \in \mathbb{N}$ such that if $n > N$, then $\left| \frac{1}{n} - 0 \right| < \varepsilon$. Since $\left| \frac{1}{n} - 0 \right| = \frac{1}{n}$, we thus need an $N \in \mathbb{N}$ so that $\frac{1}{n} < \varepsilon$ when $n > N$. Solving the inequality $\frac{1}{n} < \varepsilon$ for n, we conclude that $n > \frac{1}{\varepsilon}$. So if we take $N > \frac{1}{\varepsilon}$, then we can prove the desired result (N exists by the Archimedean property, as stated in Theorem 2.4.1). We now present a logically correct proof that follows the structure outlined in proof strategy 3.1.2.

Proof. Let $\varepsilon > 0$. Let $N > \frac{1}{\varepsilon}$ be a natural number. Let $n \in \mathbb{N}$ be such that $n > N$. We prove $\left|\frac{1}{n} - 0\right| < \varepsilon$ as follows:

$$
\begin{aligned}
\left|\frac{1}{n} - 0\right| &= \left|\frac{1}{n}\right| && \text{by algebra} \\
&= \frac{1}{n} && \text{because } \frac{1}{n} > 0 \\
&< \frac{1}{N} && \text{because } n > N \\
&< \frac{1}{\frac{1}{\varepsilon}} && \text{because } N > \frac{1}{\varepsilon} \\
&= \varepsilon && \text{by algebra.}
\end{aligned}
$$

Therefore, $\left|\frac{1}{n} - 0\right| < \varepsilon$. □

Theorem 3.1.5. $\lim\limits_{n \to \infty} \frac{1}{\sqrt{n}} = 0$.

Proof Analysis. Let $\varepsilon > 0$. We need an $N \in \mathbb{N}$ such that if $n > N$, then $\left|\frac{1}{\sqrt{n}} - 0\right| < \varepsilon$. Since $\left|\frac{1}{\sqrt{n}} - 0\right| = \frac{1}{\sqrt{n}}$, we thus need an $N \in \mathbb{N}$ so that $\frac{1}{\sqrt{n}} < \varepsilon$ when $n > N$. Solving the inequality $\frac{1}{\sqrt{n}} < \varepsilon$ for n, we obtain $n > \frac{1}{\varepsilon^2}$. So if we take $N > \frac{1}{\varepsilon^2}$, then we can prove the theorem. We now apply proof strategy 3.1.2 and present the proof.

Proof. Let $\varepsilon > 0$. Let $N > \frac{1}{\varepsilon^2}$ be a natural number. Let $n \in \mathbb{N}$ be such that $n > N$. Thus, $\frac{1}{\sqrt{n}} < \frac{1}{\sqrt{N}}$. We prove that $\left|\frac{1}{\sqrt{n}} - 0\right| < \varepsilon$ as follows:

$$
\begin{aligned}
\left|\frac{1}{\sqrt{n}} - 0\right| &= \left|\frac{1}{\sqrt{n}}\right| && \text{by algebra} \\
&= \frac{1}{\sqrt{n}} && \text{because } \frac{1}{\sqrt{n}} > 0 \\
&< \frac{1}{\sqrt{N}} && \text{because } n > N \\
&< \frac{1}{\sqrt{\frac{1}{\varepsilon^2}}} && \text{because } N > \frac{1}{\varepsilon^2} \\
&= \varepsilon && \text{by algebra.}
\end{aligned}
$$

Therefore, $\left|\frac{1}{\sqrt{n}} - 0\right| < \varepsilon$. □

In the future, we will use N to denote a natural number.

Theorem. $\lim\limits_{n \to \infty} 1 + \frac{1}{2^n} = 1$.

Proof Analysis. Let $\varepsilon > 0$. We need an $N \in \mathbb{N}$ so that if $n > N$, then $\left|1 + \frac{1}{2^n} - 1\right| < \varepsilon$. Since $\left|1 + \frac{1}{2^n} - 1\right| = \frac{1}{2^n}$, we must find an $N \in \mathbb{N}$ such that if $n > N$, then $\frac{1}{2^n} < \varepsilon$. Solving the inequality $\frac{1}{2^n} < \varepsilon$ for n is difficult. So we take a different approach. By Exercise 2 on page 26, $n < 2^n$ and so $\frac{1}{2^n} < \frac{1}{n}$, when $n \geq 1$. By solving the inequality $\frac{1}{n} < \varepsilon$ for n, we obtain $n > \frac{1}{\varepsilon}$. So if we take $N > \frac{1}{\varepsilon}$, then we can prove the theorem. We now present a proof which is guided by proof strategy 3.1.2.

Proof. Let $\varepsilon > 0$. Let $N > \frac{1}{\varepsilon}$ be a natural number. Let $n \in \mathbb{N}$ be such that $n > N$. We prove that $\left|(1 + \frac{1}{2^n}) - 1\right| < \varepsilon$ as follows:

$$\left|\left(1 + \frac{1}{2^n}\right) - 1\right| = \left|\frac{1}{2^n}\right| \qquad \text{by algebra}$$

$$= \frac{1}{2^n} \qquad \text{because } \frac{1}{2^n} > 0$$

$$< \frac{1}{n} \qquad \text{because } n < 2^n$$

$$< \frac{1}{N} \qquad \text{because } n > N$$

$$< \frac{1}{\frac{1}{\varepsilon}} \qquad \text{because } N > \frac{1}{\varepsilon}$$

$$= \varepsilon \qquad \text{by algebra.}$$

Therefore, $\left|1 + \frac{1}{2^n} - 1\right| < \varepsilon$. □

In the proof analysis for each of the above three theorems, we were able to find the required N without much difficulty. Specific sequences with more complicated definitions may require more work to find N. This work can be reduced by applying the Quotient Principles of Inequality 3.1.3 and the Sum and Product Principles of Inequality 1.1.8. In particular, when dealing with a ratio of polynomials in n, these two principles justify the techniques presented in the following remark.

Remark 3.1.6. Given a ratio of polynomials, we illustrate a method for finding a larger and simpler ratio. The first item below discusses how to find a larger numerator, while the second item considers how to find an eventually smaller denominator.

1. (Larger Numerator) Given a polynomial in n, say $cn^k + 6n^3 - 10$, with highest power n^k where $c > 0$, by dropping the negative terms, we obtain a larger value. For example, $cn^k + 6n^3 - 10 \leq cn^k + 6n^3$. Now since $n^3 \leq n^k$, we see that $cn^k + 6n^3 \leq cn^k + 6n^k$. Thus, $cn^k + 6n^3 - 10 \leq (c + 6)n^k$ for all $n \in \mathbb{N}$.

2. (Smaller Denominator) Given a polynomial in n, say $cn^k + 6n^3 - 10$, with highest power n^k where $c > 0$, by dropping the positive terms with lower powers, we get a smaller value. For example, $cn^k - 10 \leq cn^k + 6n^3 - 10$. As $cn^k - 10$ contains a negative term, let $0 < s < c$ and find an $m \in \mathbb{N}$ such that such that $sn^k \leq cn^k - 10$ when $n \geq m$. Hence, $sn^k \leq cn^k + 6n^3 - 10$ for all $n \geq m$.

We will apply Remark 3.1.6 in the proof analysis of the following theorem.

Theorem. $\lim\limits_{n \to \infty} \frac{n^2 + 2n - 3}{n^2 - n - 5} = 1$.

Proof Analysis. Let $\varepsilon > 0$. We must find a natural number N such that if $n > N$, then $\left|\frac{n^2 + 2n - 3}{n^2 - n - 5} - 1\right| < \varepsilon$. By algebra, we obtain

$$\left|\frac{n^2 + 2n - 3}{n^2 - n - 5} - 1\right| = \left|\frac{3n + 2}{n^2 - n - 5}\right|$$

We see that $3n + 2 > 0$ and $n^3 - n^2 - 5 > 0$ when $n \geq 3$. Thus, if $n \geq 3$, we have

$$\left| \frac{3n + 2}{n^2 - n - 5} \right| = \frac{3n + 2}{n^2 - n - 5}.$$

We now need to find an $N \geq 3$ such that if $n > N$, then $\frac{3n+2}{n^2-n-5} < \varepsilon$. Solving the inequality $\frac{3n+2}{n^2-n-5} < \varepsilon$ for n is difficult. So we shall take advantage of Remark 3.1.6. First, we will get a real number $b > 0$ such that

$$3n + 2 \leq bn, \text{ for all } n \in \mathbb{N}. \tag{3.2}$$

Then we will get a real number $s > 0$ such that

$$sn^2 \leq n^2 - n - 5, \text{ for all "large" } n \in \mathbb{N}. \tag{3.3}$$

To find the b in (3.2), notice that $3n + 2 \leq 3n + 2n = 5n$ for all $n \in \mathbb{N}$. So, we shall let $b = 5$. We now need to identify an s such that (3.3) holds. Let $s = 1/2$ (see Remark 3.1.6(2)). We must find m so that if $n \geq m$, then $(\star)$ $\frac{1}{2}n^2 \leq n^2 - n - 5$. This latter inequality is equivalent to $n^2 \leq 2n^2 - 2n - 10$, which is equivalent to $10 \leq n^2 - 2n = n(n - 2)$. Thus, when $n \geq 5$ we have that $(\star)$ holds. Therefore, for all $n \geq 5$, we have that

$$\left| \frac{3n + 2}{n^2 - n - 5} \right| = \frac{3n + 2}{n^2 - n - 5} \leq \frac{5n}{\frac{1}{2}n^2} = \frac{10}{n}.$$

Solving the inequality $\frac{10}{n} < \varepsilon$ for n, we see that we must let $N > \max\{\frac{10}{\varepsilon}, 5\}$.

Proof. Let $\varepsilon > 0$. Let $N > \max\{\frac{10}{\varepsilon}, 5\}$ be in $\mathbb{N}$. Let $n \in \mathbb{N}$ be such that $n > N$. Since $n > 3$, we have that

$$\left| \frac{n^2 + 2n - 3}{n^2 - n - 5} - 1 \right| = \frac{3n + 2}{n^2 - n - 5}$$

because $3n+2 > 0$ and $n^3 - n^2 - 5 > 0$. In addition, as $n > 5$, we have that $3n + 2 \leq 5n$ and $\frac{1}{2}n^2 \leq n^2 - n - 5$. Therefore, since $n > N > \max\{\frac{10}{\varepsilon}, 5\}$, we conclude that

$$\left| \frac{n^2 + 2n - 3}{n^2 - n - 5} - 1 \right| = \frac{3n + 2}{n^2 - n - 5} \leq \frac{5n}{\frac{1}{2}n^2} = \frac{10}{n} < \frac{10}{N} < \frac{10}{\frac{10}{\varepsilon}} = \varepsilon.$$

This completes the proof. □

Suppose in a proof that you are assuming that a given sequence converges. Our next strategy will be useful when dealing with such an assumption.

Assumption Strategy 3.1.7. If you are assuming that $\lim\limits_{n \to \infty} s_n = \ell$, then for any $\varepsilon > 0$, there is an $N \in \mathbb{N}$ such that $|s_n - \ell| < \varepsilon$ for all $n > N$.

So, in a proof, when you are assuming that $\lim\limits_{n \to \infty} s_n = \ell$, you can conclude that for any positive value $v > 0$, there is an N such that $|s_n - \ell| < v$ for all $n > N$. We shall express this observation as "we can make $|s_n - \ell|$ as small as we want." We now apply this idea to prove the following theorem.

Theorem 3.1.8. If $\lim_{n\to\infty} s_n = \ell$, $\ell > 0$ and $s_n \geq 0$ for all $n \geq 1$, then $\lim_{n\to\infty} \sqrt{s_n} = \sqrt{\ell}$.

Proof Analysis. Suppose $\lim_{n\to\infty} s_n = \ell$. We need to prove that $\lim_{n\to\infty} \sqrt{s_n} = \sqrt{\ell}$, where $\ell > 0$ and $s_n \geq 0$ for all $n \geq 1$. Our proof will have the following logical structure:

Assume $\lim_{n\to\infty} s_n = \ell$.
Let $\varepsilon > 0$ be an arbitrary real number.
Let $N =$ (the natural number you found).
Let $n > N$ be an arbitrary natural number.
Prove $\left|\sqrt{s_n} - \sqrt{\ell}\right| < \varepsilon$.

We must find a natural number N such that if $n > N$, then $\left|\sqrt{s_n} - \sqrt{\ell}\right| < \varepsilon$. Here is the basic plan that we will apply to get N.

Using algebra and properties of inequality on the expression $\left|\sqrt{s_n} - \sqrt{\ell}\right|$, we "extract out" a larger value that contains $|s_n - \ell|$ and no other occurrences of s_n.

Since $\lim_{n\to\infty} s_n = \ell$, we can make $|s_n - \ell|$ "as small as we want." We should then be able to make $\left|\sqrt{s_n} - \sqrt{\ell}\right| < \varepsilon$ and find N. Let us now execute this plan! First we start with $\left|\sqrt{s_n} - \sqrt{\ell}\right|$ and extract out $|s_n - \ell|$ as follows:

$$\left|\sqrt{s_n} - \sqrt{\ell}\right| = \left|\frac{\left(\sqrt{s_n} - \sqrt{\ell}\right)}{1}\frac{\left(\sqrt{s_n} + \sqrt{\ell}\right)}{\left(\sqrt{s_n} + \sqrt{\ell}\right)}\right| \quad \text{rationalizing the numerator}$$

$$= \left|\frac{s_n - \ell}{\sqrt{s_n} + \sqrt{\ell}}\right| \quad \text{by algebra}$$

$$= \frac{|s_n - \ell|}{\sqrt{s_n} + \sqrt{\ell}} \quad \text{because } \sqrt{s_n} + \sqrt{\ell} > 0$$

$$\leq \frac{|s_n - \ell|}{\sqrt{\ell}} \quad \text{because } \sqrt{\ell} \leq \sqrt{s_n} + \sqrt{\ell}.$$

Hence, $\left|\sqrt{s_n} - \sqrt{\ell}\right| \leq \frac{|s_n-\ell|}{\sqrt{\ell}}$ and we extracted out the larger value $\frac{|s_n-\ell|}{\sqrt{\ell}}$ that contains $|s_n - \ell|$ and no other occurrences of s_n. So if $\frac{|s_n-\ell|}{\sqrt{\ell}} < \varepsilon$, then we will have that $\left|\sqrt{s_n} - \sqrt{\ell}\right| < \varepsilon$. How small must $|s_n - \ell|$ be in order to ensure that $\frac{|s_n-\ell|}{\sqrt{\ell}} < \varepsilon$? By solving this latter inequality for $|s_n - \ell|$, we obtain $|s_n - \ell| < \varepsilon\sqrt{\ell}$. Hence, we need an N so that $|s_n - \ell| < \varepsilon\sqrt{\ell}$ when $n > N$. Since $\lim_{n\to\infty} s_n = \ell$, there is such an N. This is the value for N that we will use in the proof.

Proof of Theorem 3.1.8. Assume that $\lim_{n\to\infty} s_n = \ell$ where $\ell > 0$ and $s_n \geq 0$ for all $n \geq 1$. Let $\varepsilon > 0$. As $\lim_{n\to\infty} s_n = \ell$, there is an $N \in \mathbb{N}$ such that (▲) $|s_n - \ell| < \varepsilon\sqrt{\ell}$ for all $n > N$. Now let $n > N$. Thus,

$$\left|\sqrt{s_n} - \sqrt{\ell}\right| = \left|\frac{\left(\sqrt{s_n} - \sqrt{\ell}\right)\left(\sqrt{s_n} + \sqrt{\ell}\right)}{1}\right| \quad \text{rationalizing the numerator}$$

$$= \left|\frac{s_n - \ell}{\sqrt{s_n} + \sqrt{\ell}}\right| \qquad\qquad \text{by algebra}$$

$$= \frac{|s_n - \ell|}{\sqrt{s_n} + \sqrt{\ell}} \qquad\qquad \text{because } \sqrt{s_n} + \sqrt{\ell} > 0$$

$$\leq \frac{|s_n - \ell|}{\sqrt{\ell}} \qquad\qquad \text{because } \sqrt{\ell} \leq \sqrt{s_n} + \sqrt{\ell}$$

$$< \frac{\varepsilon\sqrt{\ell}}{\sqrt{\ell}} = \varepsilon \qquad\qquad \text{by (▲).}$$

Therefore, $\left|\sqrt{s_n} - \sqrt{\ell}\right| < \varepsilon$ and this completes the proof. □

Theorem 3.1.8 also holds if $\ell = 0$ (see Exercise 16). It may seem obvious that a convergent sequence cannot have more than one limit; but, this requires a proof, which is given below. In this proof, we will use two tools that are commonly used in real analysis; namely, the triangle inequality and the algebraic "trick" of adding and subtracting the same value.

Theorem 3.1.9 (Uniqueness of the Limit). If the sequence $\langle s_n \rangle$ converges, then the sequence has only one limit.

Proof. Assume that $\langle s_n \rangle$ converges. Suppose, for a contradiction, that $\ell \neq \ell'$ are both limits of the sequence $\langle s_n \rangle$. Thus, $\varepsilon = |\ell - \ell'| > 0$. Since $\langle s_n \rangle$ converges to ℓ there is an $N \in \mathbb{N}$ such that for all $n > N$, $|s_n - \ell| < \frac{\varepsilon}{2}$. Also, since $\langle s_n \rangle$ converges to ℓ' there is an $N' \in \mathbb{N}$ such that for all $n > N'$, $|s_n - \ell'| < \frac{\varepsilon}{2}$. Therefore, for all $n > \max\{N, N'\}$, we have that

$$|\ell - \ell'| = |(\ell - s_n) + (s_n - \ell')| \leq |\ell - s_n| + |s_n - \ell'| < \frac{\varepsilon}{2} + \frac{\varepsilon}{2} = \varepsilon.$$

Hence, $|\ell - \ell'| < \varepsilon$. But this contradicts the fact that $\varepsilon = |\ell - \ell'|$. □

We next present an equivalence that is sometimes useful.

Theorem 3.1.10. Let $\langle s_n \rangle$ be a sequence and $\ell \in \mathbb{R}$. Then $\lim_{n \to \infty} s_n = \ell$ if and only if $\lim_{n \to \infty} (s_n - \ell) = 0$.

Proof. Assume that $\lim_{n \to \infty} s_n = \ell$. Let $\varepsilon > 0$. Since $\lim_{n \to \infty} s_n = \ell$, there is an $N \in \mathbb{N}$ where $|s_n - \ell| < \varepsilon$ for all $n > N$. Let $n > N$. Thus, $|(s_n - \ell) - 0| = |s_n - \ell| < \varepsilon$. Hence, $\lim_{n \to \infty} (s_n - \ell) = 0$. The converse follows similarly. □

When assuming that a given sequence converges and you want to prove that another sequence converges, then assumption strategy 3.1.7 can be useful. Many times we will be assuming that $\lim_{n\to\infty} s_n = \ell$ and will be working with $\varepsilon > 0$. Using assumption strategy 3.1.7, we can conclude that for any positive $\varepsilon' < \varepsilon$ (e.g., $\varepsilon' = \frac{\varepsilon}{2}$), there is an N' such that for all $n > N'$, we we have $|s_n - \ell| < \varepsilon'$.

Theorem 3.1.11. Let $\langle s_n \rangle$, $\langle a_n \rangle$ be sequences and let $\ell \in \mathbb{R}$. If

(1) $|s_n - \ell| \leq k\,|a_n|$ for all $n \geq m$, where $k > 0$ and $m \in \mathbb{N}$,

(2) $\lim_{n\to\infty} a_n = 0$,

then $\lim_{n\to\infty} s_n = \ell$.

Proof Analysis. Assume (1) and (2). Thus, from (2), we can make $|a_n| = |a_n - 0|$ "as small as we want." Let $\varepsilon > 0$. We need an $N \in \mathbb{N}$ such that if $n > N$, then $|s_n - \ell| < \varepsilon$. As $|s_n - \ell| \leq k\,|a_n|$ when $n \geq m$, we need an $N \geq m$ such that if $n > N$, then $k\,|a_n| < \varepsilon$. Solving this latter inequality for $|a_n|$, we obtain $|a_n| < \frac{\varepsilon}{k}$. Since $\lim_{n\to\infty} a_n = 0$, there is an N' such that whenever $n > N'$ we have $|a_n - 0| = |a_n| < \frac{\varepsilon}{k}$. So we will use $N = \max\{N', m\}$. So if $n > N$, then $n > N'$ and $n > m$.

Proof. Assuming (1) and (2) we shall prove that $\lim_{n\to\infty} s_n = \ell$. To do this, let $\varepsilon > 0$. By (2), $\lim_{n\to\infty} a_n = 0$. Thus, there is an $N' \in \mathbb{N}$ such that $(\star)\ |a_n - 0| = |a_n| < \frac{\varepsilon}{k}$ for all $n > N'$. By (1), we have that $(\star\star)\ |s_n - \ell| \leq k\,|a_n|$ for all $n \geq m$. Let $N = \max\{m, N'\}$. Let $n > N$. Thus,

$$
\begin{aligned}
|s_n - \ell| &\leq k\,|a_n| && \text{by } (\star\star) \text{ because } n > N \geq m \\
&< k\left(\frac{\varepsilon}{k}\right) && \text{by } (\star) \text{ because } n > N \geq N' \\
&= \varepsilon && \text{by algebra.}
\end{aligned}
$$

Therefore, $|s_n - \ell| < \varepsilon$. This completes the proof of the theorem. $\qquad\square$

For a simple application of Theorem 3.1.11, let $\ell \in \mathbb{N}$ and consider the sequence $\left\langle \frac{k}{n^\ell} \right\rangle$ where $k \neq 0$. Since $\left| \frac{k}{n^\ell} - 0 \right| \leq |k|\frac{1}{n}$ for all $n \in \mathbb{N}$, Theorems 3.1.4 and 3.1.11 imply that $\lim_{n\to\infty} \frac{k}{n^\ell} = 0$. Theorem 3.1.11 can also be used to reduce the amount of work used to prove certain limit equations.

Example. We will use Theorem 3.1.11 to prove that $\lim_{n\to\infty} \frac{9n^2-n+1}{3n^2+n-10} = 3$. To apply this theorem, we must find an upper bound for

$$
\left| \frac{9n^2 - n + 1}{3n^2 + n - 10} - 3 \right| = \left| \frac{-4n + 31}{3n^2 + n - 10} \right|
$$

for sufficiently large n. Clearly, $|-4n + 31| \leq 4n + 31 \leq 4n + 31n = 35n$ for all n. For the denominator, we will determine when $n^2 \leq 3n^2 - 10$ (see Remark 3.1.6(2)). This latter inequality is equivalent to $10 \leq 2n^2$ which holds when $n \geq 3$. Thus, if $n \geq 3$, we have that

$$
\left| \frac{9n^2 - n + 1}{3n^2 + n - 10} - 3 \right| = \left| \frac{-4n + 31}{3n^2 + n - 10} \right| \leq \frac{35n}{n^2} = 35\frac{1}{n}.
$$

Since $\lim_{n\to\infty} \frac{1}{n} = 0$, Theorem 3.1.11 implies that $\lim_{n\to\infty} \frac{9n^2-n+1}{3n^2+n-10} = 3$.

Theorem 3.1.11 implies the following two useful corollaries.

Corollary 3.1.12. Let x be a such that $|x| < 1$. Then $\lim_{n \to \infty} x^n = 0$.

Proof. If $x = 0$, then clearly $\lim_{n \to \infty} x^n = 0$. Assume that $0 < |x| < 1$. Since $\frac{1}{|x|} > 1$, there is a $c > 0$ such that $1 + c = \frac{1}{|x|}$. By Bernoulli's inequality (Exercise 5 on page 27), $(1 + c)^n \geq 1 + nc$, for every $n \in \mathbb{N}$. Hence, $\frac{1}{|x|^n} = (1 + c)^n \geq 1 + nc > nc$, for all $n \in \mathbb{N}$. So
$$|x^n - 0| = |x|^n < \frac{1}{nc},$$
for all $n \geq 1$. As $\lim_{n \to \infty} \frac{1}{n} = 0$ and $\frac{1}{c} > 0$, Theorem 3.1.11 implies that $\lim_{n \to \infty} x^n = 0$. □

The next corollary can sometimes be used to prove that a particular sequence $\langle u_n \rangle$ converges to a specific real number r.

Corollary 3.1.13. Suppose that $\lim_{n \to \infty} s_n = \ell$. Let $r \in \mathbb{R}$ and $\langle u_n \rangle$ be such that $|u_n - r| \leq k|s_n - \ell|$ for all $n \geq m$, for some $k > 0$ and $m \in \mathbb{N}$. Then $\lim_{n \to \infty} u_n = r$.

Proof. By Theorem 3.1.10 we have that $\lim_{n \to \infty}(s_n - \ell) = 0$. Since $|u_n - r| \leq k|s_n - \ell|$ for all $n \geq m$, where $k > 0$ and $m \in \mathbb{N}$, Theorem 3.1.11 implies that $\lim_{n \to \infty} u_n = r$. □

If $\langle s_n \rangle$ converges to ℓ, then for $\varepsilon > 0$ there is an $N \in \mathbb{N}$ such that $|s_n - \ell| < \varepsilon$ for all $n > N$. So, for any $K > N$, we have $|s_n - \ell| < \varepsilon$ for all $n > K$ as well.

Remark 3.1.14. Let $\langle s_n \rangle$ be a sequence and let $\ell \in \mathbb{R}$. What does it mean to say that $\langle s_n \rangle$ does **not** converge to ℓ? By taking the negation of the logical form (3.1), we see that "the sequence $\langle s_n \rangle$ does **not** converge to ℓ" means the following:

There is an $\varepsilon > 0$ such that for all $N \in \mathbb{N}$, there is an $n > N$ so that $|s_n - \ell| \geq \varepsilon$.

Neighborhoods

Definition 3.1.15. Let $x \in \mathbb{R}$ and let $\varepsilon > 0$. The open interval $(x - \varepsilon, x + \varepsilon)$, centered at x, is said to be a **neighborhood** of x, and is denoted by U_ε^x.

Let x be a real number and let U_ε^x be the neighborhood $(x - \varepsilon, x + \varepsilon)$ of x where $\varepsilon > 0$. Then for any real number s, we have that $s \in U_\varepsilon^x$ if and only if $|s - x| < \varepsilon$. We will write U^x to denote a neighborhood of x when it is not important to specify ε. The following theorem just states that the notion of convergence can be expressed in terms of neighborhoods (see Figure 3.3).

Theorem 3.1.16. Let $\langle s_n \rangle$ be a sequence and ℓ be a real number. Then the following are equivalent:

1. The sequence $\langle s_n \rangle$ converges to ℓ.
2. For all $\varepsilon > 0$ there is an $N \in \mathbb{N}$ such that for all $n \in \mathbb{N}$, if $n > N$, then $|s_n - \ell| < \varepsilon$.
3. For every neighborhood U^ℓ there is an $N \in \mathbb{N}$ such that for all $n \in \mathbb{N}$, if $n > N$, then $s_n \in U^\ell$.

Figure 3.3: For all $n > N$, we have $|s_n - \ell| < \varepsilon$.

Corollary 3.1.17. Let $\langle s_n \rangle$ be a sequence of distinct points and suppose that $\langle s_n \rangle$ converges to ℓ. Then every neighborhood of ℓ contains an infinite number of points from the sequence $\langle s_n \rangle$.

Proof. Let U^ℓ be any neighborhood of ℓ. Since the sequence $\langle s_n \rangle$ converges to ℓ, Theorem 3.1.16 states that there is an $N \in \mathbb{N}$ such that for all $n \in \mathbb{N}$, if $n > N$, then $s_n \in U^\ell$. Thus, an infinite number of points from the sequence $\langle s_n \rangle$ are in U^ℓ. □

Given a sequence $\langle s_n \rangle$ that converges to ℓ. Theorem 3.1.16 implies that for every neighborhood U^ℓ of ℓ there exists an $N \in \mathbb{N}$ such that $s_n \in U^\ell$ for all $n > N$. We can say, in this case, that the terms of the sequence $\langle s_n \rangle$ are *eventually* in U^ℓ.

Our next result implies that every real number is the limit of a sequence of rational numbers.

Lemma 3.1.18. Let $D \subseteq \mathbb{R}$ be dense in $\mathbb{R}$. Let x be any real number. Then there is a sequence $\langle d_n \rangle$ that converges to x where $d_n \in D$ for all $n \geq 1$.

Proof. Since $D \subseteq \mathbb{R}$ is dense in $\mathbb{R}$, for each $n \in \mathbb{N}$, there is a $d_n \in D$ such that d_n is in the interval $(x - \frac{1}{n}, x + \frac{1}{n})$ and so, $|x - d_n| < \frac{1}{n}$. Since $\lim\limits_{n \to \infty} \frac{1}{n} = 0$, Theorem 3.1.11 implies that $\lim\limits_{n \to \infty} d_n = x$. □

Bounded Sequences

We now consider some other properties that a sequence $\langle s_n \rangle$ may possess. These properties are not really new, as they are very similar to the definitions on bounded sets discussed in Chapter 2.

Definition 3.1.19. A sequence $\langle s_n \rangle$ is **bounded above** if there is a real number M so that $s_n \leq M$ for all $n \in \mathbb{N}$.

Proposition 3.1.20. Let $\langle s_n \rangle$ be a sequence that is not bounded above. For every real number B and natural number m, there exists an $n > m$ such that $s_n > B$.

Proof. Suppose $\langle s_n \rangle$ is not bounded above, and let $B \in \mathbb{R}$ and $m \in \mathbb{N}$. Now let $M = \max\{s_1, s_2, \ldots, s_m, B\}$. Since $\langle s_n \rangle$ is not bounded above, there is an $n \in \mathbb{N}$ such that $s_n > M$. Because $M \geq B$ we have that $s_n > B$. Furthermore, because $s_n > M \geq s_i$ for all $i \leq m$, we must have that $n > m$. □

Definition 3.1.21. A sequence $\langle s_n \rangle$ is **bounded below** if there is a real number M so that $M \leq s_n$ for all $n \in \mathbb{N}$.

A proof similar to that of Proposition 3.1.20 will establish the following result.

Proposition 3.1.22. Let $\langle s_n \rangle$ be a sequence that is not bounded below. For every real number B and natural number m, there exists an $n > m$ such that $s_n < B$.

Definition 3.1.23. A sequence $\langle s_n \rangle$ is **bounded** if there is are real numbers a and b such that $a \leq s_n \leq b$ for all $n \in \mathbb{N}$.

Remark. A sequence $\langle s_n \rangle$ is bounded if and only if there is an $M > 0$ so that $|s_n| \leq M$ for all $n \in \mathbb{N}$ (see Theorem 2.3.2). Thus, a sequence is *unbounded* if for every $M > 0$, there is an $n \in \mathbb{N}$ such that $|s_n| > M$

Here are some examples that illustrate the above definitions.

1. The sequence $\langle n^2 \rangle$ is bounded below but it is not bounded above.
2. The sequence $\langle (-1)^n \rangle$ is bounded but it does not converge.
3. The sequence $\left\langle \frac{(-1)^n}{n} \right\rangle$ is bounded and it converges.

Let $\langle s_n \rangle$ be a sequence. Suppose there is a $B > 0$ and an N such that $|s_n| \leq B$ for all $n > N$. This leaves a finite number of terms of the sequence that may not be similarly bounded by B. But a finite set of terms is always bounded. Thus, one can conclude that the entire sequence is bounded; that is, that there is an $M > 0$ such that $|s_n| \leq M$ for all $n \geq 1$. The following example affirms this fact.

Example. Let $\langle s_n \rangle$ be the sequence where $s_n = 2 + \frac{1}{n}$ where $n \geq 1$. Let $B = 2 + \frac{1}{10}$. Notice that for all $n > 10$, we have $s_n \leq B$. But all of the values $s_1, \ldots, s_{10}$ are greater than B. Let $M = \max\{s_1, \ldots, s_{10}, B\} = 3$. We now have that $s_n \leq 3$ for all $n \geq 1$. Thus, the sequence $\langle s_n \rangle$ is bounded.

Theorem 3.1.24. If the sequence $\langle s_n \rangle$ converges, then $\langle s_n \rangle$ is bounded.

Proof. Suppose that $\lim_{n \to \infty} s_n = \ell$. Corresponding to $\varepsilon = 1$ there is an $N \in \mathbb{N}$ such that $|s_n - \ell| < 1$ for all $n > N$. Thus, $|s_n| - |\ell| \leq |s_n - \ell| < 1$ for all $n > N$. Hence, $|s_n| < |\ell| + 1$ for all $n > N$. Let $M = \max\{|s_1|, \ldots, |s_N|, |\ell| + 1\}$. Thus, $|s_n| \leq M$ for all $n \in \mathbb{N}$. Therefore, $\langle s_n \rangle$ is a bounded sequence. $\square$

By expressing Theorem 3.1.24 in terms of its contrapositive, we conclude that if a sequence $\langle s_n \rangle$ is unbounded, then the sequence $\langle s_n \rangle$ diverges.

Exercises 3.1 ──────────────────────────────────────

1. Let $a \in \mathbb{R}$. Prove that the sequence $\left\langle a + (-1)^n \frac{2n+1}{n} \right\rangle$ is bounded.

2. Let $k \neq 0$. Use Definition 3.1.1 to prove that $\lim_{n \to \infty} \frac{k}{n} = 0$.

3. Use Definition 3.1.1 to prove that $\lim_{n \to \infty} \frac{n+1}{n+2} = 1$.

4. Use Definition 3.1.1 to prove that $\lim_{n \to \infty} \frac{3n}{n+2} = 3$.

5. Use Definition 3.1.1 to prove that $\lim_{n \to \infty} \frac{6n-7}{3n-2} = 2$.

6. Use Definition 3.1.1 to prove that $\lim_{n \to \infty} \frac{6n-7}{2n-7} = 3$.

7. Use Theorem 3.1.11 to prove that $\lim\limits_{n\to\infty} \frac{1}{2n-7} = 0$ and $\lim\limits_{n\to\infty} \frac{6n^2+3n}{2n^2-5} = 3$.

8. Prove that the limits given in Exercises 3–6 hold by applying Theorems 3.1.11 and 3.1.4.

9. Prove Theorem 3.1.8 using Corollary 3.1.13.

10. Let $c \in \mathbb{R}$ be constant. Prove that $\lim\limits_{n\to\infty} c = c$.

*11. Suppose $\lim\limits_{n\to\infty} s_n = \ell$ and let $c \in \mathbb{R}$ be a constant. Prove that $\lim\limits_{n\to\infty} (c+s_n) = c+\ell$.

*12. Let $\langle s_n \rangle$ be a convergent sequence. Suppose $\lim\limits_{n\to\infty} s_n = \ell$ and let $c \in \mathbb{R}$ be a nonzero constant. Prove that $\lim\limits_{n\to\infty} (cs_n) = c\ell$.

13. Suppose that $\lim\limits_{n\to\infty} s_n = \ell$. Prove that $\lim\limits_{n\to\infty} |s_n| = |\ell|$.

14. Suppose that $\lim\limits_{n\to\infty} |s_n| = 0$. Prove that $\lim\limits_{n\to\infty} s_n = 0$.

15. Suppose that $\lim\limits_{n\to\infty} s_n = \ell$ and $|s_n| \le M$ for all $n \ge 1$, where $M > 0$. Prove that $\lim\limits_{n\to\infty} s_n^2 = \ell^2$.

*16. Suppose $\lim\limits_{n\to\infty} s_n = 0$ where $s_n \ge 0$ for all $n \ge 1$. Prove that $\lim\limits_{n\to\infty} \sqrt{s_n} = 0$.

17. Let $\langle x_n \rangle$ and $\langle y_n \rangle$ be two convergent sequences. Prove that there exists an $M > 0$ such that $|x_n| \le M$ and $|y_n| \le M$ for all $n \ge 1$.

18. Let $\langle s_n \rangle$ be a convergent sequence. Suppose that $\lim\limits_{n\to\infty} s_n = \ell$. Prove that there exists an $M > 0$ such that $|s_n + \ell| \le M$ for all $n \ge 1$.

19. Use Theorems 3.1.11 and 3.1.4 to prove that $\lim\limits_{n\to\infty} \frac{\sin(n)}{n} = 0$.

20. Use Theorem 3.1.11 and Theorem 3.1.5 to prove that $\lim\limits_{n\to\infty} (\sqrt{n+1} - \sqrt{n}) = 0$.

21. Suppose that $\lim\limits_{n\to\infty} s_n = \ell$. Prove that $\lim\limits_{n\to\infty} s_n^2 = \ell^2$.

22. Prove Proposition 3.1.22.

23. Let $\langle s_n \rangle$ be an unbounded sequence and let $k \ne 0$. Prove that $\langle s_n^2 \rangle$ and $\langle ks_n \rangle$ are unbounded sequences.

24. Let x be a such that $|x| > 1$.

   (a) Prove that the sequence $\langle x^n \rangle$ is unbounded.

   (b) Let $k \ne 0$. Conclude that $\langle kx^n \rangle$ is unbounded.

Exercise Notes: For Exercise 5, observe that $n \le 3n-2$ when $n \ge 1$. For Exercise 6, observe that $n < |2n - 7|$ when $n > 7$. In this case, we would need N to be at least 7 and so, $N > \max\{7, \frac{14}{\varepsilon}\}$ could be used in the proof. For Exercise 20, show that $\left|\sqrt{n+1} - \sqrt{n}\right| = \frac{1}{\sqrt{n+1}+\sqrt{n}} \le \frac{1}{2\sqrt{n}}$. For Exercise 24(a), modify the proof of Corollary 3.1.12.

3.2   LIMIT THEOREMS FOR SEQUENCES

In this section, we will derive some of the standard results that concern convergent sequences. In particular, we will show that the algebraic operations and the order relation $\leq$ are compatible with the limit operation.

3.2.1   Algebraic Limit Theorems

An algebraic limit theorems states that if you know the limits of some given sequences, then you can determine the limit of a new sequence that is an algebraic combination of the given sequences. Such limit theorems typically have the following form:

Theorem. If $\lim_{n \to \infty} s_n = s$ and $\lim_{n \to \infty} t_n = t$, then $\lim_{n \to \infty} u_n = u$ where $\langle u_n \rangle$ is constructed from the sequences $\langle s_n \rangle$ and $\langle t_n \rangle$, and u is a function of s and t.

Given a theorem of this form, how does one prove that $\lim_{n \to \infty} u_n = u$?

Proof Strategy 3.2.1. To prove that $\lim_{n \to \infty} u_n = u$, apply the following diagram:

Assume $\lim_{n \to \infty} s_n = s$.
Assume $\lim_{n \to \infty} t_n = t$.
Let $\varepsilon > 0$ be a real number.
  Let $N = $ (the natural number you found).
  Let $n > N$ be a natural number.
    Prove $|u_n - u| < \varepsilon$.

So given $\varepsilon > 0$, one must find a natural number N such that if $n > N$, then $|u_n - u| < \varepsilon$. Here is the basic idea that one can apply to get N.

Using algebra and properties of inequality, from the expression $|u_n - u|$ "extract out" a larger value containing $|s_n - s|$ and $|t_n - t|$, and no other occurrences of s_n or t_n.

Since $\lim_{n \to \infty} s_n = s$ and $\lim_{n \to \infty} t_n = t$, we can make $|s_n - s|$ and $|t_n - t|$ "as small as we want." One should then be able to find the desired N. We will apply this technique in our proof analysis of the next two theorems.

Given two convergent sequences $\langle s_n \rangle$ and $\langle t_n \rangle$, we will next prove that the *sum sequence* $\langle s_n + t_n \rangle$ converges. We first present an analysis that demonstrates how to employ proof strategy 3.2.1 and the above discussion.

Theorem 3.2.2. If $\lim_{n \to \infty} s_n = s$ and $\lim_{n \to \infty} t_n = t$, then $\lim_{n \to \infty} (s_n + t_n) = s + t$.

Proof Analysis. Assume that $\lim_{n \to \infty} s_n = s$ and $\lim_{n \to \infty} t_n = t$. So we can make $|s_n - s|$ and $|t_n - t|$ as small as we want. Let $\varepsilon > 0$. We need to ensure that

$$|(s_n + t_n) - (s + t)| < \varepsilon$$

for sufficiently large n. Using algebra and properties of inequality on the expression $|(s_n + t_n) - (s + t)|$ we extract out $|s_n - s|$ and $|t_n - t|$ as follows:

$$
\begin{aligned}
|(s_n + t_n) - (s + t)| &= |(s_n - s) + (t_n - t)| && \text{by algebra} \\
&\leq |s_n - s| + |t_n - t| && \text{by triangle inequality.}
\end{aligned}
$$

So, if $|s_n - s| < \frac{\varepsilon}{2}$ and $|t_n - t| < \frac{\varepsilon}{2}$, then $|(s_n + t_n) - (s + t)| < \varepsilon$. Since $\lim_{n\to\infty} s_n = s$, there is a natural number N_s such that $|s_n - s| < \frac{\varepsilon}{2}$ when $n > N_s$. Similarly, there is a $N_t \in \mathbb{N}$ such that $|t_n - t| < \frac{\varepsilon}{2}$ when $n > N_t$. So we will let $N = \max\{N_s, N_t\}$, which will ensure that $n > N_s$ and $n > N_t$, when $n > N$. We now present a valid proof that is guided by proof strategy 3.2.1.

Proof. Suppose $\lim_{n\to\infty} s_n = s$ and $\lim_{n\to\infty} t_n = t$. To prove that $\lim_{n\to\infty} (s_n + t_n) = s + t$, let $\varepsilon > 0$. Since $\lim_{n\to\infty} s_n = s$, there is an $N_s \in \mathbb{N}$ such that

$$|s_n - s| < \frac{\varepsilon}{2} \text{ for all } n > N_s. \tag{3.4}$$

Since $\lim_{n\to\infty} t_n = t$, there is an $N_t \in \mathbb{N}$ such that

$$|t_n - t| < \frac{\varepsilon}{2} \text{ for all } n > N_t. \tag{3.5}$$

Let $N = \max\{N_s, N_t\}$. Thus, for $n > N$ we have that

$$
\begin{aligned}
|(s_n + t_n) - (s + t)| &= |(s_n - s) + (t_n - t)| && \text{by algebra} \\
&\leq |s_n - s| + |t_n - t| && \text{by triangle inequality} \\
&< \frac{\varepsilon}{2} + \frac{\varepsilon}{2} && \text{by (3.4) and (3.5)} \\
&= \varepsilon && \text{by algebra.}
\end{aligned}
$$

Therefore, $|(s_n + t_n) - (s + t)| < \varepsilon$ for all $n > N$. $\qquad\square$

In the proof of Theorem 3.2.2 we were able to "cleanly" extract out $|s_n - s|$ and $|t_n - t|$, that is, there were no additional factors. This may not be the case when applying strategy 3.2.1 to prove other such theorems. There will be times, after extracting out $|s_n - s|$ or $|t_n - t|$, that "unwanted" factors involving n, s_n, or t_n will emerge. When such factors do appear, we will have to find an appropriate upper bound for these factors. This is done in the proofs of Theorems 3.2.3 and 3.2.6 below. In our proof analysis of Theorem 3.2.3 we obtain the unwanted factor $|t_n|$. As the sequence $\langle t_n \rangle$ converges, there is a $K > 0$ such that $|t_n| \leq K$ for all $n \geq 1$. So we will have a desired upper bound. Actually, in our proof analysis, we tacitly use the value $M = \max\{K, |s|\}$ for the upper bound because it simplifies the proof slightly.

Theorem 3.2.3. If $\lim_{n\to\infty} s_n = s$ and $\lim_{n\to\infty} t_n = t$, then $\lim_{n\to\infty} (s_n t_n) = st$.

Proof Analysis. Assume that $\lim\limits_{n\to\infty} s_n = s$ and $\lim\limits_{n\to\infty} t_n = t$. Since $\langle t_n \rangle$ converges, we know by Theorem 3.1.24 that $\langle t_n \rangle$ is bounded. Thus, there is an $M > 0$ such that

$$|s| \le M \text{ and } |t_n| \le M \text{ for all } n \ge 1. \tag{3.6}$$

We now use algebra and properties of inequality on the expression $|s_n t_n - st|$ to extract out $|s_n - s|$ and $|t_n - t|$ as follows:

$$
\begin{aligned}
|s_n t_n - st| &= |s_n t_n - s t_n + s t_n - st| && \text{by algebra} \\
&= |t_n(s_n - s) + s(t_n - t)| && \text{by algebra} \\
&\le |t_n(s_n - s)| + |s(t_n - t)| && \text{by triangle inequality} \\
&= |t_n| \, |s_n - s| + |s| \, |t_n - t| && \text{by property of } |\,| \\
&\le M |s_n - s| + M |t_n - t| && \text{by (3.6)}.
\end{aligned}
$$

Thus, if $|s_n - s| < \frac{\varepsilon}{2M}$ and $|t_n - t| < \frac{\varepsilon}{2M}$, then we can conclude that $|s_n t_n - st| < \varepsilon$. One can now compose a correct proof using this analysis as a guide.

Proof. See Exercise 10. □

Our next lemma shows that if the limit of a sequence is nonzero, then the terms of the sequence are eventually nonzero as well.

Lemma 3.2.4. If $\lim\limits_{n\to\infty} t_n = t$ and $t \neq 0$, then there is a real number $\delta > 0$ and an $N \in \mathbb{N}$ such that $\delta < |t|$ and $\delta < |t_n|$ for all $n > N$.

Proof. Assume that $\lim\limits_{n\to\infty} t_n = t$ and $t \neq 0$. Let $\delta = \frac{|t|}{2}$. Clearly, $0 < \delta < |t|$. Now, because $\lim\limits_{n\to\infty} t_n = t$, there is an $N \in \mathbb{N}$ such that $|t - t_n| < \delta$ for all $n > N$. Let $n > N$. Hence, $|t| - |t_n| \le |t - t_n| < \delta$, that is, $|t| - |t_n| < \delta$. Thus, $|t| - \delta < |t_n|$ and so, $|t| - \frac{|t|}{2} < |t_n|$. As $|t| - \frac{|t|}{2} = \frac{|t|}{2} = \delta$, we see that $\delta < |t_n|$. □

The following lemma shows that if all the terms of a convergent sequence are nonzero and the limit is also nonzero, then there is a single positive number γ smaller than the absolute values of the limit and all the terms of the sequence.

Lemma 3.2.5. If $\lim\limits_{n\to\infty} t_n = t$ where $t \neq 0$ and $t_n \neq 0$ for all $n \ge 1$, then there is a $\gamma > 0$ such that $\gamma \le |t|$ and $\gamma \le |t_n|$ for all $n \ge 1$.

Proof. Assume that $\lim\limits_{n\to\infty} t_n = t$, $t \neq 0$, and $t_n \neq 0$ for all $n \in \mathbb{N}$. By Lemma 3.2.4 there is a $\delta > 0$ and an $N \in \mathbb{N}$ such that $\delta < |t|$ and $\delta < |t_n|$ for all $n > N$. Let $\gamma = \min\{|t_1|, |t_2|, \dots, |t_N|, \delta\}$. Clearly, $\gamma > 0$, $\gamma \le |t|$, and $\gamma \le |t_n|$ for all $n \ge 1$. □

Our next theorem shows that if a convergent sequence $\langle t_n \rangle$ satisfies the conditions of Lemma 3.2.5, then the reciprocal sequence $\left\langle \frac{1}{t_n} \right\rangle$ also converges.

Theorem 3.2.6. Let $t \neq 0$ and $t_n \neq 0$ for all $n \ge 1$. If $\lim\limits_{n\to\infty} t_n = t$, then $\lim\limits_{n\to\infty} \frac{1}{t_n} = \frac{1}{t}$.

Proof Analysis. Assume that $\lim_{n\to\infty} t_n = t$, $t \neq 0$, and $t_n \neq 0$ for all $n \in \mathbb{N}$. Lemma 3.2.5 implies that there is a $\gamma > 0$ such that $(\star)$ $\gamma \leq |t|$ and $(\star\star)$ $\gamma \leq |t_n|$ for all $n \geq 1$. We can now use algebra and properties of inequality on the expression $\left|\frac{1}{t_n} - \frac{1}{t}\right|$ to extract out $|t_n - t|$ as follows:

$$\left|\frac{1}{t_n} - \frac{1}{t}\right| = \left|\frac{t - t_n}{t_n t}\right| \quad \text{by algebra}$$

$$= \frac{|t_n - t|}{|t_n||t|} \quad \text{by property of } |\ |$$

$$\leq \frac{|t_n - t|}{\gamma^2} \quad \text{by } (\star) \text{ and } (\star\star).$$

Thus, if we have that $|t_n - t| < \gamma^2 \varepsilon$, we can conclude that $\left|\frac{1}{t_n} - \frac{1}{t}\right| < \varepsilon$. Guided by this analysis and proof strategy 3.2.1, we can now present a logically correct proof.

Proof. Assume that $t \neq 0$, $t_n \neq 0$ for all $n \geq 1$, and $\lim_{n\to\infty} t_n = t$. By Lemma 3.2.5, there is a real number $\gamma > 0$ such that $\gamma \leq |t|$ and $\gamma \leq |t_n|$ for all $n \geq 1$. Thus,

$$\gamma^2 \leq |t_n t| \text{ for all } n \geq 1. \tag{3.7}$$

To prove that $\lim_{n\to\infty} \frac{1}{t_n} = \frac{1}{t}$, let $\varepsilon > 0$. Since $\lim_{n\to\infty} t_n = t$, there is an $N \in \mathbb{N}$ such that

$$|t_n - t| < \varepsilon\gamma^2 \text{ for all } n > N. \tag{3.8}$$

Thus, for $n > N$ we have that

$$\left|\frac{1}{t_n} - \frac{1}{t}\right| = \left|\frac{t - t_n}{t_n t}\right| \quad \text{by algebra}$$

$$= \frac{|t - t_n|}{|t_n||t|} \quad \text{by property of absolute value}$$

$$\leq \frac{|t - t_n|}{\gamma^2} \quad \text{by (3.7)}$$

$$< \frac{\varepsilon\gamma^2}{\gamma^2} = \varepsilon \quad \text{by (3.8)}.$$

Therefore, $\left|\frac{1}{t_n} - \frac{1}{t}\right| < \varepsilon$ for all $n > N$. This completes the proof of the theorem. $\square$

Theorem 3.2.7. If $\lim_{n\to\infty} s_n = s$ and $\lim_{n\to\infty} t_n = t$ where $t \neq 0$ and $t_n \neq 0$ for all $n \geq 1$, then $\lim_{n\to\infty} \frac{s_n}{t_n} = \frac{s}{t}$.

Proof. Assume that $\lim_{n\to\infty} s_n = s$ and $\lim_{n\to\infty} t_n = t$ where $t \neq 0$ and $t_n \neq 0$ for all $n \geq 1$. Since $\frac{s_n}{t_n} = s_n\left(\frac{1}{t_n}\right)$, Theorems 3.2.6 and 3.2.3 imply the desired conclusion. $\square$

3.2.2 The Squeeze Theorem

Suppose that two convergent sequences have the same limit and the terms of a third sequence are eventually between the corresponding terms of the convergent sequences. We now show that this third sequence converges to the same limit as well.

Theorem 3.2.8 (Squeeze Theorem). Let $\langle s_n \rangle$ and $\langle t_n \rangle$ be convergent sequences such that $\lim\limits_{n \to \infty} s_n = \ell$ and $\lim\limits_{n \to \infty} t_n = \ell$. If $\langle y_n \rangle$ is a sequence satisfying $s_n \le y_n \le t_n$ for all $n \ge m$ where $m \in \mathbb{N}$, then $\lim\limits_{n \to \infty} y_n = \ell$.

Proof. Let $\langle s_n \rangle$, $\langle y_n \rangle$, $\langle t_n \rangle$, ℓ, m be as stated. Let $\varepsilon > 0$. Since $\lim\limits_{n \to \infty} s_n = \ell$, there is an $N_s \in \mathbb{N}$ such that

$$|s_n - \ell| < \varepsilon \text{ for all } n > N_s. \tag{3.9}$$

Since $\lim\limits_{n \to \infty} t_n = \ell$, there is an $N_t \in \mathbb{N}$ such that

$$|t_n - \ell| < \varepsilon \text{ for all } n > N_t. \tag{3.10}$$

Let $N = \max\{m, N_s, N_t\}$. Let $n > N$. We now prove that $|y_n - \ell| < \varepsilon$. Since $n > m$, we conclude that $s_n \le y_n \le t_n$. Thus, $s_n - \ell \le y_n - \ell \le t_n - \ell$. By Lemma 2.2.8, we infer that $|y_n - \ell| \le \max\{|s_n - \ell|, |t_n - \ell|\}$. Since $n > N_s$ and $n > N_t$, from (3.9) and (3.10) we see that $\max\{|s_n - \ell|, |t_n - \ell|\} < \varepsilon$. Hence, $|y_n - \ell| < \varepsilon$. □

Theorem 3.2.8 will allow us to establish the limit of two important sequences. First, for $c > 0$, it is not obvious that the sequence $\langle \sqrt[n]{c} \rangle = \langle c, \sqrt[2]{c}, \sqrt[3]{c}, \sqrt[4]{c}, \ldots \rangle$ converges, and if it does, then one could surmise that the limit depends on c. Our next result shows that the sequence does converge and that its limit is independent of c. Recall that $\sqrt[n]{c} = c^{\frac{1}{n}}$, for all $n \in \mathbb{N}$.

Corollary 3.2.9. Let $c > 0$. Then $\lim\limits_{n \to \infty} c^{\frac{1}{n}} = 1$.

Proof. First suppose that $c \ge 1$. Let $d \ge 0$ be such that $1 + d = c$ and let $n \ge 1$ be any natural number. Since $d \ge 0$, we see that $\frac{d}{n} > -1$. By Bernoulli's inequality (see Exercise 5 on page 27), we have that

$$\left(1 + \frac{d}{n}\right)^n \ge 1 + n\frac{d}{n} = 1 + d = c \ge 1.$$

Hence, $\left(1 + \frac{d}{n}\right) \ge c^{\frac{1}{n}} \ge 1$, for all $n \ge 1$. Since $\lim\limits_{n \to \infty} \left(1 + \frac{d}{n}\right) = 1$ and $\lim\limits_{n \to \infty} 1 = 1$, Theorem 3.2.8 implies that $\lim\limits_{n \to \infty} c^{\frac{1}{n}} = 1$. Now assume that $0 < c < 1$. So $\frac{1}{c} > 1$. The previous argument shows that $\lim\limits_{n \to \infty} \frac{1}{c^{\frac{1}{n}}} = 1$. Thus, $\lim\limits_{n \to \infty} c^{\frac{1}{n}} = 1$ by Theorem 3.2.6. □

We know that the sequence $\langle n \rangle = \langle 1, 2, 3, 4, \ldots \rangle$ diverges. So, one would suspect that the sequence $\langle \sqrt[n]{n} \rangle = \langle 1, \sqrt[2]{2}, \sqrt[3]{3}, \sqrt[4]{4}, \ldots \rangle$ also diverges. However, it converges.

Corollary 3.2.10. $\lim\limits_{n \to \infty} n^{\frac{1}{n}} = 1$.

Proof. Define the sequence $\langle b_n \rangle$ by $b_n = n^{\frac{1}{n}} - 1$, for all $n \in \mathbb{N}$. As $\left\langle n^{\frac{1}{n}} \right\rangle = \langle b_n + 1 \rangle$, to prove that $\lim_{n \to \infty} n^{\frac{1}{n}} = 1$, it is sufficient to show that $\lim_{n \to \infty} b_n = 0$ by Theorem 3.2.2. Since $n^{\frac{1}{n}} \geq 1$, we see that $b_n \geq 0$ for each $n \in \mathbb{N}$. Let $n \in \mathbb{N}$. Since $n^{\frac{1}{n}} = b_n + 1$, we have $n = (1 + b_n)^n$. Thus, by Corollary 1.4.12, for all $n \geq 2$

$$n = (1 + b_n)^n \geq 1 + \frac{n(n-1)}{2} b_n^2.$$

Thus, $\frac{2}{n} \geq b_n^2$ and $\frac{\sqrt{2}}{\sqrt{n}} \geq b_n \geq 0$ for all $n \geq 2$. Since $\lim_{n \to \infty} \frac{\sqrt{2}}{\sqrt{n}} = 0$ and $\lim_{n \to \infty} 0 = 0$, Theorem 3.2.8 implies that $\lim_{n \to \infty} b_n = 0$. Hence, $\lim_{n \to \infty} n^{\frac{1}{n}} = 1$, by Theorem 3.2.2. □

We now present another "squeeze" result that will be applied later in the book.

Lemma 3.2.11. Suppose that $\lim_{n \to \infty} x_n = c$. If a sequence $\langle y_n \rangle$ satisfies $|x_n - y_n| < \frac{1}{n}$ for all $n \geq 1$, then $\lim_{n \to \infty} y_n = c$.

Proof. See Exercise 4. □

3.2.3 Order Limit Theorems

We next consider how an inequality that is satisfied by every term in a convergent sequence affects the limit of the sequence. We present several proofs in this section that show that the limit operation preserves the inequality relation $\leq$. Moreover, these proofs only involve the definition of convergence and some standard properties of inequality.

Theorem 3.2.12. If $\lim_{n \to \infty} t_n = t$ where $t_n \geq 0$ for all $n \in \mathbb{N}$, then $t \geq 0$.

Proof. Suppose $\lim_{n \to \infty} t_n = t$ and (▲) $t_n \geq 0$ for all $n \in \mathbb{N}$. We will prove that $t \geq 0$. Suppose, for a contradiction, that $t < 0$. Let $\varepsilon = -t > 0$. Since $\lim_{n \to \infty} t_n = t$, there is an $N \in \mathbb{N}$ such that $|t_n - t| < \varepsilon$ for all $n > N$. Let $n > N$. So $t_n - t \leq |t_n - t| < \varepsilon$. Thus, $t_n - t < \varepsilon = -t$. Hence, $t_n < t - t = 0$ which contradicts (▲). □

Theorem 3.2.12 asserts that if a sequence with nonnegative terms converges, then its limit is also nonnegative. This fact implies the remaining results of this section.

Theorem 3.2.13. If $\lim_{n \to \infty} s_n = s$ and $\lim_{n \to \infty} t_n = t$ where $s_n \leq t_n$ for all $n \in \mathbb{N}$, then $s \leq t$.

Proof. Suppose that $\lim_{n \to \infty} s_n = s$, $\lim_{n \to \infty} t_n = t$ and that $s_n \leq t_n$ for all $n \in \mathbb{N}$. Hence, $0 \leq (t_n - s_n)$ for all $n \in \mathbb{N}$. Exercise 1 implies that $\lim_{n \to \infty} (t_n - s_n) = t - s$. Thus, $0 \leq t - s$ by Theorem 3.2.12. Therefore, $s \leq t$. □

Theorem 3.2.14. Suppose that $\lim_{n \to \infty} s_n = c$. Let a and b be real numbers.

1. If $s_n \leq b$ for all $n \in \mathbb{N}$, then $c \leq b$.
2. If $a \leq s_n$ for all $n \in \mathbb{N}$, then $a \leq c$.

Proof. Suppose $\lim_{n\to\infty} s_n = c$. To prove item 1, suppose that $s_n \leq b$ for all $n \in \mathbb{N}$. Since the constant sequence $\langle b \rangle$ converges to b and $s_n \leq b$ for all $n \in \mathbb{N}$, Theorem 3.2.13 implies that $c \leq b$. A similar argument establishes item 2. $\qquad\square$

Corollary 3.2.15. Let $\langle s_n \rangle$ be a sequence whose terms are all in $[a, b]$. If $\lim_{n\to\infty} s_n = c$, then c is in $[a, b]$.

Corollary 3.2.16. Let $\langle s_n \rangle$ be a sequence such that $\lim_{n\to\infty} s_n = c$ and let $M \geq 0$. If $|s_n| \leq M$ for all $n \in \mathbb{N}$, then $|c| \leq M$.

Proof. If $|s_n| \leq M$ for all $n \in \mathbb{N}$, then $-M \leq s_n \leq M$ for all $n \in \mathbb{N}$. Theorem 3.2.14 implies that $-M \leq c \leq M$, that is, $|c| \leq M$. $\qquad\square$

It is important to realize that strict inequalities are not necessarily preserved under the limit operation. For example, $\frac{1}{n} > 0$ for all $n \in \mathbb{N}$, but $\lim_{n\to\infty} \frac{1}{n} = 0$.

Exercises 3.2

*1. Suppose $\lim_{n\to\infty} s_n = s$ and $\lim_{n\to\infty} t_n = t$. Using proof strategy 3.2.1 as a guide, prove that $\lim_{n\to\infty} (s_n - t_n) = s - t$.

2. Suppose $\lim_{n\to\infty} s_n = s$ and $\lim_{n\to\infty} t_n = t$. Let $a, b \in \mathbb{R}$ be nonzero. Using proof strategy 3.2.1 as a guide, prove that $\lim_{n\to\infty} (as_n + bt_n) = as + bt$.

3. Let $\langle s_n \rangle$ and $\langle t_n \rangle$ be sequences. Suppose that $\langle s_n \rangle$ converges and $\gamma > 0$ is such that $\gamma \leq |t_n|$ for all $n \geq 1$. Prove that the sequence $\left\langle \frac{s_n}{t_n} \right\rangle$ is bounded.

*4. Prove Lemma 3.2.11. (Hint: $|y_n - c| = |y_n - x_n + x_n - c|$.)

5. Let $\lim_{n\to\infty} (a_n + b_n) = \ell$ and $\lim_{n\to\infty} (a_n - b_n) = m$ for $\ell, m \in \mathbb{R}$. Use Theorem 3.2.2 and Exercise 12 on page 67 to show that $\langle a_n \rangle$ converges and evaluate its limit.

6. Given that $\lim_{n\to\infty} (a_n + b_n) = \ell$ and $\lim_{n\to\infty} (a_n - b_n) = m$ for $\ell, m \in \mathbb{R}$, use Exercise 1 and Exercise 12 on page 67 to prove that $\langle b_n \rangle$ converges and evaluate its limit.

7. Let $\langle a_n \rangle$ be a bounded sequence and suppose that $\lim_{n\to\infty} b_n = 0$. Prove that $\lim_{n\to\infty} a_n b_n = 0$.

8. Let $0 < r < 1$. Show that $\lim_{n\to\infty} \sqrt[n]{1 + r^n} = 1$. Conclude that if $x > 1$, then $\lim_{n\to\infty} \sqrt[n]{1 + x^n} = x$.

9. Let $\gamma \in \mathbb{R}$ and $K \in \mathbb{N}$. Suppose that $\lim_{n\to\infty} s_n = \ell$ and $s_n \geq \gamma$ for all $n \geq K$. Prove that $\ell \geq \gamma$.

*10. Prove Theorem 3.2.3.

11. Let $\langle a_n \rangle$ and $\langle b_n \rangle$ be sequences, and let $\ell \in \mathbb{R}$. Suppose that $a_n \leq \ell \leq b_n$ for all $n \geq 1$, and $\lim_{n\to\infty} (a_n - b_n) = 0$. Prove that $\lim_{n\to\infty} a_n = \ell$ and $\lim_{n\to\infty} b_n = \ell$.

12. Let $k \in \mathbb{N}$. Show that $\lim\limits_{n \to \infty} \left(\frac{n+1}{n}\right)^k = 1$.

13. Prove that $\lim\limits_{n \to \infty} (n+1)^{\frac{1}{n}} = 1$.

14. Prove that $\lim\limits_{n \to \infty} (n-1)^{\frac{1}{n}} = 1$.

15. Prove that $\lim\limits_{n \to \infty} (n^2-1)^{\frac{1}{n}} = 1$.

16. Suppose $\lim\limits_{n \to \infty} a_n = \ell > \rho$. Prove that there is a $k \in \mathbb{N}$ such that $a_n > \rho$ for all $n \geq k$.

17. Let $\langle a_n \rangle$ be a sequence of positive terms such that $\lim\limits_{n \to \infty} \frac{a_{n+1}}{a_n} > 1$. Let $\rho \in \mathbb{R}$ be such that $\lim\limits_{n \to \infty} \frac{a_{n+1}}{a_n} > \rho > 1$.

    (a) Show that there exists a $k \in \mathbb{N}$ such that $a_{n+1} > \rho a_n$ for all $n \geq k$. (Use Exercise 16.)

    (b) Prove, by induction, that $a_{k+n} > \rho^n a_k$ for all $n \geq 1$.

    (c) Conclude that $\langle a_n \rangle$ is unbounded.

18. Using Exercise 17 and Theorem 3.4.8, show that the sequence $\left\langle \frac{n^n}{n!} \right\rangle$ is unbounded.

19. Suppose that $\lim\limits_{n \to \infty} s_n = \ell$. Prove that $\lim\limits_{n \to \infty} \frac{1}{n} \sum\limits_{k=1}^{n} s_k = \ell$.

Exercise Notes: For Exercise 8, $1 \leq 1 + r^n \leq (1 + r^n)^n$ (see Exercise 1 on page 26). For Exercise 10, use proof strategy 3.2.1 and the proof analysis on page 70. For Exercise 17, use Exercise 24 on page 67. For Exercise 19, let N_1 be so that $|s_n - \ell| < \frac{\varepsilon}{2}$ when $n > N_1$, and let $N > N_1$ be such that $\frac{1}{n} \sum\limits_{k=1}^{N_1} |s_k - \ell| < \frac{\varepsilon}{2}$, when $n > N$.

3.3 SUBSEQUENCES

Given a sequence, we can create a new sequence, called a subsequence, by choosing certain terms of the given sequence and retaining the same order as in the original sequence. We now present a more formal definition of a subsequence.

Definition 3.3.1. Let $\langle s_n \rangle_{n=1}^{\infty}$ be a sequence. Let $n_1 < n_2 < n_3 < \cdots < n_k < \cdots$ be a strictly increasing infinite sequence of natural numbers. The sequence $\langle s_{n_k} \rangle_{k=1}^{\infty}$ is a **subsequence** of $\langle s_n \rangle$.

Example. Consider the sequence $\langle s_n \rangle = \langle s_1, s_2, s_3, \ldots \rangle$. Let $n_1 = 2$, $n_2 = 5$, $n_3 = 6$, $n_4 = 9$, and $n_4 < n_5 < \cdots < n_k < \cdots$ continue to be a strictly increasing infinite sequence of natural numbers. We can now list the sequence $\langle s_n \rangle$ and the subsequence $\langle s_{n_k} \rangle = \langle s_{n_1}, s_{n_2}, s_{n_3}, \ldots \rangle$ as follows:

$$s_1, \quad s_2, \quad s_3, \quad s_4, \quad s_5, \quad s_6, \quad s_7, \quad s_8, \quad s_9, \quad \cdots$$
$$s_{n_1}, \qquad\qquad\quad s_{n_2}, \quad s_{n_3}, \qquad\qquad\quad s_{n_4}, \quad \cdots$$

Example. For the sequence $\langle \frac{1}{n} \rangle = \langle 1, \frac{1}{2}, \frac{1}{3}, \frac{1}{4}, \ldots \rangle$, let $n_1 < n_2 < \cdots < n_k < \cdots$ be the strictly increasing infinite sequence of natural numbers defined by $n_k = 2k$. Thus, $n_1 = 2$, $n_2 = 4$, $n_3 = 6$, and so on. Thus, $s_{n_1} = s_2 = \frac{1}{2}$, $s_{n_2} = s_4 = \frac{1}{4}$, $s_{n_3} = s_6 = \frac{1}{6}$, and so on. The resulting subsequence $\langle s_{n_k} \rangle_{k=1}^{\infty}$ is $\langle \frac{1}{2}, \frac{1}{4}, \frac{1}{6}, \ldots \rangle$.

Remark 3.3.2. One can give an alternative definition of a subsequence. A function $\sigma \colon \mathbb{N} \to \mathbb{N}$ is *strictly increasing* if and only if for all $m, n \in \mathbb{N}$, if $m < n$, then $\sigma(m) < \sigma(n)$. Let $\langle s_n \rangle$ be a sequence and let $\sigma \colon \mathbb{N} \to \mathbb{N}$ be a strictly increasing function. The range of σ produces the strictly increasing infinite sequence of natural numbers $\sigma(1) < \sigma(2) < \cdots < \sigma(n) < \cdots$. Thus, $\langle s_{\sigma(n)} \rangle$ is a subsequence of $\langle s_n \rangle$. Moreover, if $\langle s_{n_k} \rangle$ is a subsequence of $\langle s_n \rangle$ as in Definition 3.3.1, then the function $\sigma \colon \mathbb{N} \to \mathbb{N}$ defined by $\sigma(k) = n_k$ is strictly increasing and $\langle s_{n_k} \rangle = \langle s_{\sigma(k)} \rangle$. In addition, $\langle s_{\sigma(k)} \rangle = \langle s_{\sigma(n)} \rangle$, that is, $\langle s_{\sigma(k)} \rangle_{k=1}^{\infty} = \langle s_{\sigma(n)} \rangle_{n=1}^{\infty}$.

Lemma 3.3.3. Let $n_1 < n_2 < n_3 < \cdots < n_k < \cdots$ be a strictly increasing sequence of natural numbers; that is, $n_k < n_{k+1}$ for all $k \in \mathbb{N}$. Then for all $k \in \mathbb{N}$, $k \leq n_k$.

Theorem 3.3.4. If the sequence $\langle s_n \rangle$ converges to s, then every subsequence of $\langle s_n \rangle$ also converges to s.

Proof. Suppose that $\lim_{n \to \infty} s_n = s$ and let $\langle s_{n_k} \rangle$ be a subsequence of $\langle s_n \rangle$. To prove that $\lim_{k \to \infty} s_{n_k} = s$, let $\varepsilon > 0$. Since $\lim_{n \to \infty} s_n = s$, there is an $N \in \mathbb{N}$ such that

$$|s_n - s| < \varepsilon \text{ for all } n > N. \tag{3.11}$$

Let $k > N$. By Lemma 3.3.3, $n_k \geq k > N$. Thus, (3.11) implies that $|s_{n_k} - s| < \varepsilon$. $\square$

The above theorem offers two methods for showing that a sequence diverges.

Corollary 3.3.5. If a subsequence of $\langle s_n \rangle$ diverges, then $\langle s_n \rangle$ diverges.

Example. Consider the sequence $\langle s_n \rangle = \langle 1, 2, 1, 3, 1, 4, 1, 5, 1, 6, \ldots \rangle$. This sequence has the divergent subsequence $\langle 2, 3, 4, 5, 6, \ldots \rangle$. Thus, $\langle s_n \rangle$ diverges.

Corollary 3.3.6. If a subsequence of $\langle s_n \rangle$ converges to x and another subsequence of $\langle s_n \rangle$ converges to y where $x \neq y$, then the sequence $\langle s_n \rangle$ diverges.

Example. Consider sequence $\langle s_n \rangle = \langle 1, \frac{1}{2}, 1, \frac{1}{3}, 1, \frac{1}{4}, 1, \frac{1}{5}, 1, \frac{1}{6}, \ldots \rangle$. This sequence has the constant subsequence $\langle 1, 1, 1, 1, 1, \ldots \rangle$ that converges to 1, and the subsequence $\langle \frac{1}{2}, \frac{1}{3}, \frac{1}{4}, \frac{1}{5}, \ldots \rangle$ that converges to 0. Thus, $\langle s_n \rangle$ diverges.

The above example shows that if a sequence $\langle s_n \rangle$ has a convergent subsequence, then one cannot conclude that $\langle s_n \rangle$ converges. However, there is a condition on such a subsequence which does imply that $\langle s_n \rangle$ converges.

Lemma 3.3.7. Let $\langle s_{n_k} \rangle$ be a subsequence of $\langle s_n \rangle$ that converges to a real number ℓ. Suppose that for all $k \in \mathbb{N}$ and $n \in \mathbb{N}$, if $n_k \leq n \leq n_{k+1}$, then either $s_{n_k} \leq s_n \leq s_{n_{k+1}}$ or $s_{n_{k+1}} \leq s_n \leq s_{n_k}$. Then $\langle s_n \rangle$ converges to ℓ.

Proof. Let $\langle s_{n_k} \rangle$ and $\langle s_n \rangle$ be as stated. Let $\varepsilon > 0$. Since $\langle s_{n_k} \rangle$ converges ℓ, there exists a $K \in \mathbb{N}$ such that

$$|s_{n_k} - \ell| < \varepsilon \text{ for all } k \geq K. \tag{3.12}$$

Let $n > n_K$. Now let $k \geq K$ be such that $n_k \leq n \leq n_{k+1}$. So either $s_{n_k} \leq s_n \leq s_{n_{k+1}}$ or $s_{n_{k+1}} \leq s_n \leq s_{n_k}$. Since $k + 1 > k \geq K$, (3.12) implies that $|s_{n_k} - \ell| < \varepsilon$ and $|s_{n_{k+1}} - \ell| < \varepsilon$. Lemma 2.2.11 implies that $|s_n - \ell| < \varepsilon$. Hence, $\lim_{n \to \infty} s_n = \ell$. $\qquad \square$

If a sequence $\langle s_n \rangle$ does not converge to a real number ℓ, then there is an $\varepsilon > 0$ such that all $N \in \mathbb{N}$ there is an $n > N$ where $|s_n - \ell| \geq \varepsilon$ (see Remark 3.1.14). In fact, the following proposition shows that there is a subsequence $\langle s_{n_k} \rangle$ such that $|s_{n_k} - \ell| \geq \varepsilon$ for *all* of the terms in $\langle s_{n_k} \rangle$. This result can be used to show that if a bounded sequence does not converge, then it has two subsequences that converge to different limits (see Exercise 2 on page 86).

Proposition 3.3.8. Suppose that the sequence $\langle s_n \rangle$ does not converge to the real number ℓ. Then there is an $\varepsilon > 0$ and a subsequence $\langle s_{n_k} \rangle$ such that $|s_{n_k} - \ell| \geq \varepsilon$ for all $k \geq 1$. Hence, no subsequence of $\langle s_{n_k} \rangle$ can converge to ℓ.

Proof. Suppose that $\langle s_n \rangle$ does not converge to ℓ. Thus, there is an $\varepsilon > 0$ such that for all $N \in \mathbb{N}$, there is an $n > N$ such that $|s_n - \ell| \geq \varepsilon$. For $N = 1$ let $n_1 \geq 1$ be such that $|s_{n_1} - \ell| \geq \varepsilon$; for $N = n_1$, let $n_2 > n_1$ be such that $|s_{n_2} - \ell| \geq \varepsilon$; continuing with this method, we obtain a subsequence $\langle s_{n_k} \rangle$ such that $|s_{n_k} - \ell| \geq \varepsilon$ for all $k \geq 1$. It follows that no subsequence of $\langle s_{n_k} \rangle$ can converge to ℓ (see Exercise 11). $\qquad \square$

In the next proposition, we dovetail two sequences together to obtain a new sequence and provide a condition under which the new sequence will converge.

Proposition 3.3.9. Let $\langle s_n \rangle$, $\langle t_n \rangle$ be sequences. Then $\langle s_n \rangle$ and $\langle t_n \rangle$ both converge to ℓ if and only if the sequence $\langle s_1, t_1, s_2, t_2, \ldots, s_n, t_n, \ldots \rangle$ converges to ℓ.

Proof. Let $\langle s_n \rangle$ and $\langle t_n \rangle$ be sequences.

($\Rightarrow$). Assume that $\langle s_n \rangle$ and $\langle t_n \rangle$ converge to ℓ. Let $\varepsilon > 0$. Since $\langle s_n \rangle$ converges to ℓ, there is an $N_1 \in \mathbb{N}$ such that $|s_n - \ell| < \varepsilon$ for all $n > N_1$. As $\langle t_n \rangle$ converges to ℓ, there is an $N_2 \in \mathbb{N}$ such that $|t_n - \ell| < \varepsilon$ for all $n > N_2$. Let $N = \max\{N_1, N_2\}$. For all $n > N$, we have that $|s_n - \ell| < \varepsilon$ and $|t_n - \ell| < \varepsilon$. Thus, $\langle s_1, t_1, s_2, t_2, \ldots, s_n, t_n, \ldots \rangle$ converges to ℓ (see Exercise 13).

($\Leftarrow$). If $\langle s_1, t_1, s_2, t_2, \ldots, s_n, t_n, \ldots \rangle$ converges to ℓ, then Theorem 3.3.4 implies that $\langle s_n \rangle$ and $\langle t_n \rangle$ also converge to ℓ. $\qquad \square$

With respect to convergence, the terms at the beginning of a sequence are of little importance. This observation motivates our next definition.

Definition 3.3.10. Let $\langle s_n \rangle_{n=1}^{\infty}$ be a sequence and let $m \geq 0$ be an integer. The subsequence $\langle s_{m+n} \rangle_{n=1}^{\infty} = \langle s_n \rangle_{n=m+1}^{\infty}$ is called a **tail-end** of the sequence $\langle s_n \rangle$.

Example. Consider the sequence $\langle \frac{1}{n} \rangle = \langle 1, \frac{1}{2}, \frac{1}{3}, \frac{1}{4}, \ldots \rangle$. Let $m = 3$. Then we see that the subsequence $\langle \frac{1}{3+n} \rangle = \langle \frac{1}{4}, \frac{1}{5}, \frac{1}{6}, \frac{1}{7}, \ldots \rangle$ is a tail-end of the sequence $\langle \frac{1}{n} \rangle$.

Since the convergence of a sequence does not depend on the terms at the beginning of the sequence, we have the following equivalence.

Corollary 3.3.11. Let $\langle s_n \rangle$ be a sequence and let $m \in \mathbb{N}$. Then $\langle s_n \rangle$ converges to ℓ if and only if $\langle s_{m+n} \rangle$ converges to ℓ.

Proof. Let $m \in \mathbb{N}$. Assume that $\langle s_n \rangle$ converges to ℓ. Since $\langle s_{m+n} \rangle$ is a subsequence of $\langle s_n \rangle$, Theorem 3.3.4 implies that $\langle s_{m+n} \rangle$ converges to ℓ. The converse follows directly from Definition 3.1.1. □

Clearly, every tail-end of a sequence is a subsequence, but not every subsequence is a tail-end.

Exercises 3.3

1. Prove Lemma 3.3.3 by induction on k.

2. Give an example of a sequence that is not bounded above but has a convergent subsequence.

3. Give an example of a sequence that is not bounded below but has a convergent subsequence.

4. Give an example of a bounded sequence that does not converge.

5. Suppose that $\langle s_n \rangle$ is a sequence that is not bounded above. Show that there is a subsequence $\langle s_{n_k} \rangle_{k=1}^{\infty}$ such that $s_{n_k} > k$ for all $k \in \mathbb{N}$.

6. Suppose that $\langle s_n \rangle$ is a sequence that is not bounded below. Show that there is a subsequence $\langle s_{n_k} \rangle_{k=1}^{\infty}$ such that $s_{n_k} < -k$ for all $k \in \mathbb{N}$.

7. Suppose that a sequence $\langle s_n \rangle$ converges to $s \in \mathbb{R}$ and a sequence $\langle t_n \rangle$ converges to $t \in \mathbb{R}$, where $s \neq t$. Explain why the sequence $\langle s_1, t_1, s_2, t_2, \ldots, s_n, t_n, \ldots \rangle$ does not converge.

8. Let $\langle s_n \rangle$ be a sequence and let $m \in \mathbb{N}$. Suppose that the tail-end sequence $\langle s_{m+n} \rangle$ converges to s. Prove that $\langle s_n \rangle$ converges to s.

9. Suppose that $\langle s_n \rangle$ has a subsequence that converges to $x \in \mathbb{R}$ and another subsequence that converges to $y \in \mathbb{R}$, where $x \neq y$. Explain why the sequence $\langle s_n \rangle$ does not converge.

10. Suppose that the sequence $\langle s_n \rangle$ is bounded above and is also bounded below. Show that there is a real number $K > 0$ such that $|s_n| \leq K$ for all $n \in \mathbb{N}$.

*11. Let $\varepsilon > 0$ and let $\ell \in \mathbb{R}$. Suppose that a sequence $\langle s_n \rangle$ satisfies $|s_n - \ell| \geq \varepsilon$ for all $n \geq 1$. Prove that no subsequence of $\langle s_n \rangle$ can converge to ℓ.

12. Suppose $\lim_{n \to \infty} s_n = c$. Let $\sigma \colon \mathbb{N} \to \mathbb{N}$ be one-to-one. Prove that $\lim_{n \to \infty} s_{\sigma(n)} = c$.

*13. Suppose that $\lim_{n \to \infty} s_n = \ell$ and $\lim_{n \to \infty} t_n = \ell$. Define the sequence $\langle u_n \rangle$ by

$$u_n = \begin{cases} s_{\frac{n+1}{2}}, & \text{if } n \text{ is odd;} \\ t_{\frac{n}{2}}, & \text{if } n \text{ is even.} \end{cases}$$

Using Definition 3.1.1, prove that $\lim\limits_{n\to\infty} u_n = \ell$.

*14. Let $\langle s_n \rangle$ be a sequence. Prove that $\langle s_n \rangle$ converges to ℓ if and only if the two subsequences $\langle s_{2k} \rangle$ and $\langle s_{2k-1} \rangle$ converge to ℓ.

15. Suppose $\lim\limits_{n\to\infty} a_n = \ell < \rho$. Prove that there is a $k \in \mathbb{N}$ such that $a_n < \rho$ for all $n \geq k$.

*16. Let $\langle a_n \rangle$ be a sequence of positive terms such that $\lim\limits_{n\to\infty} \frac{a_{n+1}}{a_n} < 1$. Let $\rho \in \mathbb{R}$ be such that $\lim\limits_{n\to\infty} \frac{a_{n+1}}{a_n} < \rho < 1$.

   (a) Show that there exists a $k \in \mathbb{N}$ such that $a_{n+1} < \rho a_n$ for all $n \geq k$. (Use Exercise 15.)
   (b) Prove, by induction, that $a_{k+n} < \rho^n a_k$ for all $n \geq 1$.
   (c) Conclude that $\lim\limits_{n\to\infty} a_n = 0$.

17. Let $0 < |r| < 1$ and $k \in \mathbb{N}$. Use Exercise 16 to show that the sequences converge to 0.

   (a) $\left\langle \frac{2^n}{n!} \right\rangle$
   (b) $\left\langle \frac{n^k}{n!} \right\rangle$
   (c) $\langle n^k r^n \rangle$.

18. Let $\sigma: \mathbb{N} \to \mathbb{N}$ and $\tau: \mathbb{N} \to \mathbb{N}$ be strictly increasing. Prove that $(\sigma \circ \tau): \mathbb{N} \to \mathbb{N}$ is strictly increasing. Let $\langle s_n \rangle$ be a sequence. Using Remark 3.3.2, conclude that (a) $\langle s_{\sigma(\tau(n))} \rangle$ is a subsequence of $\langle s_n \rangle$, and (b) $\langle s_{\sigma(\tau(n))} \rangle$ is a subsequence of $\langle s_{\sigma(n)} \rangle$.

Exercise Notes: For Exercise 5, Proposition 3.1.20 allows one to construct the required subsequence as follows: For $k = 1$ let n_1 be such that $s_{n_1} > 1$; for $k = 2$ let $n_2 > n_1$ be such that $s_{n_2} > 2$; etc. For 5(b), Proposition 3.1.22 allows one to construct the desired subsequence as well. For Exercise 12, observe that for any $N \in \mathbb{N}$, the set $\{n \in \mathbb{N} : \sigma(n) \leq N\}$ is finite. This exercise implies that if a sequence $\langle s_n \rangle$ converges to c, then any sequence obtained by reordering the terms in a subsequence of $\langle s_n \rangle$ also converges to c. For Exercises 13 and 14, given the appropriate N_1 and N_2, let $N = \max\{2N_1 - 1, 2N_2\}$. Note: n is odd iff $n = 2k - 1$ for some $k \in \mathbb{N}$. For Exercise 17(c), see Corollary 3.1.12. For Exercise 18(b), let $\langle t_n \rangle = \langle s_{\sigma(n)} \rangle$, where $t_n = s_{\sigma(n)}$ for all $n \in \mathbb{N}$.

3.4 MONOTONE SEQUENCES

A sequence is *increasing* if each term is less than or equal to the term after it. A sequence is *decreasing* if each term is greater than or equal to the term after it.

Definition 3.4.1. A sequence $\langle s_n \rangle$ is an **increasing sequence** if for all $n \in \mathbb{N}$, we have that $s_n \leq s_{n+1}$. A sequence $\langle s_n \rangle$ is a **decreasing sequence** if for all $n \in \mathbb{N}$, we have that $s_n \geq s_{n+1}$. A sequence is **monotone** if it is either increasing or decreasing.

Thus, if sequence $\langle s_n \rangle$ is increasing, then $s_n \leq s_m$ whenever $n \leq m$, and if $\langle s_n \rangle$ is decreasing, then $s_n \geq s_m$ when $n \leq m$.

Example. The sequences $\langle 1 - \frac{1}{n} \rangle$, $\langle 3n \rangle$, and $\langle 1, 1, 2, 2, 3, 3, \ldots \rangle$ are increasing. The sequence $\langle \frac{1}{n} \rangle$ is decreasing, and the sequences $\langle (-1)^n \rangle$ and $\left\langle \frac{(-1)^n}{n} \right\rangle$ are not monotone.

An increasing sequence that is not bounded above does not converge; for example, the sequence $\langle 3n \rangle$ diverges. However, if a sequence is increasing and bounded above, then the terms of the sequence eventually cluster together with the supremum of the set of these terms (see Figure 3.4).

Lemma 3.4.2. If $\langle s_n \rangle$ is increasing, bounded above, and $\beta = \sup\{s_n : n \in \mathbb{N}\}$, then $\lim\limits_{n \to \infty} s_n = \beta$.

Proof. Assume that $\langle s_n \rangle$ is increasing and bounded above. Let $S = \{s_n : n \in \mathbb{N}\}$ and $\beta = \sup(S)$. Thus, $s_n \leq \beta$ for all $n \in \mathbb{N}$. To prove that $\lim\limits_{n \to \infty} s_n = \beta$, let $\varepsilon > 0$. As $\beta - \varepsilon < \beta$, there is an $N \in \mathbb{N}$ such that $\beta - \varepsilon < s_N$. Thus, (▲) $\beta - s_N < \varepsilon$. Let $n > N$. So $s_N \leq s_n$ and (◆) $\beta - s_n \leq \beta - s_N$. We now prove that $|s_n - \beta| < \varepsilon$ as follows:

$$|s_n - \beta| = \beta - s_n \qquad \text{because } s_n \leq \beta$$
$$\leq \beta - s_N < \varepsilon \quad \text{by (▲) and (◆).} \qquad \qquad \square$$

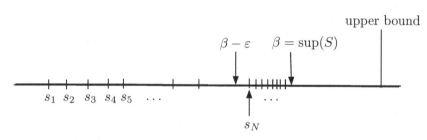

Figure 3.4: Representation of the proof of Lemma 3.4.2, where $S = \{s_n : n \in \mathbb{N}\}$.

Corollary 3.4.3. If $\langle s_n \rangle$ is increasing and $\lim\limits_{n \to \infty} s_n = s$, then $s_n \leq s$ for all $n \geq 1$.

Proof. Suppose that $\langle s_n \rangle$ is increasing and $\lim\limits_{n \to \infty} s_n = s$. Then $s = \sup\{s_n : n \in \mathbb{N}\}$ by Lemma 3.4.2. Thus, $s_n \leq s$ for all $n \geq 1$. $\qquad \square$

Our next lemma (see Exercise 4) shows that if a sequence is decreasing and bounded below, then the limit of the sequence is the infimum of the set of its terms.

Lemma 3.4.4. If $\langle s_n \rangle$ is decreasing, bounded below, and $\alpha = \inf\{s_n : n \in \mathbb{N}\}$, then $\lim\limits_{n \to \infty} s_n = \alpha$.

Corollary 3.4.5. If $\langle s_n \rangle$ is decreasing and $\lim\limits_{n \to \infty} s_n = s$, then $s_n \geq s$ for all $n \geq 1$.

We can now show that a bounded monotone sequence converges.

Theorem 3.4.6 (Monotone Convergence Theorem). Suppose that $\langle s_n \rangle$ is a monotone sequence. Then $\langle s_n \rangle$ is convergent if and only if $\langle s_n \rangle$ is bounded.

Proof. Let $\langle s_n \rangle$ be a monotone sequence. If $\langle s_n \rangle$ is convergent, then Theorem 3.1.24 implies that the sequence is bounded. If $\langle s_n \rangle$ is bounded, then either Lemma 3.4.2 or Lemma 3.4.4 implies $\langle s_n \rangle$ is convergent. □

Problem 3.4.7. Consider the sequence $\langle s_n \rangle$ which is inductively defined by $s_1 = 2$ and (▲) $s_{n+1} = \sqrt{6 + s_n}$ for all $n \geq 1$. Prove by induction that the sequence is monotone and bounded. Using the Monotone Convergence Theorem show that the sequence $\langle s_n \rangle$ converges and then evaluate its limit.

Solution. We begin by proving the following proposition.

Proposition. For every natural number $n \geq 1$, $0 < s_n \leq s_{n+1} \leq 10$.

Proof. We use mathematical induction.

Base step: For $n = 1$, we have that $s_1 = 2$ and $s_2 = \sqrt{6 + s_1} = \sqrt{8} = 2\sqrt{2}$. Thus, $0 < s_1 \leq s_2 \leq 10$.

Inductive step: Let $n \geq 1$ and assume the induction hypothesis that

$$0 < s_n \leq s_{n+1} \leq 10. \tag{IH}$$

We prove that $0 < s_{n+1} \leq s_{n+2} \leq 10$. Note that (▲) implies $s_{n+1} = \sqrt{6 + s_n}$ and $s_{n+2} = \sqrt{6 + s_{n+1}}$. Thus,

$$
\begin{aligned}
&0 < s_n \leq s_{n+1} \leq 10 && \text{by (IH)} \\
&0 < 6 + s_n \leq 6 + s_{n+1} \leq 6 + 10 && \text{by prop. of inequality} \\
&0 < \sqrt{6 + s_n} \leq \sqrt{6 + s_{n+1}} \leq \sqrt{16} && \text{by prop. of inequality} \\
&0 < s_{n+1} \leq s_{n+2} \leq 4 && \text{by (▲).}
\end{aligned}
$$

Hence, $0 < s_{n+1} \leq s_{n+2} \leq 10$. This completes the proof of the proposition. □

Thus, $\langle s_n \rangle$ is monotone and bounded and so, $\lim_{n \to \infty} s_n = s$ exists by Theorem 3.4.6. Since $s_{n+1} = \sqrt{6 + s_n}$, we conclude that (▼) $s_{n+1}^2 = 6 + s_n$. Moreover, $\lim_{n \to \infty} s_{n+1} = s$ by Theorem 3.3.4. Hence, $\lim_{n \to \infty} s_{n+1}^2 = s^2$ by Theorem 3.2.3. Thus,

$$
\begin{aligned}
s^2 = \lim_{n \to \infty} s_{n+1}^2 &= \lim_{n \to \infty} (6 + s_n) && \text{by (▼)} \\
&= 6 + \lim_{n \to \infty} s_n && \text{by Exercise 11 on page 67} \\
&= 6 + s && \text{because } \lim_{n \to \infty} s_n = s.
\end{aligned}
$$

Therefore, $s^2 - s - 6 = 0$. The roots of this equation are $s = -2, 3$. Since $s_n \geq 0$ for all $n \geq 1$, Theorem 3.2.12 implies that $s = 3$. So, $\lim_{n \to \infty} s_n = 3$.

Consider the sequence $\langle (1 + \frac{1}{n})^n \rangle$, called *Euler's sequence*. Does it converge? The answer to this question is not at all obvious. In fact, the sequence does converge and this will be verified in the following proof by appealing to the Monotone Convergence Theorem 3.4.6, a very important theoretical tool.

Theorem 3.4.8. The sequence $\langle (1 + \frac{1}{n})^n \rangle$ converges to a real number ℓ. Moreover, $2 \leq \ell \leq 3$.

Proof. We shall show that $\langle (1 + \frac{1}{n})^n \rangle$ is increasing, bounded above by 3, and bounded below by 2. By Theorem 1.4.11, for all natural numbers $m \geq 3$, we have that

$$\left(1 + \frac{1}{m}\right)^m = \sum_{k=0}^{m} \binom{m}{k} \frac{1}{m^k} = 1 + 1 + \sum_{k=2}^{m} \binom{m}{k} \frac{1}{m^k}$$

$$= 2 + \left(\sum_{k=2}^{m-1} \binom{m}{k} \frac{1}{m^k}\right) + \frac{1}{m^m}. \tag{3.13}$$

One can check that ($\blacktriangle$) $2 = (1 + \frac{1}{1})^1 < (1 + \frac{1}{2})^2 < (1 + \frac{1}{3})^3$. We will now show that $(1 + \frac{1}{n})^n < (1 + \frac{1}{n+1})^{n+1}$ for all $n \geq 3$, as follows:

$$\left(1 + \frac{1}{n}\right)^n = 1 + 1 + \sum_{k=2}^{n} \binom{n}{k} \frac{1}{n^k} \qquad \text{by (3.13)}$$

$$= 2 + \sum_{k=2}^{n} \frac{1}{k!} \left(1 - \frac{1}{n}\right)\left(1 - \frac{2}{n}\right)\cdots\left(1 - \frac{k-1}{n}\right) \qquad \text{by Exercise 8}$$

$$< 2 + \sum_{k=2}^{n} \frac{1}{k!} \left(1 - \frac{1}{n+1}\right)\left(1 - \frac{2}{n+1}\right)\cdots\left(1 - \frac{k-1}{n+1}\right) \qquad \text{Quo. Prin. 3.1.3}$$

$$= 2 + \sum_{k=2}^{n} \binom{n+1}{k} \frac{1}{(n+1)^k} \qquad \text{by Exercise 8}$$

$$< 2 + \left(\sum_{k=2}^{n} \binom{n+1}{k} \frac{1}{(n+1)^k}\right) + \frac{1}{(n+1)^{n+1}} \qquad \frac{1}{(n+1)^{n+1}} > 0$$

$$= 1 + 1 + \sum_{k=2}^{n+1} \binom{n+1}{k} \frac{1}{(n+1)^k} \qquad \text{by (3.13)}$$

$$= \left(1 + \frac{1}{n+1}\right)^{n+1} \qquad \text{by (3.13).}$$

Therefore, $(1 + \frac{1}{n})^n < (1 + \frac{1}{n+1})^{n+1}$. Hence, the sequence is increasing. Moreover, ($\blacktriangle$), (3.13), and Exercise 8 imply that for $n \geq 3$,

$$2 < \left(1 + \frac{1}{n}\right)^n = \sum_{k=0}^{n} \binom{n}{k} \frac{1}{n^k} = 2 + \sum_{k=2}^{n} \binom{n}{k} \frac{1}{n^k} \leq 2 + \sum_{k=2}^{n} \frac{1}{2^{k-1}} < 3$$

(see Theorem 1.4.7 and the Shift Rule 1.4.8). Theorems 3.4.6 and 3.2.14 now imply that $\langle (1 + \frac{1}{n})^n \rangle$ converges to a limit ℓ where $2 \leq \ell \leq 3$. $\square$

3.4.1 The Monotone Subsequence Theorem

In this section we shall prove that every sequence of real numbers has a monotone subsequence. First we define what it means for a term in a sequence to be a "peak."

Definition 3.4.9. Let $\langle s_n \rangle$ be a sequence. Let s_m be the mth term of this sequence. We say that s_m is a **peak** if $s_m > s_n$ for all $n > m$ (see Figure 3.5).

Let s_m be the mth term of the sequence $\langle s_n \rangle$. Then s_m is a peak if it is strictly greater than every term that follows it. On the other hand, s_m is not a peak if s_m is less than or equal to some subsequent term; that is, $s_m \leq s_n$ for some $n > m$.

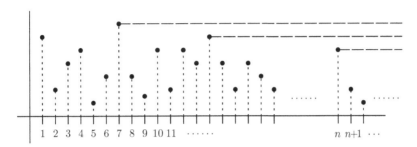

Figure 3.5: s_7, s_{14}, and s_n are peaks; s_4 and s_8 are not a peaks.

Example. Consider the sequence $\langle s_n \rangle = \left\langle \frac{(-1)^n}{n} \right\rangle$. So, $\langle s_n \rangle = \langle -1, \frac{1}{2}, -\frac{1}{3}, \frac{1}{4}, -\frac{1}{5}, \dots \rangle$. Thus, $s_4 = \frac{1}{4}$ is a peak and $s_5 = -\frac{1}{5}$ is not a peak because $s_5 = -\frac{1}{5} < \frac{1}{6} = s_6$. Let $P = \{ s_m : s_m \text{ is a peak} \}$. Then $P = \{ \frac{1}{2}, \frac{1}{4}, \frac{1}{6}, \dots \}$ and P is infinite.

Example. Consider the sequence $\langle s_n \rangle = \langle 2, 1, \frac{1}{2}, \frac{2}{3}, \frac{3}{4}, \dots \rangle$. Thus, $s_1 = 2$ is a peak and $s_3 = \frac{1}{2}$ is not a peak since $s_3 = \frac{1}{2} < \frac{3}{4} = s_5$. Let $P = \{ s_m : s_m \text{ is a peak} \}$. Then $P = \{ 2, 1 \}$ and P is finite.

Theorem 3.4.10 (Monotone Subsequence Theorem). Every sequence of real numbers has a monotone subsequence.

Proof. Let $\langle s_n \rangle$ be a sequence of real numbers. Let $P = \{ s_m : s_m \text{ is a peak} \}$. There are two cases to consider: Either P is infinite or P is finite.

CASE 1: P is infinite. Since P is infinite, we can construct a subsequence of $\langle s_n \rangle$, consisting of peaks, as follows: Let m_1 be the first index such that s_{m_1} is a peak. Let m_2 be the smallest natural number larger than m_1 such that s_{m_2} is a peak. Since P is infinite, we can continue in this manner obtaining $m_1 < m_2 < m_3 < \cdots < m_k < \cdots$ where the subsequence $\langle s_{m_k} \rangle$ is such that each s_{m_k} is a peak. Thus, as each s_{m_k} is a peak, we conclude that $s_{m_1} > s_{m_2} > s_{m_3} > \cdots > s_{m_k} > \cdots$ and thus, we have constructed a monotone subsequence.

CASE 2: P is finite. Since P is finite, let $P = \{ s_{m_1}, s_{m_2}, \dots, s_{m_r} \}$ be a finite listing of *all* the peaks. We can construct a subsequence of $\langle s_n \rangle$, consisting of terms that are not peaks, as follows: Let n_1 be larger than all of natural numbers in $\{ m_1, m_2, \dots, m_r \}$. Thus, s_{n_1} is not a peak and that s_k is not a peak for all $k \geq n_1$. Since s_{n_1} is not a peak, there is a natural number $n_2 > n_1$ where $s_{n_1} \leq s_{n_2}$. Now, since s_{n_2} is not a peak, there is a natural number $n_3 > n_2$ such that $s_{n_2} \leq s_{n_3}$. We can continue in this manner obtaining $n_1 < n_2 < n_3 < \cdots < n_k < \cdots$ where the subsequence $\langle s_{n_k} \rangle$ is such that each term s_{n_k} is not a peak and $s_{n_1} \leq s_{n_2} \leq s_{n_3} \leq \cdots \leq s_{n_k} \leq \cdots$. Hence, we have a monotone subsequence. □

Exercises 3.4

1. Give an example of a sequence that converges and is not monotone.

2. Prove that each of the following sequences is monotone and bounded:

    (a) $\left\langle \frac{2^n}{n!} \right\rangle$ (b) $\left\langle \frac{n+3}{6n-1} \right\rangle$ (c) $\left\langle \sqrt[n]{4} \right\rangle$.

3. Show that the sequence $\left\langle n + \frac{(-1)^n}{n} \right\rangle$ is increasing and does not converge.

*4. Prove Lemma 3.4.4.

5. Prove Corollary 3.4.5.

6. Inductively define the sequence $\langle s_n \rangle$ by $s_1 = 1$ and $s_{n+1} = \frac{1}{4}(2s_n + 3)$ for all $n \geq 1$. Show that $\langle s_n \rangle$ converges and then find its limit. (See Problem 3.4.7.)

7. Inductively define the sequence $\langle s_n \rangle$ by $s_1 = 1$ and $s_{n+1} = 3 - \frac{1}{s_n}$ for all $n \geq 1$. Show that $\langle s_n \rangle$ converges and then find its limit. (See Problem 3.4.7.)

*8. Let $m \in \mathbb{N}$ and $k \in \mathbb{N}$ be such that $m \geq k \geq 2$. Using algebra, show that

$$\binom{m}{k} \frac{1}{m^k} = \frac{1}{k!} \left(1 - \frac{1}{m}\right) \left(1 - \frac{2}{m}\right) \cdots \left(1 - \frac{k-1}{m}\right).$$

    Now conclude that $\binom{m}{k} \frac{1}{m^k} < \frac{1}{k!} \leq \frac{1}{2^{k-1}}$.

9. Let $\langle s_n \rangle$ be a monotone sequence with a subsequence $\langle s_{n_k} \rangle$ that converges to ℓ. Prove that $\langle s_n \rangle$ converges to ℓ.

10. Let $A \subseteq \mathbb{R}$ be nonempty and bounded. Let $\beta = \sup(A)$. Thus, for each $n \in \mathbb{N}$, there is a $b_n \in A$ such that $\beta - \frac{1}{n} < b_n$. By Theorem 3.4.10 the sequence $\langle b_n \rangle$ has a monotone subsequence $\langle b_{n_k} \rangle$.

    (a) Show that $\lim_{n\to\infty} b_n = \beta$.
    (b) Prove that $\lim_{k\to\infty} b_{n_k} = \beta$.
    (c) Suppose that $\beta \notin A$. Prove that $\langle b_{n_k} \rangle$ must be an increasing sequence.

11. Let $A \subseteq \mathbb{R}$ be nonempty and bounded. Let $\alpha = \inf(A)$. Thus, for each $n \in \mathbb{N}$, there is an $a_n \in A$ such that $a_n < \alpha + \frac{1}{n}$. By Theorem 3.4.10 the sequence $\langle a_n \rangle$ has a monotone subsequence $\langle a_{n_k} \rangle$.

    (a) Show that $\lim_{n\to\infty} a_n = \alpha$.
    (b) Prove that $\lim_{k\to\infty} a_{n_k} = \alpha$.
    (c) Suppose that $\alpha \notin A$. Prove that $\langle a_{n_k} \rangle$ must be a decreasing sequence.

3.5   BOLZANO–WEIERSTRASS THEOREMS

Bolzano–Weierstrass Theorem for sequences is a fundamental result which states that each bounded sequence has a convergent subsequence. This theorem is named after the mathematicians Bernard Bolzano and Karl Weierstrass. It was first proved by Bolzano, but his proof was lost. It was re-proven by Weierstrass and has become an important centerpiece of analysis.

Theorem 3.5.1 (Bolzano–Weierstrass Theorem for Sequences). If the sequence $\langle s_n \rangle$ is bounded, then $\langle s_n \rangle$ has a convergent subsequence.

Proof. Let $\langle s_n \rangle$ be a bounded sequence. By Theorem 3.4.10, there exists a monotone subsequence $\langle s_{n_k} \rangle$. Since $\langle s_n \rangle$ is bounded, it follows that $\langle s_{n_k} \rangle$ is bounded. As $\langle s_{n_k} \rangle$ is a bounded monotone sequence, Theorem 3.4.6 implies that $\langle s_{n_k} \rangle$ converges.   □

Given a point $x \in \mathbb{R}$ and a set $S \subseteq \mathbb{R}$, item 1 of our next definition can be viewed as addressing the question: What does it mean to say that x is "close" to S?

Definition 3.5.2. Let S be a subset of $\mathbb{R}$.

1. A point $x \in \mathbb{R}$ is an **accumulation point** of S if every neighborhood of x contains an infinite number of points from S; that is, if U^x is any neighborhood of x, then $S \cap U^x$ is infinite.
2. A point $x \in \mathbb{R}$ is an **isolated point of** S if $x \in S$ and x is not an accumulation point of S; that is, there is a neighborhood U^x of x such that $S \cap U^x = \{x\}$.

A point $x \in \mathbb{R}$ is an *accumulation point* of a set S if for every neighborhood of x (no matter how small), there are an infinite number of points from the set S that are in the neighborhood. So if I is an interval and either $x \in I$ or x is an endpoint of I, then x is an accumulation point of I.

A point $x \in S$ is an *isolated point* of S if there is a neighborhood of x in which there are no other points from the set S (x is all alone; i.e., x is the only point from S living in this neighborhood).

Remark. An accumulation point of S may be in the set S or it may not be in S. On the other hand, an isolated point must be in S.

Example. For $S = [0, 3) \cup \{4\}$, we see that every point $x \in [0, 3]$ is an accumulation point of S whereas, 4 is an isolated point of S. For $S = \mathbb{Q}$, as $\mathbb{Q}$ is dense in $\mathbb{R}$, we see that every point $x \in \mathbb{R}$ is an accumulation of S.

We will next show that a bounded infinite set of real numbers has at least one accumulation point.

Theorem 3.5.3 (Bolzano–Weierstrass Theorem for Sets). Let $S \subseteq \mathbb{R}$ be infinite. If S is bounded, then there is a point $x \in \mathbb{R}$ such that x is an accumulation point of S.

Proof. Let $S \subseteq \mathbb{R}$ be infinite and bounded. As S is infinite, there is a sequence $\langle s_n \rangle$ of distinct points in S, that is, $s_n \in S$ for all $n \in \mathbb{N}$. Since S is bounded, the sequence $\langle s_n \rangle$ is also bounded. Theorem 3.5.1 implies that $\langle s_n \rangle$ has a convergent subsequence $\langle s_{n_k} \rangle$. Let x be the limit of this subsequence $\langle s_{n_k} \rangle$. Corollary 3.1.17 implies that every neighborhood of x contains an infinite number of points from the subsequence $\langle s_{n_k} \rangle$. Since each $s_{n_k} \in S$, it follows that x is an accumulation point of S. □

Theorem 3.5.4. Let $S \subseteq \mathbb{R}$ and x be an accumulation point of S. Then there is a sequence of distinct points $\langle s_n \rangle$ in S that converges to x.

Proof. Let x be an accumulation point of S. Recall that for any $n \geq 1$, a real number a is in the neighborhood $(x - \frac{1}{n}, x + \frac{1}{n})$ if and only if $|a - x| < \frac{1}{n}$. Now, for each $n \in \mathbb{N}$, we have that the $(x - \frac{1}{n}, x + \frac{1}{n})$ contains an infinite number of points in S. We now define a sequence with distinct points as follows: For $n = 1$ choose $s_1 \in S$ such that $|s_1 - x| < \frac{1}{1}$. For $n = 2$ choose an $s_2 \in S$ so that $s_2 \neq s_1$ and $|s_2 - x| < \frac{1}{2}$. For $n = 3$ choose an $s_3 \in S$ so that $s_3 \neq s_1, s_2$ and $|s_3 - x| < \frac{1}{3}$. Continuing in this manner, we obtain a sequence $\langle s_n \rangle$ of distinct points in S such that $|s_n - x| < \frac{1}{n}$ for all $n \geq 1$. Theorem 3.1.11 implies that the sequence $\langle s_n \rangle$ converges to x. □

Exercises 3.5

1. Can you find a sequence $\langle s_n \rangle$ such that $1 \leq s_n \leq 5$ for all $n \geq 1$ where $\langle s_n \rangle$ has no convergent subsequence?

*2. Let $\langle s_n \rangle$ be a bounded sequence that does not converge. By Theorem 3.5.1 there is a subsequence $\langle s_{n_i} \rangle$ that converges to some real number ℓ. Show that there is another subsequence of $\langle s_n \rangle$ that converges to a real number different than ℓ.

3. Let $\langle s_n \rangle$ be a bounded sequence. Suppose that every convergent subsequence of $\langle s_n \rangle$ converges to the same value ℓ. Explain why Exercise 2 implies that $\langle s_n \rangle$ must converge.

4. For each of the following subsets S of $\mathbb{R}$, find some accumulation points (if any) and find some isolated points (if any).

   (a) $S = [0, 3)$. (c) $S = \{\frac{1}{n} : n \in \mathbb{N}\}$.
   (b) $S = \mathbb{N}$. (d) $S = \{q \in \mathbb{Q} : 0 < q < 1\}$.

5. Let $\mathbb{I} \subseteq \mathbb{R}$ be the set of irrational numbers. Find the set of all accumulation points of $\mathbb{I}$.

6. Let $S = \{x \in \mathbb{Q} : x < 1\}$.

   (a) Show that 1 is an accumulation point of S.
   (b) Define a sequence $\langle x_n \rangle$ of points in S that converges to 1.

7. Let $S = \{x\sqrt{2} : x \in \mathbb{Q} \text{ and } x < 1\}$.

   (a) Show that $\sqrt{2}$ is an accumulation point of S.
   (b) Define a sequence $\langle x_n \rangle$ of points in S that converges to $\sqrt{2}$.

8. Let $S \subseteq \mathbb{R}$. Suppose that $\langle s_n \rangle$ is a sequence of distinct points in S that converges to x. Show that x is an accumulation point of S.

9. Let $\langle s_n \rangle$ be a sequence satisfying $|s_n - s_m| < M$ for all $n, m \geq 1$, where $M > 0$. Prove that $\langle s_n \rangle$ has a convergent subsequence.

10. Let $\langle s_n \rangle$ be a sequence such that $a \leq s_n \leq b$ for infinitely many $n \in \mathbb{N}$, where a and b are real numbers. Show that $\langle s_n \rangle$ has a convergent subsequence.

11. Let S be a subset of $\mathbb{R}$, and let $x \in \mathbb{R}$. Suppose that every neighborhood of x contains a point from S that is different from x. Prove that every neighborhood of x contains an infinite number of points from S that are all different from x.

Exercise Notes: For Exercise 2, Proposition 3.3.8 implies that there is a subsequence $\langle s_{n_k} \rangle$ that does not converge to ℓ. This subsequence is bounded. For Exercise 11, let U_ε^x be a neighborhood of x, where $\varepsilon > 0$. Suppose that $(U_\varepsilon^x \setminus \{x\}) \cap S = \{u_1, u_2, \ldots, u_n\}$ is finite. Let $\varepsilon^* = \min\{|u_1 - x|, |u_2 - x|, \ldots, |u_n - x|\} > 0$. Show that $U_{\varepsilon^*}^x \subseteq U_\varepsilon^x$ and each $u_i \notin U_{\varepsilon^*}^x$. Derive a contradiction.

3.6 CAUCHY SEQUENCES

One of the problems with deciding if a sequence converges is that we need to have a purported limit before we can apply the limit definition. Augustin Cauchy found a way around this problem, called the Cauchy Convergence Criterion. A sequence that satisfies this criterion is said to be a Cauchy sequence. The formal definition of a Cauchy sequence, below, makes no reference to a limit; but, it does require that the terms of the sequence become arbitrarily close to each other as the sequence progresses.

Definition 3.6.1. A sequence $\langle s_n \rangle$ is a **Cauchy Sequence** if for every $\varepsilon > 0$ there exists an $N \in \mathbb{N}$ such that for all $m, n \in \mathbb{N}$, if $m, n > N$, then $|s_n - s_m| < \varepsilon$.

Proof Strategy 3.6.2. To prove that a sequence $\langle s_n \rangle$ is Cauchy, use the proof diagram:

> Let $\varepsilon > 0$ be an arbitrary real number.
> Let $N = $ (the natural number you found).
> Let $m, n > N$ be arbitrary natural numbers.
> Prove $|s_n - s_m| < \varepsilon$.

Lemma 3.6.3. Every convergent sequence is a Cauchy sequence.

Proof. Let $\langle s_n \rangle$ be a convergent sequence. Let $\lim_{n \to \infty} s_n = s$. To prove that $\langle s_n \rangle$ is a Cauchy sequence, let $\varepsilon > 0$. Since $\lim_{n \to \infty} s_n = s$, there is an $N \in \mathbb{N}$ such that

$$\text{for all } n \in \mathbb{N}, \text{ if } n > N, \text{ then } |s_n - s| < \frac{\varepsilon}{2}. \tag{3.14}$$

Now let $m, n > N$. We shall prove that $|s_n - s_m| < \varepsilon$ as follows

$$
\begin{aligned}
|s_n - s_m| &= |(s_n - s) + (s - s_m)| &\text{by algebra}\\
&= |s_n - s| + |s - s_m| &\text{by triangle inequality}\\
&< \frac{\varepsilon}{2} + \frac{\varepsilon}{2} = \varepsilon &\text{by (3.14).}
\end{aligned}
$$

Thus, the sequence $\langle s_n \rangle$ is a Cauchy sequence. □

We know by Theorem 3.1.24 that a convergent sequence is bounded. The same is true of a Cauchy sequence.

Lemma 3.6.4. Every Cauchy sequence is bounded.

Proof. Let $\langle s_n \rangle$ be a Cauchy sequence. Let $\varepsilon = 1$ and $N \in \mathbb{N}$ be so that $|s_n - s_m| < 1$ holds for all $m, n > N$. Thus,

$$
|s_n| - |s_m| \le |s_n - s_m| < 1
$$

for all $m, n > N$. Let m_0 be any fixed natural number $m_0 > N$. Hence, $|s_n| < |s_{m_0}| + 1$ for all $n > N$. Let $M = \max\{|s_1|, \ldots, |s_N|, |s_{m_0}| + 1\}$. We see that $|s_n| \le M$ for all $n \in \mathbb{N}$. Therefore, $\langle s_n \rangle$ is a bounded sequence. □

In the proof of the following theorem, Lemma 3.6.3 is used to conclude that a convergent sequence is a Cauchy sequence. Lemma 3.6.4 and the Bolzano–Weierstrass Theorem 3.5.1 are then used to prove the converse.

Theorem 3.6.5 (Cauchy Convergence Criterion). Let $\langle s_n \rangle$ be a sequence. Then $\langle s_n \rangle$ is convergent if and only if $\langle s_n \rangle$ is a Cauchy sequence.

Proof. If $\langle s_n \rangle$ is convergent, then Lemma 3.6.3 implies that $\langle s_n \rangle$ is a Cauchy sequence. To prove the converse, assume that $\langle s_n \rangle$ is a Cauchy sequence. Lemma 3.6.4 implies that $\langle s_n \rangle$ is a bounded sequence. Theorem 3.5.1 implies that $\langle s_n \rangle$ has a convergent subsequence $\langle s_{n_k} \rangle$. Let x be the limit of this subsequence $\langle s_{n_k} \rangle$. We now prove that the sequence $\langle s_n \rangle$ also converges to x. To do this, let $\varepsilon > 0$. Since $\langle s_n \rangle$ is a Cauchy sequence, there is an $N \in \mathbb{N}$ such that

$$
\text{for all } m, n \in \mathbb{N}, \text{ if } m, n > N, \text{ then } |s_n - s_m| < \frac{\varepsilon}{2}. \tag{3.15}
$$

Let $n > N$. Because x is the limit of the subsequence $\langle s_{n_k} \rangle$, there is a natural number $n_k > N$ such that (▲) $|s_{n_k} - x| < \frac{\varepsilon}{2}$. Therefore,

$$
\begin{aligned}
|s_n - x| &= |(s_n - s_{n_k}) + (s_{n_k} - x)| &\text{by algebra}\\
&\le |s_n - s_{n_k}| + |s_{n_k} - x| &\text{by triangle inequality}\\
&< \frac{\varepsilon}{2} + \frac{\varepsilon}{2} = \varepsilon &\text{by (3.15) and (▲).}
\end{aligned}
$$

Thus, $|s_n - x| < \varepsilon$. Therefore, $\langle s_n \rangle$ converges. □

Theorem 3.6.5 allows one to confirm that a sequence converges without knowing the value of its limit. Since the proof of Theorem 3.6.5 applies Theorem 3.5.1, the Cauchy convergence criterion follows from the completeness axiom.

The following lemma can be a used to show that certain sequences are Cauchy.

Lemma 3.6.6. Let $\langle s_n \rangle$ be a sequence and let $a > 0$. Suppose $0 < r < 1$ is such that

$$|s_{n+1} - s_n| \le ar^n \text{ for all } n \ge 1. \tag{3.16}$$

Then $\langle s_n \rangle$ is a Cauchy sequence and hence, converges.

Proof. Let $\langle s_n \rangle$, a, and r be as stated in the lemma. Corollary 3.1.12 implies that $\lim_{n \to \infty} ar^n = a \lim_{n \to \infty} r^n = 0$. By Theorem 3.6.5, $\langle ar^n \rangle$ is a Cauchy sequence. Now let $\varepsilon > 0$. Since $\langle ar^n \rangle$ is a Cauchy sequence, there is a natural number N such that (▲) $|ar^n - ar^m| < \varepsilon(1-r)$ for all natural numbers $m, n > N$. Let $m, n > N$. We can assume that $m > n$. We show that $|s_m - s_n| < \varepsilon$ as follows:

$$
\begin{aligned}
|s_m - s_n| &= |(s_{n+1} - s_n) + (s_{n+2} - s_{n+1}) + \cdots + (s_m - s_{m-1})| && \text{by algebra}^1 \\
&\le |s_{n+1} - s_n| + |s_{n+2} - s_{n+1}| + \cdots + |s_m - s_{m-1}| && \text{by triangle inequality} \\
&\le ar^n + ar^{n+1} + \cdots + ar^{m-1} && \text{by (3.16)} \\
&= ar^n(1 + r + r^2 + \cdots + r^{m-n-1}) && \text{by algebra} \\
&= ar^n \left(\frac{1 - r^{m-n}}{1 - r} \right) && \text{by Theorem 1.4.7} \\
&= \frac{ar^n - ar^m}{1 - r} && \text{by algebra} \\
&< \frac{\varepsilon(1 - r)}{1 - r} = \varepsilon && \text{by (▲).} \qquad \square
\end{aligned}
$$

Problem. Let $\langle a_k \rangle$ be a bounded sequence and let $x \in \mathbb{R}$ be such that $0 < |x| < 1$. Consider the sequence $\langle s_n \rangle$ where $s_n = \sum_{k=1}^{n} a_k x^k$. Prove that $\langle s_n \rangle$ converges.

Proof. We show that $\langle s_n \rangle$ is a Cauchy sequence by applying Lemma 3.6.6. Since $\langle a_k \rangle$ is bounded, there is an $M > 0$ such that $|a_k| \le M$ for all $k \in \mathbb{N}$. Let $n \in \mathbb{N}$. Then

$$|s_{n+1} - s_n| = \left| \sum_{k=1}^{n+1} a_k x^k - \sum_{k=1}^{n} a_k x^k \right| = \left| a_{n+1} x^{n+1} \right| = |a_{n+1}| |x|^{n+1} \le (M |x|) |x|^n .$$

So, by Lemma 3.6.6, $\langle s_n \rangle$ is a Cauchy sequence and converges by Theorem 3.6.5. $\quad \square$

In the above proof, upon showing that $\langle s_n \rangle$ is a Cauchy sequence, Theorem 3.6.5 allowed us to infer that $\langle s_n \rangle$ converges without knowing its limit! Cauchy sequences play an important role in real analysis and they will surface again in the text.

---

[1] For example, $s_7 - s_3 = (s_4 - s_3) + (s_5 - s_4) + (s_6 - s_5) + (s_7 - s_6)$.

Exercises 3.6

1. Using Definition 3.6.1, prove that $\left\langle \frac{n}{n+3} \right\rangle$ is a Cauchy sequence.

2. Prove that any subsequence of a Cauchy sequence is also a Cauchy sequence.

3. Let $\langle s_n \rangle$ be a Cauchy sequence. Let $\langle s_{n_k} \rangle$ be a subsequence that converges to ℓ. Prove that $\langle s_n \rangle$ also converges to ℓ.

4. Let $\langle s_n \rangle$ be a Cauchy sequence and let $k > 0$. Let $\langle t_n \rangle$ be a sequence satisfying $|t_n - t_m| \le k |s_n - s_m|$ for all $n, m \ge 1$. Prove that $\langle t_n \rangle$ is a Cauchy sequence.

5. Suppose that the sequence $\langle s_n \rangle$ is such that $|s_n - s_m| \le \frac{1}{mn}$ for all $m, n \in \mathbb{N}$.

   (a) Prove that $\langle s_n \rangle$ is a Cauchy sequence.
   (b) Prove that $\langle s_n \rangle$ is a constant sequence.

6. Suppose that $\langle s_n \rangle$ and $\langle t_n \rangle$ are Cauchy sequences. Using Definition 3.6.1, prove that $\langle s_n + t_n \rangle$ is a Cauchy sequence.

7. Suppose that the sequence $\langle s_n \rangle$ satisfies $|s_{n+1} - s_n| \le \frac{1}{(n+1)!}$ for all $n \ge 1$. Show that $\langle s_n \rangle$ is a Cauchy sequence.

8. Consider the sequence $\langle s_n \rangle$ where $s_n = \sum_{k=1}^{n} \frac{1}{k!} = 1 + \frac{1}{2!} + \frac{1}{3!} + \cdots + \frac{1}{n!}$. Using Exercise 7 show that the sequence $\langle s_n \rangle$ converges.

9. Let $\langle s_n \rangle$ be a sequence where $s_1 \ne s_2$. Let $0 < r < 1$ and suppose that

$$|s_{n+2} - s_{n+1}| \le r |s_{n+1} - s_n| \text{ for all } n \ge 1.$$

   Prove the following:

   (a) $|s_{n+1} - s_n| \le r^{n-1} |s_2 - s_1|$ for all $n \ge 1$, by induction.
   (b) $\langle s_n \rangle$ is a Cauchy sequence.

10. Inductively define the sequence $\langle s_n \rangle$ by $s_1 = c \ge 1$ and $s_{n+1} = \frac{1}{2+s_n}$ for all $n \ge 1$. Observe that $s_n > 0$ for all $n \ge 1$.

    (a) Using Exercise 9, prove $\langle s_n \rangle$ is a Cauchy sequence.
    (b) Evaluate $\lim_{n \to \infty} s_n$.

Exercise Notes: For Exercise 1, $\left| \frac{n}{n+3} - \frac{m}{m+3} \right| = \left| \frac{n}{n+3} - 1 + 1 - \frac{m}{m+3} \right| \le \left| \frac{n}{n+3} - 1 \right| + \left| 1 - \frac{m}{m+3} \right|$. For Exercises 2 and 3, use Lemma 3.3.3. For part (b) of Exercise 5, let ℓ be the limit of the sequence. Show that $s_n = \ell$ for all n, by first showing that $|s_n - \ell| - |s_m - \ell| \le |s_n - s_m|$ for all $m, n \in \mathbb{N}$. For Exercise 7, first show that $\frac{1}{(n+1)!} \le \frac{1}{2^n}$ for all $n \ge 1$.

3.7 INFINITE LIMITS

An unbounded sequence does not converge, but some unbounded sequences "take off" to infinity; that is, the terms of the sequence become, progressively, larger and larger without bound. In other words, there are unbounded sequences $\langle s_n \rangle$ whose terms s_n become arbitrarily large as n becomes large. For example, consider the unbounded sequences

$$\langle 1, 2, 3, 4, 5, 6, 7, \ldots \rangle, \quad \langle 2, 1, 1, 4, 3, 6, 5, 8, 7, \ldots \rangle, \quad \text{and} \quad \langle 1, 0, 2, 0, 3, 0, 4, \ldots \rangle.$$

In the first two sequences, the terms are eventually all getting larger and larger without bound, whereas this is not the case for the third sequence. Here is a precise definition of this notion.

Definition 3.7.1. Let $\langle s_n \rangle$ be a sequence.

- We say that $\langle s_n \rangle$ **diverges to** ∞ provided that for every $M > 0$ there exists an $N \in \mathbb{N}$ such that for all $n \in \mathbb{N}$, if $n > N$, then $s_n > M$. In this case, we shall write $\lim_{n \to \infty} s_n = \infty$.

- We say that $\langle s_n \rangle$ **diverges to** $-\infty$ if for every $M < 0$ there exists an $N \in \mathbb{N}$ such that for all $n \in \mathbb{N}$, if $n > N$, then $s_n < M$. When this is the case, we will write $\lim_{n \to \infty} s_n = -\infty$.

Problem. Prove that $\lim_{n \to \infty} (n^2 - 3n + 2) = \infty$.

Proof Analysis. Let $M > 0$. We must find an N such that $n^2 - 3n + 2 > M$ for all $n > N$. Note that $n^2 - 3n + 2 > n^2 - 3n = n(n-3) \geq n - 3$, when $n - 3 > 0$. So we need $n - 3 > M$. Solving for n, we need $n > M + 3$. Thus, we let $N > M + 3$. We can now give a proof.

Proof. We prove that $\lim_{n \to \infty} (n^2 - 3n + 2) = \infty$. To do this, let $M > 0$. Let $N \in \mathbb{N}$ be such that $N > M + 3$. Let $n > N$ be a natural number. Then

$$
\begin{aligned}
n^2 - 3n + 2 &> n^2 - 3n && \text{by property of inequality} \\
&= n(n-3) && \text{by factoring} \\
&\geq n - 3 && \text{because } n \geq 4 \\
&> (M+3) - 3 && \text{because } n > M + 3 \\
&= M && \text{by algebra.}
\end{aligned}
$$

Therefore, $n^2 - 3n + 2 > M$, for all natural numbers $n > N$. □

Theorem 3.7.2. Let $\langle s_n \rangle$ and $\langle t_n \rangle$ be sequences. Suppose that $K \in \mathbb{N}$ is such that $s_n \leq t_n$ for all $n \geq K$. Then the following hold:

(a) If $\lim_{n \to \infty} s_n = \infty$, then $\lim_{n \to \infty} t_n = \infty$.
(b) If $\lim_{n \to \infty} t_n = -\infty$, then $\lim_{n \to \infty} s_n = -\infty$.

Proof. Let $\langle s_n \rangle$ and $\langle t_n \rangle$ be sequences such that $s_n \leq t_n$ for all $n \geq K$. To prove (a), assume that $\lim_{n \to \infty} s_n = \infty$. Let $M > 0$. Since $\lim_{n \to \infty} s_n = \infty$, there exists an $N' \in \mathbb{N}$ such that $s_n > M$ for all $n > N'$. Let $N = \max\{N', K\}$. As $s_n \leq t_n$ for all $n \geq K$, we conclude that $M < s_n \leq t_n$, for all $n > N$. So $t_n > M$ for all $n > N$ and thus, $\lim_{n \to \infty} t_n = \infty$. The proof of (b) is similar. □

Theorem 3.7.3. Let $\langle s_n \rangle$ is a sequence of positive terms. Then $\lim_{n \to \infty} s_n = \infty$ if and only if $\lim_{n \to \infty} \frac{1}{s_n} = 0$.

Corollary 3.7.4. Let $x > 1$. Then $\lim_{n \to \infty} x^n = \infty$.

Proof. Let $x > 1$. Thus, $0 < \frac{1}{x} < 1$. Corollary 3.1.12 implies that $\lim_{n \to \infty} \frac{1}{x^n} = 0$. Thus, Theorem 3.7.3 implies that $\lim_{n \to \infty} x^n = \infty$. □

Exercises 3.7

1. Prove that $\lim_{n \to \infty} \sqrt{n} = \infty$.

2. Prove that $\lim_{n \to \infty} \frac{\sqrt{n^2+1}}{\sqrt{n}} = \infty$. [Hint: Show that $\frac{\sqrt{n^2+1}}{\sqrt{n}} \geq \sqrt{n}$ when $n \geq 1$.]

3. Prove that $\lim_{n \to \infty} -n + \sin(2n) = -\infty$.

4. Prove that $\lim_{n \to \infty} n - 4\sqrt{n} = \infty$. [Hint: Show that $n - 4\sqrt{n} \geq \sqrt{n}$ when $n \geq 25$.]

5. Let $c > 0$. Suppose that $\lim_{n \to \infty} s_n = \infty$. Prove that $\lim_{n \to \infty} c s_n = \infty$.

6. Suppose that $\lim_{n \to \infty} s_n = \infty$ and $\lim_{n \to \infty} t_n = \infty$. Prove that $\lim_{n \to \infty} (s_n + t_n) = \infty$.

7. Prove Theorem 3.7.2(b).

8. Prove Theorem 3.7.3.

9. Let $\langle s_n \rangle$ be bounded below. Prove that if $\lim_{n \to \infty} t_n = \infty$, then $\lim_{n \to \infty} (s_n + t_n) = \infty$.

*10. Let $\langle a_n \rangle$ be a sequence of positive terms. Suppose $\lim_{n \to \infty} \frac{a_{n+1}}{a_n} = \infty$.

   (a) Prove that there exists a $K \in \mathbb{N}$ such that $a_{n+1} > a_n$ for all $n \geq K$.

   (b) Show that $a_n > a_K$ for all $n \geq K+1$.

   (c) Let $c = \min\{a_1, a_2, \ldots a_K\}$. Conclude that $c > 0$ and $a_n \geq c$ for all $n \in \mathbb{N}$.

   (d) Show that there is a real number $\alpha > 0$ such that $c^{\frac{1}{n}} \geq \alpha$ for all $n \in \mathbb{N}$.

   (e) Let $\rho > 0$. Show that there is a $N \in \mathbb{N}$ such that $a_{n+1} > (\frac{\rho}{\alpha}) a_n$ for all $n \geq N$.

   (f) Prove, by induction, that $a_n > (\frac{\rho}{\alpha})^n a_N$ for all $n \geq N+1$.

   (g) Infer that $a_n > (\frac{\rho}{\alpha})^n c$ for all $n \geq N+1$.

   (h) Conclude that $\lim_{n \to \infty} (a_n)^{\frac{1}{n}} = \infty$.

3.8 LIMIT SUPERIOR AND LIMIT INFERIOR

A bounded sequence may not converge, but we can use such a sequence together with the supremum and infimum operations to define two sequences that do converge.

3.8.1 The Limit Superior of a Bounded Sequence

Let $\langle x_n \rangle$ be a bounded sequence. Thus, the set $\{x_k : k \geq 1\}$ is bounded. Let a be a lower bound and let b be an upper bound for the set $\{x_k : k \geq 1\}$. So a and b are, respectively, a lower bound and an upper bound for any subset of $\{x_k : k \geq 1\}$. For each $n \in \mathbb{N}$, consider the set $\{x_k : k \geq n\}$. Clearly, we have that

$$\{x_k : k \geq 1\} \supseteq \{x_k : k \geq 2\} \supseteq \{x_k : k \geq 3\} \supseteq \cdots \supseteq \{x_k : k \geq n\} \supseteq \cdots \quad (3.17)$$

where $A \supseteq B$ means that $B \subseteq A$. As $\{x_k : k \geq 1\}$ is a bounded set, it follows from (3.17) that each of the sets $\{x_k : k \geq n\}$ are bounded. Moreover, (3.17) implies (see Exercise 4 on page 46) that

$$\sup\{x_k : k \geq 1\} \geq \sup\{x_k : k \geq 2\} \geq \cdots \geq \sup\{x_k : k \geq n\} \geq \cdots . \quad (3.18)$$

Thus, using the bounded sequence $\langle x_n \rangle$, we can construct a decreasing sequence $\langle \beta_n \rangle$ by defining

$$\beta_n = \sup\{x_k : k \geq n\}$$

for all $n \geq 1$. Hence, $\beta_1 \geq \beta_2 \geq \beta_3 \geq \cdots$. Since a is a lower bound for each set $\{x_k : k \geq n\}$, it follows that $\beta_n \geq a$ for all $n \geq 1$. Lemma 3.4.4 thus implies that the sequence $\langle \beta_n \rangle$ converges. We now introduce notation which is used to identify the limit of the sequence $\langle \beta_n \rangle$.

Definition 3.8.1. Let $\langle x_n \rangle$ be a bounded sequence. Let $\beta_n = \sup\{x_k : k \geq n\}$ for each $n \geq 1$. The **limit superior** of $\langle x_n \rangle$, denoted by $\limsup\limits_{n \to \infty} x_n$, is defined by $\limsup\limits_{n \to \infty} x_n = \lim\limits_{n \to \infty} \beta_n$, that is,

$$\limsup\limits_{n \to \infty} x_n = \lim\limits_{n \to \infty} (\sup\{x_k : k \geq n\}).$$

The sequential limit theorems covered in Section 3.2 may not hold for the limit superior operation. Compare the following theorem with Theorem 3.2.2.

Theorem 3.8.2. Let $\langle x_n \rangle$ and $\langle y_n \rangle$ be bounded sequences. Then

$$\limsup\limits_{n \to \infty}(x_n + y_n) \leq \limsup\limits_{n \to \infty} x_n + \limsup\limits_{n \to \infty} y_n.$$

Proof. Let $\langle x_n \rangle$ and $\langle y_n \rangle$ be bounded. So $\langle x_n + y_n \rangle$ is bounded. For each $n \in \mathbb{N}$, let

$$\sigma_n = \sup\{x_k + y_k : k \geq n\}, \quad \beta_n = \sup\{x_k : k \geq n\}, \quad \gamma_n = \sup\{y_k : k \geq n\}.$$

Let $n \in \mathbb{N}$. As $x_k \leq \beta_n$ and $y_k \leq \gamma_n$ for all $k \geq n$, we see that $x_k + y_k \leq \beta_n + \gamma_n$ for all $k \geq n$. So $\beta_n + \gamma_n$ is an upper bound for the set $\{x_k + y_k : k \geq n\}$. Thus, $\sigma_n \leq \beta_n + \gamma_n$ for all $n \geq 1$. Hence, $\lim\limits_{n \to \infty} \sigma_n \leq \lim\limits_{n \to \infty} \beta_n + \lim\limits_{n \to \infty} \gamma_n$ by Theorems 3.2.13 and 3.2.2. Now, by Definition 3.8.1, we conclude that $\limsup\limits_{n \to \infty}(x_n + y_n) \leq \limsup\limits_{n \to \infty} x_n + \limsup\limits_{n \to \infty} y_n$. □

There are bounded sequences where the inequality in Theorem 3.8.2 is strict (see Exercise 19). Thus, the properties that we established for the limit of a sequence (e.g., Theorem 3.2.2) do not necessarily hold for the limit superior of a bounded sequence. But the sequential limit theorems can be used to prove limit superior theorems.

Theorem 3.8.3. If $\langle x_n \rangle$ is a bounded sequence and $\langle x_{n_i} \rangle$ is a convergent subsequence of $\langle x_n \rangle$, then $\lim_{i \to \infty} x_{n_i} \leq \limsup_{n \to \infty} x_n$.

Proof. Let $\langle x_n \rangle$ be bounded and let $\langle x_{n_i} \rangle$ be a convergent subsequence of $\langle x_n \rangle$. For each $n \geq 1$, let $\beta_n = \sup\{x_k : k \geq n\}$. So $\lim_{n \to \infty} \beta_n = \limsup_{n \to \infty} x_n$. Observe that

$$x_{n_i} \leq \sup\{x_k : k \geq n_i\} = \beta_{n_i}, \text{ for all } i \geq 1.$$

Thus, $x_{n_i} \leq \beta_{n_i}$ for all $i \geq 1$. Therefore,

$$\lim_{i \to \infty} x_{n_i} \leq \lim_{i \to \infty} \beta_{n_i} = \lim_{n \to \infty} \beta_n$$

by Theorems 3.2.13 and 3.3.4, respectively. Hence, $\lim_{i \to \infty} x_{n_i} \leq \limsup_{n \to \infty} x_n$. □

Let $\langle x_n \rangle$ be a bounded sequence. For each $n \geq 1$, let $\beta_n = \sup\{x_k : k \geq n\}$. Then for every $n \geq 1$ and $\varepsilon > 0$, there exists a $k \geq n$ such that $\beta_n - \varepsilon < x_k$ (see Exercise 7 on page 46). This observation will be applied in the following proof.

Theorem 3.8.4. If $\langle x_n \rangle$ is bounded, then $\langle x_n \rangle$ has a convergent subsequence $\langle x_{n_i} \rangle$ such that $\lim_{i \to \infty} x_{n_i} = \limsup_{n \to \infty} x_n$.

Proof. Let $\langle x_n \rangle$ be a bounded sequence and for each $n \geq 1$, let $\beta_n = \sup\{x_k : k \geq n\}$. Thus, $\limsup_{n \to \infty} x_n = \lim_{n \to \infty} \beta_n$ and $\langle \beta_n \rangle$ is a decreasing sequence. We will now define a subsequence $\langle x_{n_i} \rangle$ of $\langle x_n \rangle$, by induction on i, such that

$$\beta_{n_i} - \frac{1}{i} < x_{n_i}, \text{ for all } i \geq 1. \tag{3.19}$$

(1) Since $\beta_1 = \sup\{x_k : k \geq 1\}$, let $n_1 \geq 1$ be such that $\beta_1 - 1 < x_{n_1}$. As $\beta_{n_1} \leq \beta_1$, it follows that $\beta_{n_1} - 1 < x_{n_1}$.

(2) Let $i \geq 1$ and assume that the natural number n_i has been defined. Since

$$\beta_{n_i + 1} = \sup\{x_k : k \geq n_i + 1\},$$

let $n_{i+1} \geq n_i + 1$ be such that $\beta_{n_i + 1} - \frac{1}{i+1} < x_{n_{i+1}}$. Since $n_{i+1} \geq n_i + 1$, we have that $\beta_{n_{i+1}} \leq \beta_{n_i + 1}$. Hence,

$$\beta_{n_{i+1}} - \frac{1}{i+1} \leq \beta_{n_i + 1} - \frac{1}{i+1} < x_{n_{i+1}}.$$

Thus, $\beta_{n_{i+1}} - \frac{1}{i+1} < x_{n_{i+1}}$. This completes the inductive definition of $\langle x_{n_i} \rangle$.

Theorem 3.3.4 implies that $\lim\limits_{i\to\infty} \beta_{n_i} = \lim\limits_{n\to\infty} \beta_n$. As $\beta_{n_i} \geq x_{n_i}$, we see from (3.19) that

$$|\beta_{n_i} - x_{n_i}| = \beta_{n_i} - x_{n_i} < \frac{1}{i}, \text{ for all } i \geq 1.$$

Lemma 3.2.11 implies that $\lim\limits_{i\to\infty} x_{n_i} = \lim\limits_{i\to\infty} \beta_{n_i}$. Thus, $\lim\limits_{i\to\infty} x_{n_i} = \limsup\limits_{n\to\infty} x_n$. □

Corollary 3.8.5. Let $\langle x_n \rangle$ be a convergent sequence. Then $\lim\limits_{n\to\infty} x_n = \limsup\limits_{n\to\infty} x_n$.

Our next theorem will be used in Chapter 8 to derive results on power series (see Theorem 8.3.5).

Theorem 3.8.6. Suppose that $\lim\limits_{n\to\infty} x_n > 0$ where $x_n \neq 0$ for all $n \geq 1$. If $\langle y_n \rangle$ is a bounded sequence, then $\limsup\limits_{n\to\infty}(x_n y_n) = \lim\limits_{n\to\infty} x_n \cdot \limsup\limits_{n\to\infty} y_n$.

Proof. Suppose that $\lim\limits_{n\to\infty} x_n = \ell > 0$ where $x_n \neq 0$ for all $n \geq 1$. Let $\langle y_n \rangle$ be a bounded sequence. Since $\langle x_n \rangle$ converges, Theorem 3.1.24 implies that $\langle x_n \rangle$ is bounded. Thus, $\langle x_n y_n \rangle$ is bounded. Let $p = \limsup\limits_{n\to\infty}(x_n y_n)$ and $y = \limsup\limits_{n\to\infty} y_n$. We will show that $p = \ell y$. By Theorem 3.8.4 there is a convergent subsequence $\langle y_{n_i} \rangle$ of $\langle y_n \rangle$ such that $\lim\limits_{i\to\infty} y_{n_i} = y$. Theorem 3.3.4 implies that $\lim\limits_{i\to\infty} x_{n_i} = \ell$. Thus, by Theorem 3.2.3, we conclude that $\lim\limits_{i\to\infty} x_{n_i} y_{n_i} = \ell y$. Therefore, $\ell y \leq p$ by Theorem 3.8.3.

Since $p = \limsup\limits_{n\to\infty}(x_n y_n)$, there is a convergent subsequence $\langle x_{n_i} y_{n_i} \rangle$ of $\langle x_n y_n \rangle$ such that $\lim\limits_{i\to\infty} x_{n_i} y_{n_i} = p$, by Theorem 3.8.4. Since $\lim\limits_{i\to\infty} x_{n_i} = \ell$, Theorem 3.2.7 implies that

$$\lim\limits_{i\to\infty} \frac{x_{n_i} y_{n_i}}{x_{n_i}} = \lim\limits_{i\to\infty} y_{n_i} = \frac{p}{\ell}.$$

Thus, $\lim\limits_{i\to\infty} y_{n_i} = \frac{p}{\ell}$. Hence, $\frac{p}{\ell} \leq y$ by Theorem 3.8.3. So $p \leq \ell y$. Therefore, $p = \ell y$. □

Definition 3.8.7. Let $\langle x_n \rangle$ be a bounded sequence. A **subsequential limit** of $\langle x_n \rangle$ is any real number that is the limit of some convergent subsequence of $\langle x_n \rangle$.

If $\langle x_n \rangle$ is a bounded sequence of real numbers, then by Theorem 3.5.1 we know that $\langle x_n \rangle$ has a convergent subsequence. It is sometimes useful to know the maximum limit that such a subsequence can have. Our next theorem shows that the limit superior of $\langle x_n \rangle$ is, in fact, this maximum subsequential limit (see Definition 2.3.5).

Theorem 3.8.8. Let $\langle x_n \rangle$ be a bounded sequence and also let S be the set of all subsequential limits of $\langle x_n \rangle$. Then $\limsup\limits_{n\to\infty} x_n \in S$ and $\limsup\limits_{n\to\infty} x_n = \sup(S)$.

Proof. Let $\langle x_n \rangle$ be bounded and let S be the set of all subsequential limits of $\langle x_n \rangle$. Theorem 3.8.4 implies that $\limsup\limits_{n\to\infty} x_n \in S$. Also, $\limsup\limits_{n\to\infty} x_n$ is an upper bound for S, by Theorem 3.8.3. Hence, $\limsup\limits_{n\to\infty} x_n = \sup(S)$. □

3.8.2 The Limit Inferior of a Bounded Sequence

Let $\langle x_n \rangle$ be a bounded sequence. Thus, the set $\{x_k : k \geq 1\}$ is bounded. Let a be a lower bound and let b be an upper bound for the set $\{x_k : k \geq 1\}$. So a and b are, respectively, a lower bound and an upper bound for any subset of $\{x_k : k \geq 1\}$. For each $n \in \mathbb{N}$, consider the set $\{x_k : k \geq n\}$. Clearly, we have that

$$\{x_k : k \geq 1\} \supseteq \{x_k : k \geq 2\} \supseteq \{x_k : k \geq 3\} \supseteq \cdots \supseteq \{x_k : k \geq n\} \supseteq \cdots \quad (3.20)$$

where $A \supseteq B$ means that $B \subseteq A$. As $\{x_k : k \geq 1\}$ is a bounded set, it follows from (3.20) that each of the sets $\{x_k : k \geq n\}$ are bounded. Moreover, (3.20) implies (see Exercise 4 on page 46) that

$$\inf\{x_k : k \geq 1\} \leq \inf\{x_k : k \geq 2\} \leq \inf\{x_k : k \geq 3\} \leq \cdots \leq \inf\{x_k : k \geq n\} \leq \cdots.$$

Thus, using the bounded sequence $\langle x_n \rangle$, we can construct an increasing sequence $\langle \alpha_n \rangle$ by defining

$$\alpha_n = \inf\{x_k : k \geq n\}$$

for all $n \geq 1$. Hence, $\alpha_1 \leq \alpha_2 \leq \alpha_3 \leq \cdots$. Since b is an upper bound for each set $\{x_k : k \geq n\}$, it follows that $\alpha_n \leq b$ for all $n \geq 1$. Lemma 3.4.2 thus implies that the sequence $\langle \alpha_n \rangle$ converges. The next definition introduces notation that is used to identify the limit of the sequence $\langle \alpha_n \rangle$.

Definition 3.8.9. Let $\langle x_n \rangle$ be a bounded sequence. Let $\alpha_n = \inf\{x_k : k \geq n\}$ for each $n \geq 1$. The **limit inferior** of $\langle x_n \rangle$, denoted by $\liminf\limits_{n \to \infty} x_n$, is defined by $\liminf\limits_{n \to \infty} x_n = \lim\limits_{n \to \infty} \alpha_n$, that is,

$$\liminf_{n \to \infty} x_n = \lim_{n \to \infty} \left(\inf\{x_k : k \geq n\} \right).$$

The proofs of the following five results, from Theorem 3.8.10 to Theorem 3.8.14, can be adapted from the proofs of the corresponding theorems and corollary in the previous section.

Theorem 3.8.10. Let $\langle x_n \rangle$ and $\langle y_n \rangle$ be a bounded sequences. Then

$$\liminf_{n \to \infty} x_n + \liminf_{n \to \infty} y_n \leq \liminf_{n \to \infty} (x_n + y_n).$$

Theorem 3.8.11. Let $\langle x_n \rangle$ be a bounded sequence and let $\langle x_{n_i} \rangle$ be a convergent subsequence of $\langle x_n \rangle$. Then $\liminf\limits_{n \to \infty} x_n \leq \lim\limits_{i \to \infty} x_{n_i}$.

Theorem 3.8.12. Let $\langle x_n \rangle$ be a bounded sequence. Then $\langle x_n \rangle$ has a convergent subsequence $\langle x_{n_i} \rangle$ such that $\lim\limits_{i \to \infty} x_{n_i} = \liminf\limits_{n \to \infty} x_n$.

Corollary 3.8.13. Let $\langle x_n \rangle$ be a convergent sequence. Then $\lim\limits_{n \to \infty} x_n = \liminf\limits_{n \to \infty} x_n$.

If $\langle x_n \rangle$ is a bounded sequence, then our next result shows that the limit inferior of $\langle x_n \rangle$ is equal to the minimum subsequential limit of $\langle x_n \rangle$ (see Definition 2.3.5).

Theorem 3.8.14. Let $\langle x_n \rangle$ be a bounded sequence and also let S be the set of all subsequential limits of $\langle x_n \rangle$. Then $\liminf\limits_{n \to \infty} x_n \in S$ and $\liminf\limits_{n \to \infty} x_n = \inf(S)$.

3.8.3 Connections and Relations

We now focus on some relationships between the limit superior and limit inferior of a bounded sequence. We will also derive some useful tools, which will be applied in Chapter 7, for dealing with the convergence of infinite series. In particular, these tools will be used to derive a ratio test and a root test.

Theorem 3.8.15. Let $\langle x_n \rangle$ be a bounded sequence. Then $\langle x_n \rangle$ converges if and only if $\liminf\limits_{n\to\infty} x_n = \limsup\limits_{n\to\infty} x_n$.

Proof. Let $\langle x_n \rangle$ be bounded. Corollary 3.8.5 and Corollary 3.8.13 imply that if $\langle x_n \rangle$ converges, then $\liminf\limits_{n\to\infty} x_n = \limsup\limits_{n\to\infty} x_n$. The converse also holds (see Exercise 21). □

Theorem 3.8.16. Let $\langle x_n \rangle$ be a bounded sequences. Then

1. $\liminf\limits_{n\to\infty} x_n \leq \limsup\limits_{n\to\infty} x_n$.
2. If $\limsup\limits_{n\to\infty} x_n < b$, then there is an $N \in \mathbb{N}$ such that $x_n < b$ for all $n \geq N$.
3. If $\liminf\limits_{n\to\infty} x_n > a$, then there is an $N \in \mathbb{N}$ such that $x_n > a$ for all $n \geq N$.

Proof. See Exercises 3, 4, and 5. □

Theorem 3.8.17. Let $\langle x_n \rangle$ be a sequence of positive terms where $\left\langle \frac{x_{n+1}}{x_n} \right\rangle$ is bounded. Then $\left\langle (x_n)^{\frac{1}{n}} \right\rangle$ is bounded and

$$\liminf_{n\to\infty} \frac{x_{n+1}}{x_n} \leq \liminf_{n\to\infty} (x_n)^{\frac{1}{n}} \leq \limsup_{n\to\infty} (x_n)^{\frac{1}{n}} \leq \limsup_{n\to\infty} \frac{x_{n+1}}{x_n}. \qquad (3.21)$$

Proof. Suppose that $\langle x_n \rangle$ is a sequence whose terms are all positive and that $\left\langle \frac{x_{n+1}}{x_n} \right\rangle$ is bounded. By Exercise 16, $\left\langle (x_n)^{\frac{1}{n}} \right\rangle$ is bounded. The "middle" inequality in (3.21) holds by Theorem 3.8.16(1). To prove that $\limsup\limits_{n\to\infty} (x_n)^{\frac{1}{n}} \leq \limsup\limits_{n\to\infty} \frac{x_{n+1}}{x_n}$, suppose that $\limsup\limits_{n\to\infty} \frac{x_{n+1}}{x_n} < \limsup\limits_{n\to\infty} (x_n)^{\frac{1}{n}}$. Thus, there is a $b > 0$ (see Exercise 7) such that

$$\limsup_{n\to\infty} \frac{x_{n+1}}{x_n} < b < \limsup_{n\to\infty} (x_n)^{\frac{1}{n}}. \qquad (3.22)$$

Since $\limsup\limits_{n\to\infty} \frac{x_{n+1}}{x_n} < b$, there exists an $N \in \mathbb{N}$ such that $\frac{x_{n+1}}{x_n} < b$ for all $n \geq N$, by Theorem 3.8.16(2). Thus, by Exercise 16, $\limsup\limits_{n\to\infty} (x_n)^{\frac{1}{n}} \leq b$ which contradicts (3.22). A similar argument, using Exercise 18, shows that $\liminf\limits_{n\to\infty} \frac{x_{n+1}}{x_n} \leq \liminf\limits_{n\to\infty} (x_n)^{\frac{1}{n}}$. □

Corollary 3.8.18. Let $\langle x_n \rangle$ be a sequence of positive terms.

(1) If $\left\langle \frac{x_{n+1}}{x_n} \right\rangle$ converges to $\gamma \in \mathbb{R}$, then $\lim\limits_{n\to\infty} (x_n)^{\frac{1}{n}} = \lim\limits_{n\to\infty} \frac{x_{n+1}}{x_n} = \gamma$.

(2) If $\left\langle \frac{x_{n+1}}{x_n} \right\rangle$ diverges to ∞, then $\lim\limits_{n\to\infty} (x_n)^{\frac{1}{n}} = \lim\limits_{n\to\infty} \frac{x_{n+1}}{x_n} = \infty$.

Proof. Corollaries 3.8.5, 3.8.13, and Theorems 3.8.17, 3.8.15 imply (1). Exercise 10 on page 92 establishes (2). □

Exercises 3.8

1. Find an example to show that equality in Theorem 3.8.2 may not hold.

2. Let $\langle x_n \rangle$ be a bounded sequence and let $m \in \mathbb{N}$. Hence, $\langle x_{m+n} \rangle$ is also a bounded sequence (see Definition 3.3.10). Show that $\limsup_{n \to \infty} x_n = \limsup_{n \to \infty} x_{m+n}$ and $\liminf_{n \to \infty} x_n = \liminf_{n \to \infty} x_{m+n}$.

*3. Prove Theorem 3.8.16(1).

*4. Prove Theorem 3.8.16(2).

*5. Prove Theorem 3.8.16(3).

6. Let $\langle x_n \rangle$ and $\langle y_n \rangle$ be bounded sequences. Suppose that $x_n \leq y_n$ for all $n \in \mathbb{N}$. Prove that $\limsup_{n \to \infty} x_n \leq \limsup_{n \to \infty} y_n$ and $\liminf_{n \to \infty} x_n \leq \liminf_{n \to \infty} y_n$.

*7. Let $\langle x_n \rangle$ be a bounded sequence. Let N, m, and M be such that $m \leq x_n \leq M$ for all $n \geq N$. Prove that $m \leq \liminf_{n \to \infty} x_n \leq \limsup_{n \to \infty} x_n \leq M$.

8. Prove that $\limsup_{n \to \infty}(c + x_n) = c + \limsup_{n \to \infty} x_n$, when $\langle x_n \rangle$ is bounded and $c \in \mathbb{R}$.

9. Prove that $\liminf_{n \to \infty}(c + x_n) = c + \liminf_{n \to \infty} x_n$, when $\langle x_n \rangle$ is bounded and $c \in \mathbb{R}$.

*10. Prove that $\limsup_{n \to \infty}(cx_n) = c \limsup_{n \to \infty} x_n$, when $\langle x_n \rangle$ is bounded and $c > 0$.

11. Prove that $\liminf_{n \to \infty}(cx_n) = c \liminf_{n \to \infty} x_n$, when $\langle x_n \rangle$ is bounded and $c > 0$.

12. Prove that $\limsup_{n \to \infty}(cx_n) = c \liminf_{n \to \infty} x_n$, when $\langle x_n \rangle$ is bounded and $c < 0$.

13. Prove that $\liminf_{n \to \infty}(cx_n) = c \limsup_{n \to \infty} x_n$, when $\langle x_n \rangle$ is bounded and $c < 0$.

14. Prove that $\limsup_{n \to \infty}(c^{\frac{1}{n}} x_n) = \limsup_{n \to \infty} x_n$, when $\langle x_n \rangle$ is bounded and $c > 0$.

15. Prove Theorem 3.8.10.

*16. Let $\langle x_n \rangle$ be a sequence of positive terms. Let $b > 0$ and $N \in \mathbb{N}$ be such that $\frac{x_{n+1}}{x_n} < b$ for all $n \geq N$. Let $c = x_N > 0$.

   (a) Prove, by induction, that $x_{N+n} < cb^n$, for all $n \geq 1$.

   (b) Infer that $(x_{N+n})^{\frac{1}{n}} < (c)^{\frac{1}{n}} b$ for all $n \geq 1$, and that $\left\langle (x_{N+n})^{\frac{1}{n}} \right\rangle$ is bounded.

   (c) Prove that $\limsup_{n \to \infty} (x_n)^{\frac{1}{n}} \leq b$.

17. Let $\langle x_n \rangle$ be bounded with positive terms. Show that $\left\langle (x_n)^{\frac{1}{n}} \right\rangle$ is bounded.

*18. Let $\langle x_n \rangle$ be a sequence of positive terms where $\langle x_n \rangle$ and $\left\langle \frac{x_{n+1}}{x_n} \right\rangle$ are bounded. Let $a > 0$ and $N \in \mathbb{N}$ be such that $a < \frac{x_{n+1}}{x_n}$ for all $n \geq N$. Let $c = x_N > 0$.

   (a) Prove, by induction, that $ca^n < x_{N+n}$, for all $n \geq 1$.

   (b) Infer that $(c)^{\frac{1}{n}} a < (x_{N+n})^{\frac{1}{n}}$ for all $n \geq 1$.

   (c) Prove that $a \leq \liminf_{n \to \infty} (x_n)^{\frac{1}{n}}$.

*19. Consider the bounded sequences $\langle x_n \rangle = \langle (-1)^n \rangle$ and $\langle y_n \rangle = \langle (-1)^{n+1} \rangle$. Compare the values of

$$\limsup_{n \to \infty}(x_n + y_n) \text{ and } \limsup_{n \to \infty} x_n + \limsup_{n \to \infty} y_n,$$

and compare the two values

$$\limsup_{n \to \infty}(x_n \cdot y_n) \text{ and } \limsup_{n \to \infty} x_n \cdot \limsup_{n \to \infty} y_n.$$

20. Prove Corollary 3.8.5.

*21. Let $\langle x_n \rangle$ be a bounded sequence. For each $n \in \mathbb{N}$, let $\alpha_n = \inf\{x_k : k \geq n\}$ and $\beta_n = \sup\{x_k : k \geq n\}$.

(a) Show that $\alpha_n \leq x_n \leq \beta_n$ for all $n \geq 1$.

(b) Show that $\liminf_{n \to \infty} x_n \leq \limsup_{n \to \infty} x_n$.

(c) Suppose that $\liminf_{n \to \infty} x_n = \ell = \limsup_{n \to \infty} x_n$. Show that $\lim_{n \to \infty} x_n = \ell$.

22. Prove Theorem 3.8.10 by appropriately modifying the proof of Theorem 3.8.2.

23. Prove Theorem 3.8.11 by appropriately modifying the proof of Theorem 3.8.3.

24. Prove Theorem 3.8.12 by appropriately modifying the proof of Theorem 3.8.4.

25. Prove Theorem 3.8.14 by appropriately modifying the proof of Theorem 3.8.8.

26. Let $\langle x_n \rangle$ and $\langle y_n \rangle$ be a bounded sequences with positive terms.

(a) Prove that $\limsup_{n \to \infty}(x_n \cdot y_n) \leq \limsup_{n \to \infty} x_n \cdot \limsup_{n \to \infty} y_n$.

(b) Find an example to show that equality may not hold in part (a).

27. Show that there no sequence whose set of subsequential limits is $\{\frac{n}{n+1} : n \in \mathbb{N}\}$.

28. Suppose that $\lim_{n \to \infty} x_n = \ell > 0$ where $x_n \neq 0$ for all $n \geq 1$. Let $\langle y_n \rangle$ be a bounded sequence. Prove that $\liminf_{n \to \infty}(x_n y_n) = \ell \liminf_{n \to \infty} y_n$.

29. Suppose that $\lim_{n \to \infty} x_n = \ell$. Let $\langle y_n \rangle$ be a bounded sequence. Prove that

$$\limsup_{n \to \infty}(x_n + y_n) = \ell + \limsup_{n \to \infty} y_n \text{ and } \liminf_{n \to \infty}(x_n + y_n) = \ell + \liminf_{n \to \infty} y_n.$$

Exercise Notes: For Exercises 4, let $\sigma = \lim_{n \to \infty} \beta_n$ where $\beta_n = \sup\{x_k : k \geq n\}$. Let $\varepsilon = b - \sigma > 0$. For Exercises 10-13, recall Theorem 2.3.13. For Exercises 14, recall Corollary 3.2.9. For Exercises 28 and 29, review the proof of Theorem 3.8.6.

CHAPTER 4

Continuity

Virtually all the functions discussed in a calculus course are continuous. In such a course it is often said that a function is continuous if its graph is an "unbroken" curve, that is, one can pencil-sketch the graph without raising the pencil off the page. Of course, the concept of an unbroken curve is intuitive and easily understood, but this description is not mathematically precise. In the following section we shall present a mathematical rigorous definition of the continuity concept. We also show that the algebraic combination and the composition of continuous functions produces a continuous function. In the remaining sections of this chapter, we will show how continuity is connected to sequences and limits. Moreover, continuous functions have many nice properties that will be identified in theorems that will appear in this and the remaining chapters of the book.

4.1   CONTINUOUS FUNCTIONS

Let $D \subseteq \mathbb{R}$. What does it mean to say that a function $f \colon D \to \mathbb{R}$ is continuous? Intuitively, a continuous function is one for which a small change in the input results in a small change in the output. Here is the precise definition.[1]

Definition 4.1.1. Let $f \colon D \to \mathbb{R}$ and let c be in D. Then f is **continuous** at c when the following holds:

> For every $\varepsilon > 0$ there exists a $\delta > 0$ such that for all $x \in D$, if $|x - c| < \delta$, then $|f(x) - f(c)| < \varepsilon$.

We say that $f \colon D \to \mathbb{R}$ is **continuous** when f is continuous at every point $c \in D$.

The phrase "if $|x - c| < \delta$, then $|f(x) - f(c)| < \varepsilon$" in Definition 4.1.1 is illustrated in Figure 4.1. The logical form of Definition 4.1.1 can be expressed as

$$(\forall \varepsilon > 0)(\exists \delta > 0)(\forall x \in D)(|x - c| < \delta \to |f(x) - f(c)| < \varepsilon) \qquad (4.1)$$

and it is this logical form that motivates the next proof strategy.

---

[1] We will always presume that $D \subseteq \mathbb{R}$.

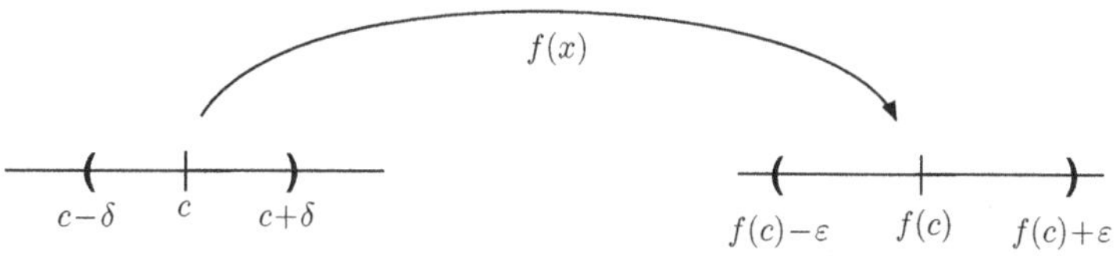

Figure 4.1: Illustration for Definition 4.1.1.

Proof Strategy 4.1.2. Let $f: D \to \mathbb{R}$ and let $c \in D$. To prove that f is continuous at c, use the proof diagram:

> Let $\varepsilon > 0$ be an arbitrary real number.
> Let $\delta =$ (the positive value you found).
> Let $x \in D$ be arbitrary.
> Assume $|x - c| < \delta$.
> Prove $|f(x) - f(c)| < \varepsilon$.

To apply proof strategy 4.1.2 to a function f that is defined by an explicit formula, first let $\varepsilon > 0$. Then we must find a $\delta > 0$ so that when $x \in D$ and $|x - c| < \delta$, we can prove that $|f(x) - f(c)| < \varepsilon$. To find the desired δ, we use algebra and properties of inequality on the expression $|f(x) - f(c)|$ to "extract out" a larger value that contains $|x - c|$ and no other occurrences of x. This process may involve finding an appropriate upper bounds for "unwanted" factors that involve x, but are different than $|x - c|$. Upon completing this process, we should then be able to find δ so that when $|x - c| < \delta$, we will have $|f(x) - f(c)| < \varepsilon$. We illustrate this technique in our proof analysis of the next theorem.

Theorem. Define $f: \mathbb{R} \to \mathbb{R}$ by $f(x) = 3x^2 + 5$. Then f is continuous at $c \in \mathbb{R}$.

Proof Analysis. Let $c \in \mathbb{R}$ and let $\varepsilon > 0$. Using algebra and properties of inequality on the expression $|f(x) - f(c)|$, we extract out $|x - c|$ as follows:

$$
\begin{aligned}
|f(x) - f(c)| &= \left| 3x^2 + 5 - (3c^2 + 5) \right| && \text{by definition of } f \\
&= \left| 3(x^2 - c^2) \right| && \text{by algebra} \\
&= 3 \, |x + c| \, |x - c| && \text{by algebra and property of } |\,|.
\end{aligned}
$$

In this case, we end up with the unwanted factor $|x + c|$, which involves x. We need to find an upper bound for this factor. We are free to make $|x - c|$ as small as is needed to obtain an upper bound for $|x + c|$. Let us start with $|x + c|$ and extract out a larger value that involves $|x - c|$ as follows:

$$
\begin{aligned}
|x + c| &= |x - c + 2c| && \text{by subtracting and adding } c \\
&\leq |x - c| + 2 \, |c| && \text{by triangle inequality.}
\end{aligned}
$$

So if $|x - c| < 1$, we have that $|x + c| < 1 + 2 \, |c|$ and $|f(x) - f(c)| \leq 3(1 + 2 \, |c|) \, |x - c|$. Solving the inequality $3(1 + 2 \, |c|) \, |x - c| < \varepsilon$ for $|x - c|$, we obtain $|x - c| < \frac{\varepsilon}{3(1 + 2|c|)}$. Thus, if $|x - c| < 1$ and $|x - c| < \frac{\varepsilon}{3(1 + 2|c|)}$, we can conclude that $|f(x) - f(c)| < \varepsilon$. So, we will let $\delta = \min\{1, \frac{\varepsilon}{3(1 + 2|c|)}\}$. We can now present a logically correct proof.

Proof. Let $f\colon \mathbb{R} \to \mathbb{R}$ be the function defined by $f(x) = 3x^2 + 5$. Let $c \in \mathbb{R}$ and $\varepsilon > 0$. Let $\delta = \min\{1, \frac{\varepsilon}{3M}\}$ where $M = 1 + 2\,|c|$. Let $x \in \mathbb{R}$ satisfy $|x - c| < \delta$. Since $|x - c| < \delta \le 1$, it follows that

$$
\begin{aligned}
|x + c| &= |x - c + 2c| && \text{by subtracting and adding } c \\
&\le |x - c| + 2\,|c| && \text{by triangle inequality} \\
&< 1 + 2\,|c| = M && \text{because } |x - c| < \delta \le 1.
\end{aligned}
$$

Thus, (▲) $|x + c| < M$. We now prove $|f(x) - f(c)| < \varepsilon$ as follows:

$$
\begin{aligned}
|f(x) - f(c)| &= \left|3x^2 + 5 - (3c^2 + 5)\right| && \text{by definition of } f \\
&= \left|3(x^2 - c^2)\right| && \text{by algebra} \\
&= 3\,|x + c|\,|x - c| && \text{by algebra and property of } |\ |. \\
&\le 3M\,|x - c| && \text{by (▲)} \\
&< 3M\frac{\varepsilon}{3M} = \varepsilon && \text{because } |x - c| < \delta \le \frac{\varepsilon}{3M}.
\end{aligned}
$$

Therefore, $|f(x) - f(c)| < \varepsilon$ and f is continuous at c. □

There may times when one is required to assume or show that a function is not continuous at a point. By evaluating the negation of (4.1), our next remark clarifies the meaning of discontinuity at a point.

Remark 4.1.3. Let $f\colon D \to \mathbb{R}$ and let $c \in D$. Then f is **not** continuous at c if and only if there exists an $\varepsilon > 0$ such that for all $\delta > 0$, there is an $x \in D$ such that $|x - c| < \delta$ and $|f(x) - f(c)| \ge \varepsilon$.

Combinations of Continuous Functions

Recall that a real-valued function is one that has the form $f\colon D \to \mathbb{R}$ where $D \subseteq \mathbb{R}$. Given two real-valued functions with the same domain, one can use the standard algebraic operations on the real numbers to define a new real-valued function.

Definition 4.1.4. Let $f\colon D \to \mathbb{R}$ and $g\colon D \to \mathbb{R}$. We define new functions as follows:

(1) The **sum** $(f + g)\colon D \to \mathbb{R}$ is defined by $(f + g)(x) = f(x) + g(x)$ for all $x \in D$.

(2) The **difference** $(f - g)\colon D \to \mathbb{R}$ is defined by $(f - g)(x) = f(x) - g(x)$ for all $x \in D$.

(3) The **product** $(fg)\colon D \to \mathbb{R}$ is defined by $(fg)(x) = f(x)g(x)$ for all $x \in D$.

(4) For $k \in \mathbb{R}$, the **constant multiple** $(kf)\colon D \to \mathbb{R}$ is defined by $(kf)(x) = kf(x)$ for all $x \in D$.

(5) Let $D^* = \{x \in D : g(x) \ne 0\}$. Then the **quotient** $\left(\frac{f}{g}\right)\colon D^* \to \mathbb{R}$ is defined by $\left(\frac{f}{g}\right)(x) = \frac{f(x)}{g(x)}$ for all $x \in D^*$.

We will show that if f and g are continuous at a point c, then all of the operations defined in Definition 4.1.4 will produce a function that is also continuous at c. In other words, the above operations (1)–(5) preserve continuity.

Preservation-of-Continuity Theorems

Consider a theorem having the following generic form.

Theorem. Let $f\colon D \to \mathbb{R}$ and $g\colon D \to \mathbb{R}$ be functions and let $c \in D$. Suppose that f is continuous at c and g is continuous at c. Then a new function $h\colon D \to \mathbb{R}$, which is defined from the given functions, is also continuous at c.

How does one prove theorems that have this form? We are assuming that f and g are continuous at c, and we must prove that h is continuous at c.

Proof Strategy 4.1.5. To prove that h is continuous at c, use the proof diagram:

> Assume f and g are continuous at c.
> Let $\varepsilon > 0$ be an arbitrary real number.
> Let $\delta =$ (the positive value you found).
> Let $x \in D$ be arbitrary.
> Assume $|x - c| < \delta$.
> Prove $|h(x) - h(c)| < \varepsilon$.

To apply proof strategy 4.1.5, let $\varepsilon > 0$. We must then find a $\delta > 0$ such that $|h(x) - h(c)| < \varepsilon$ when $|x - c| < \delta$. Since f and g are continuous at c, we can make $|f(x) - f(c)|$ and $|g(x) - g(c)|$ "as small as we want." Here is the basic idea that we will apply to get δ:

> *Using algebra and properties of inequality on the expression $|h(x) - h(c)|$, we extract out a larger value that contains $|f(x) - f(c)|$ and $|g(x) - g(c)|$, and no other occurrences of x, $f(x)$, or $g(x)$.*

Since we can make $|f(x) - f(c)|$ and $|g(x) - g(c)|$ "as small as we want," we should be able to find δ. These ideas appear in the proof analysis of our next two theorems.

Remark. If $f\colon D \to \mathbb{R}$ is continuous at c, then for any value $e > 0$, there is a $d > 0$ such that $|f(x) - f(c)| < e$ whenever $|x - c| < \delta$ and $x \in D$. This is the meaning behind the phrase "we can make $|f(x) - f(c)|$ as small as we want."

Theorem 4.1.6. Let $f\colon D \to \mathbb{R}$ and $g\colon D \to \mathbb{R}$ be functions, and let $c \in D$. Suppose that f and g are both continuous at c. Then $(f + g)$ is continuous at c.

Proof Analysis. Let f and g be continuous at c. So we can make $|f(x) - f(c)|$ and $|g(x) - g(c)|$ as small as we want. Let $\varepsilon > 0$. To ensure that

$$|(f + g)(x) - (f + g)(c)| < \varepsilon \tag{4.2}$$

when x is sufficiently closet to c, we first extract out $|f(x) - f(c)|$ and $|g(x) - g(c)|$ from the expression $|(f + g)(x) - (f + g)(c)|$ as follows:

$$
\begin{aligned}
|(f + g)(x) - (f + g)(c)| &= |(f(x) + g(x)) - (f(c) + g(c))| &&\text{by definition of } f + g \\
&= |(f(x) - f(c)) + (g(x) - g(c))| &&\text{by algebra} \\
&\leq |f(x) - f(c)| + |g(x) - g(c)| &&\text{by triangle inequality.}
\end{aligned}
$$

So if $|f(x) - f(c)| < \frac{\varepsilon}{2}$ and $|g(x) - g(c)| < \frac{\varepsilon}{2}$, then we can conclude (4.2). This analysis and proof strategy 4.1.5, guides the following logically correct proof.

Proof. Let $f: D \to \mathbb{R}$ and $g: D \to \mathbb{R}$ be continuous at $c \in D$. To prove that $f + g$ is continuous at c, let $\varepsilon > 0$. Since f is continuous at c, there is a $\delta_1 > 0$ such that for all $x \in D$,

$$|f(x) - f(c)| < \frac{\varepsilon}{2} \text{ when } |x - c| < \delta_1. \tag{4.3}$$

Also, because g is continuous at c, there is a $\delta_2 > 0$ such that for all $x \in D$,

$$|g(x) - g(c)| < \frac{\varepsilon}{2} \text{ when } |x - c| < \delta_2. \tag{4.4}$$

Let $\delta = \min\{\delta_1, \delta_2\}$. Let $x \in D$ and assume $|x - c| < \delta$. Thus, $|x - c| < \delta_1$ and $|x - c| < \delta_2$. We now prove that $|(f + g)(x) - (f + g)(c)| < \varepsilon$ as follows:

$$
\begin{aligned}
|(f + g)(x) - (f + g)(c)| &= |(f(x) + g(x)) - (f(c) + g(c))| && \text{by definition of } f + g \\
&= |(f(x) - f(c)) + (g(x) - g(c))| && \text{by algebra} \\
&\leq |f(x) - f(c)| + |g(x) - g(c)| && \text{by triangle inequality} \\
&< \frac{\varepsilon}{2} + \frac{\varepsilon}{2} = \varepsilon && \text{by (4.3) and (4.4).}
\end{aligned}
$$

Therefore, $(f + g)$ is continuous at c. $\qquad\qquad\qquad\qquad\qquad\qquad\qquad\qquad$ □

In the proof of Theorem 4.1.6, we were able to "cleanly" extract out $|f(x) - f(c)|$ and $|g(x) - g(c)|$, that is, there were no additional factors. In other situations there may be "unwanted" factors involving x, $f(x)$, or $g(x)$. You will then have to find an appropriate upper bound for these factors. This will have to be done in the proofs of Theorems 4.1.8 and 4.1.10. In our proof of Theorem 4.1.8, we obtain the unwanted factor $\frac{1}{|g(x)|}$. Our next lemma shows that if g is continuous at c and $g(c) \neq 0$, then there is a neighborhood of c and an $m > 0$ such that $m < |g(x)|$ for any x in this neighborhood. Thus, we will also have that $\frac{1}{|g(x)|} < \frac{1}{m}$. So $\frac{1}{m}$ will be our desired upper bound for the proof of Theorem 4.1.8.

Lemma 4.1.7. Let $g: D \to \mathbb{R}$ be continuous at $c \in D$ where $g(c) \neq 0$. Then there exists an $m > 0$ and a $\delta > 0$ such that for all $x \in D$, if $|x - c| < \delta$, then $m < |g(x)|$.

Proof. Let $g: D \to \mathbb{R}$ be continuous at $c \in D$ where $g(c) \neq 0$. Since $g(c) \neq 0$, consider the positive value $\frac{|g(c)|}{2}$. Since g is continuous at c, there exists a $\delta > 0$ such that $|g(x) - g(c)| < \frac{|g(c)|}{2}$ if $x \in D$ and $|x - c| < \delta$. Let $x \in D$ satisfy $|x - c| < \delta$. Thus, $|g(x) - g(c)| < \frac{|g(c)|}{2}$. So by the backward triangle inequality, $|g(c)| - |g(x)| < \frac{|g(c)|}{2}$ and $|g(c)| - \frac{|g(c)|}{2} < |g(x)|$. Thus, $\frac{|g(c)|}{2} < |g(x)|$. Let $m = \frac{|g(c)|}{2} > 0$. Hence, $m < |g(x)|$ when $x \in D$ and $|x - c| < \delta$. $\qquad\qquad\qquad\qquad\qquad\qquad\qquad$ □

Let $g: D \to \mathbb{R}$ be a function and let $D^* = \{x \in D : g(x) \neq 0\}$. We define the **reciprocal function** by $\left(\frac{1}{g}\right)(x) = \frac{1}{g(x)}$ for all $x \in D^*$. If g is continuous at $c \in D$ where $g(c) \neq 0$, then Lemma 4.1.7 will be used to prove that $\left(\frac{1}{g}\right)$ is continuous at c.

Theorem 4.1.8. Let $g: D \to \mathbb{R}$ be continuous at $c \in D$ where $g(c) \neq 0$. Then $\left(\frac{1}{g}\right)$ is continuous at c.

Proof Analysis. Let g be continuous at $c \in D$ where $g(c) \neq 0$. So we can make $|g(x) - g(c)|$ as small as we want. Let $\varepsilon > 0$. We need to make $\left| \frac{1}{g(x)} - \frac{1}{g(c)} \right| < \varepsilon$ when x is close enough to c. Since g is continuous at c, Lemma 4.1.7 implies that there is a $\delta_1 > 0$ and an $m > 0$ such that $(\star)$ $m < |g(x)|$ when $|x - c| < \delta_1$ and $x \in D$. Thus, $m < |g(c)|$ because $|c - c| = 0 < \delta_1$ and $c \in D$. Note that when $m < |g(x)|$, we have that $\frac{1}{|g(x)|} < \frac{1}{m}$. So assume that $|x - c| < \delta_1$. We can now extract out $|g(x) - g(c)|$ from the expression $\left| \frac{1}{g(x)} - \frac{1}{g(c)} \right|$ as follows:

$$\left| \frac{1}{g(x)} - \frac{1}{g(c)} \right| = \left| \frac{g(c) - g(x)}{g(x)g(c)} \right| \quad \text{by algebra}$$

$$= \frac{|g(x) - g(c)|}{|g(x)|\,|g(c)|} \quad \text{property of absolute value}$$

$$\leq \frac{|g(x) - g(c)|}{m^2} \quad \text{by } (\star).$$

Thus, if $|g(x) - g(c)| < m^2 \varepsilon$, we can conclude that $\left| \frac{1}{g(x)} - \frac{1}{g(c)} \right| < \varepsilon$. We can now present a logically correct proof which is guided by proof strategy 4.1.5.

Proof. Let $g \colon D \to \mathbb{R}$ be continuous at $c \in D$ where $g(c) \neq 0$. Lemma 4.1.7 implies that there is a $\delta_1 > 0$ and an $m > 0$ such that

$$m < |g(x)| \text{ when } |x - c| < \delta_1 \text{ and } x \in D. \tag{4.5}$$

Let $\varepsilon > 0$. Since g is continuous at c, there is a δ_2 such that

$$|g(x) - g(c)| < \varepsilon m^2 \text{ when } |x - c| < \delta_2 \text{ and } x \in D. \tag{4.6}$$

Let $\delta = \min\{\delta_1, \delta_2\}$. Let $x \in D$ be such that $|x - c| < \delta$. Thus, $|x - c| < \delta_1$ and $|x - c| < \delta_2$, and so (4.5) and (4.6) apply. We prove that $\left| \frac{1}{g(x)} - \frac{1}{g(c)} \right| < \varepsilon$ as follows:

$$\left| \frac{1}{g(x)} - \frac{1}{g(c)} \right| = \left| \frac{g(c) - g(x)}{g(x)g(c)} \right| \quad \text{by algebra}$$

$$= \frac{|g(x) - g(c)|}{|g(x)|\,|g(c)|} \quad \text{property of absolute value}$$

$$\leq \frac{|g(x) - g(c)|}{m^2} \quad \text{by (4.5)}$$

$$< \frac{\varepsilon m^2}{m^2} = \varepsilon \quad \text{by (4.6.)}$$

Therefore, $\left(\frac{1}{g} \right)$ is continuous at c. □

We next show that if a function is continuous at a point, then it is bounded near this point. This will allow us to prove that the product and quotient of two functions is continuous at a point, if each of the two functions is continuous at the point.

Lemma 4.1.9. If $f \colon D \to \mathbb{R}$ is continuous at $c \in D$, then there is a $\delta > 0$ and an $M > 0$ such that $|f(x)| \leq M$ for all $x \in D$ satisfying $|x - c| < \delta$.

Proof. Let $f \colon D \to \mathbb{R}$ be continuous at $c \in D$. Let $\varepsilon = 1$. Since f is continuous at c, there exists a $\delta > 0$ such that $|f(x) - f(c)| < 1$ whenever $x \in D$ and $|x - c| < \delta$. Let $x \in D$ satisfy $|x - c| < \delta$. Thus, $|f(x) - f(c)| < 1$. By the backward triangle inequality, we obtain $|f(x)| - |f(c)| < 1$. Hence, $|f(x)| < 1 + |f(c)|$ whenever $x \in D$ and $|x - c| < \delta$. Thus, for $M = 1 + |f(c)| > 0$ we have that $|f(x)| \leq M$ for all $x \in D$ satisfying $|x - c| < \delta$. □

Theorem 4.1.10. Let $f \colon D \to \mathbb{R}$ and $g \colon D \to \mathbb{R}$ be continuous at $c \in D$. Then fg is continuous at c.

Proof. See Exercise 15. □

Theorem 4.1.11. Let $f \colon D \to \mathbb{R}$ and $g \colon D \to \mathbb{R}$ be continuous at $c \in D$. If $g(c) \neq 0$, then $\left(\frac{f}{g}\right)$ is continuous at c.

Proof. Let $f \colon D \to \mathbb{R}$ and $g \colon D \to \mathbb{R}$ be continuous at $c \in D$ where $g(c) \neq 0$. Thus, $\left(\frac{1}{g}\right)$ is continuous at c, by Theorem 4.1.8. Since $\left(\frac{f}{g}\right) = f \cdot \left(\frac{1}{g}\right)$, Theorem 4.1.10 implies that $\left(\frac{f}{g}\right)$ is continuous at c. □

Theorems 4.1.6, 4.1.10, and 4.1.11 ensure that any function that is an algebraic combination of continuous functions is a continuous function. Thus, we have the following result.

Theorem 4.1.12. A polynomial function is continuous at every point $c \in \mathbb{R}$, and a rational function is continuous at every point c for which the function is defined.

Proof. One can show that every constant functions is continuous at each point and the function $f(x) = x$ is also continuous at each point. These two facts, together with Theorem 4.1.6, Theorem 4.1.10 and mathematical induction, show that a polynomial function is continuous at every real number. Theorem 4.1.11 now implies that rational functions are continuous at the points stated in the theorem. □

We now identify conditions which ensure that a composite function is continuous at a point.

Theorem 4.1.13. Let $f \colon D \to \mathbb{R}$ and $g \colon E \to \mathbb{R}$ be functions such that $f[D] \subseteq E$. If f is continuous at $c \in D$ and g is continuous at $f(c)$, then $(g \circ f) \colon D \to \mathbb{R}$ is continuous at c.

Proof. Let $f \colon D \to \mathbb{R}$ and $g \colon E \to \mathbb{R}$ be as stated. Let $\varepsilon > 0$. Since g is continuous at $f(c)$, there is a $\delta_1 > 0$ such that

$$|g(y) - g(f(c))| < \varepsilon \text{ for all } y \in E \text{ satisfying } |y - f(c)| < \delta_1. \tag{4.7}$$

Since f is continuous at c, there is a $\delta > 0$ such that $|f(x) - f(c)| < \delta_1$ for all $x \in D$ satisfying $|x - c| < \delta$. Now, let $x \in D$ be such that $|x - c| < \delta$. Then $f(x) \in E$ and $|f(x) - f(c)| < \delta_1$. Since $|f(x) - f(c)| < \delta_1$, we see that (4.7) implies

$$|(g \circ f)(x) - (g \circ f)(c)| = |g(f(x)) - g(f(c))| \quad \text{by definition of } g \circ f$$
$$< \varepsilon \quad \text{by (4.7).}$$

Therefore, $g \circ f$ is continuous at c. □

Our next lemma shows that if a function is continuous and positive at a point c, then the function is positive at all of the points in a particular neighborhood of c.

Lemma 4.1.14. Let $f\colon D \to \mathbb{R}$ be continuous at $c \in D$ where $f(c) \neq 0$.

(a) If $f(c) > 0$, there is a $\delta > 0$ where $f(x) > 0$ for all $x \in D$ satisfying $|x - c| < \delta$.

(b) If $f(c) < 0$, there is a $\delta > 0$ where $f(x) < 0$ for all $x \in D$ satisfying $|x - c| < \delta$.

Proof. Let $f\colon D \to \mathbb{R}$ be continuous at $c \in D$ where $f(c) \neq 0$. Let $\varepsilon = |f(c)| > 0$. Since f is continuous at c, there exists a $\delta > 0$ such that $|f(x) - f(c)| < \varepsilon$ whenever $x \in D$ and $|x - c| < \delta$. We now prove (a). Assume $f(c) > 0$. Thus, $\varepsilon = f(c)$. So we have that $|f(x) - f(c)| < \varepsilon = f(c)$ whenever $x \in D$ and $|x - c| < \delta$. So, if $x \in D$ and $|x - c| < \delta$, then

$$f(c) - f(x) \leq |f(x) - f(c)| < f(c),$$

and thus, $f(c) - f(x) < f(c)$. We conclude that $0 < f(x)$ whenever $x \in D$ and $|x - c| < \delta$. Thus, (a) holds. The proof of (b), using $\varepsilon = -f(c)$, is very similar. □

Many important theorems in real analysis assume that a function is continuous at every point in its domain. Our next strategy shows how to prove that a function is continuous.

Proof Strategy 4.1.15. Let $f\colon D \to \mathbb{R}$. To prove that f is continuous, use the proof diagram:

Let $c \in D$ be an arbitrary.
Prove f is continuous at c.

Exercises 4.1 ——————————————————————————————

1. Let m and d be real numbers where $m \neq 0$. Consider the function $f\colon \mathbb{R} \to \mathbb{R}$ defined by $f(x) = mx + d$. Using Proof Strategies 4.1.15 and 4.1.2, prove that f is continuous.

2. Let $g\colon [-1, 1] \to \mathbb{R}$ be defined by

$$g(x) = \begin{cases} 1, & \text{if } x \neq 0; \\ 0, & \text{if } x = 0. \end{cases}$$

Prove that g is not continuous at 0.

3. Find functions $f\colon [-1, 1] \to \mathbb{R}$ and $g\colon [-1, 1] \to \mathbb{R}$ that are not continuous at 0 but $f + g$ and fg are continuous at 0.

*4. Define $f\colon \mathbb{R} \to \mathbb{R}$ by $f(x) = |x|$ for all $x \in \mathbb{R}$. Using Proof Strategies 4.1.15 and 4.1.2, prove that f is continuous.

*5. Let $f\colon \mathbb{R} \to \mathbb{R}$ be defined by $f(x) = x^2$ for all $x \in \mathbb{R}$. Using Proof Strategies 4.1.15 and 4.1.2, prove that f is continuous.

6. Let $f\colon [0, \infty) \to \mathbb{R}$ be defined by $f(x) = \sqrt{x}$ for all $x \in [0, \infty)$. Using Proof Strategies 4.1.15 and 4.1.2, prove that f is continuous.

7. Let $f: [1,3] \to \mathbb{R}$ be defined by $f(x) = \frac{1}{x+2}$ for all $x \in [1,3]$. Using Proof Strategies 4.1.15 and 4.1.2, prove that f is continuous.

8. Let $f: D \to \mathbb{R}$ and $g: D \to \mathbb{R}$ be functions and let $c \in D$. Suppose that f and g are both continuous at c. Prove that there exists an $M > 0$ and a $\delta > 0$ such that $|f(x)| \le M$ and $|g(x)| \le M$ when $|x - c| < \delta$.

9. Let $f: D \to \mathbb{R}$ and $g: D \to \mathbb{R}$ be functions and let $c \in D$. Suppose that f and g are both continuous at c. Prove that $f - g$ is continuous at c.

10. Let $f: D \to \mathbb{R}$ be a function and let $c \in D$. Suppose that f is continuous at c. Define $g: D \to \mathbb{R}$ by $g(x) = |f(x)|$. Prove that g is continuous at c.

11. Let $f: D \to \mathbb{R}$ be a function and let $c \in D$. Suppose that f is continuous at c and that $k \in \mathbb{R}$ is nonzero. Prove that kf is continuous at c.

12. Let $f: D \to \mathbb{R}$ be continuous at $c \in D$. Let $D^* \subseteq D$ be such that $c \in D^*$. Define $h: D^* \to \mathbb{R}$ by $h(x) = f(x)$. Show that h is continuous at c.

13. Let $f: D \to \mathbb{R}^+$ be a function. Suppose that f is continuous at $c \in D$. Define $\sqrt{f}: D \to \mathbb{R}^+$ by $\sqrt{f}(x) = \sqrt{f(x)}$. Prove that $\sqrt{f}$ is continuous at c.

14. Let $f: D \to \mathbb{R}$ and $g: D \to \mathbb{R}$ be functions and let $c \in D$. Suppose that
    (a) g is continuous at c, and
    (b) $|f(x) - f(c)| \le |g(x)|\,|x - c|$ for all $x \in D$.
    Prove that f is continuous at c.

*15. Prove Theorem 4.1.10.

16. Prove Lemma 4.1.14(b).

17. Let $f: D \to \mathbb{R}$ be continuous at $c \in D$ and let $h \in \mathbb{R}$. Prove the following:
    (a) If $f(c) > h$, there is a $\delta > 0$ such that $f(x) > h$ for all $x \in D$ satisfying $|x - c| < \delta$.
    (b) If $f(c) < h$, there is a $\delta > 0$ such that $f(x) < h$ for all $x \in D$ satisfying $|x - c| < \delta$.

18. Let $f: D \to \mathbb{R}$ and $g: D \to \mathbb{R}$ be continuous at $c \in D$.
    (a) Let $h: D \to \mathbb{R}$ be defined by $h(x) = \max\{f(x), g(x)\}$. Prove that h is continuous at c.
    (b) Let $k: D \to \mathbb{R}$ be defined by $k(x) = \min\{f(x), g(x)\}$. Prove that k is continuous at c.

Exercise Notes: For Exercises 8 and 14, use Lemma 4.1.9. For Exercise 15, to prove fg is continuous at c, note that

$$|f(x)g(x) - f(c)g(c)| = |f(x)g(x) - f(x)g(c) + f(x)g(c) - f(c)g(c)|.$$

After "extracting out" $|g(x) - g(c)|$ and $|f(x) - f(c)|$, use Exercise 8. For the proof of Exercise 17, the function $f(x) - h$ is continuous. Apply Lemma 4.1.14. For Exercise 18, apply Exercise 16 on page 35, Theorem 4.1.6, and the above Exercises 9, 10, and 11.

4.2 CONTINUITY AND SEQUENCES

Our next theorem shows that there is another way, using sequences, to show that a function is continuous at a point. This result establishes an important relationship between continuity and convergent sequences. As the following proof uses the negation of continuity, one should review Remark 4.1.3.

Theorem 4.2.1 (Sequential Criterion for Continuity). Let $f\colon D \to \mathbb{R}$ and let $c \in D$. Then the following are equivalent:

(1) f is continuous at c.

(2) For every sequence $\langle x_n \rangle$ of points in D, if $\lim_{n\to\infty} x_n = c$, then $\lim_{n\to\infty} f(x_n) = f(c)$.

Proof. Let $f\colon D \to \mathbb{R}$ and let $c \in D$. We shall prove that f is continuous at c if and only if for every sequence $\langle x_n \rangle$ of points in D, if $\lim_{n\to\infty} x_n = c$, then $\lim_{n\to\infty} f(x_n) = f(c)$.

($\Rightarrow$). Assume that f is continuous at c. Let $\langle x_n \rangle$ be an arbitrary sequence of points in D such that $\lim_{n\to\infty} x_n = c$. We will prove that $\lim_{n\to\infty} f(x_n) = f(c)$. Let $\varepsilon > 0$. Since f is continuous at c, there is a $\delta > 0$ such that

$$\text{for all } x \in D, \text{ if } |x - c| < \delta, \text{ then } |f(x) - f(c)| < \varepsilon. \tag{4.8}$$

Now, because $\langle x_n \rangle$ converges to c, there is an $N \in \mathbb{N}$ such that for all $n > N$, $|x_n - c| < \delta$. Thus, (4.8) implies that $|f(x_n) - f(c)| < \varepsilon$ for all $n > N$. Hence, $\lim_{n\to\infty} f(x_n) = f(c)$.

($\Leftarrow$). Assume that

$$\text{for any sequence } \langle x_n \rangle \text{ of points in } D, \text{ if } \lim_{n\to\infty} x_n = c, \text{ then } \lim_{n\to\infty} f(x_n) = f(c). \tag{4.9}$$

We prove that f is continuous at c. Suppose, for a contradiction, that f is not continuous at c. Thus, there is an $\varepsilon > 0$ such that

$$\text{for all } \delta > 0, \text{ there is an } x \in D \text{ such that } |x - c| < \delta \text{ and } |f(x) - f(c)| \geq \varepsilon.$$

Thus, for each $n \in \mathbb{N}$ (letting $\delta = \frac{1}{n}$), there is an $x_n \in D$ so that $|x_n - c| < \frac{1}{n}$ and $|f(x_n) - f(c)| \geq \varepsilon$. Therefore, using this ε, we obtain a sequence $\langle x_n \rangle$ such that (i) $x_n \in D$, (ii) $|x_n - c| < \frac{1}{n}$, and (iii) $|f(x_n) - f(c)| \geq \varepsilon$ for all $n \in \mathbb{N}$. It follows, using (i) and (ii) (see Theorem 3.1.11), that

$$\langle x_n \rangle \text{ is a sequence of points in } D \text{ and } \lim_{n\to\infty} x_n = c. \tag{4.10}$$

However, (4.9) and (4.10) imply that $\lim_{n\to\infty} f(x_n) = f(c)$. So there is an n such that $|f(x_n) - f(c)| < \varepsilon$; however, this contradicts (iii). This completes the proof. $\square$

In Section 4.1, we used Definition 4.1.1 (the definition of continuity) to prove theorems about continuity. Using previously established results on sequences, we can produce new proofs of these same theorems, using Theorem 4.2.1. To illustrate this, we now restate Theorem 4.1.6 and give it a new "sequential" proof.

Theorem 4.1.6. Let $f\colon D \to \mathbb{R}$ and $g\colon D \to \mathbb{R}$ be functions, and let $c \in D$. Suppose that f and g are both continuous at c. Then $(f + g)$ is continuous at c.

Proof. Let $f\colon D \to \mathbb{R}$ and $g\colon D \to \mathbb{R}$ be continuous at $c \in D$. Let $\langle x_n \rangle$ be an arbitrary sequence of points in D such that $\lim\limits_{n \to \infty} x_n = c$. Since f and g are continuous at c, Theorem 4.2.1 implies that

$$\lim_{n \to \infty} f(x_n) = f(c) \text{ and } \lim_{n \to \infty} g(x_n) = g(c) \tag{4.11}$$

We show that $\lim\limits_{n \to \infty} (f + g)(x_n) = (f + g)(c)$ as follows:

$$
\begin{aligned}
\lim_{n \to \infty} (f + g)(x_n) &= \lim_{n \to \infty} [f(x_n) + g(x_n)] && \text{by Definition 4.1.4(1)} \\
&= \lim_{n \to \infty} f(x_n) + \lim_{n \to \infty} g(x_n) && \text{by Theorem 3.2.2} \\
&= f(c) + g(c) && \text{by (4.11)} \\
&= (f + g)(c) && \text{by Definition 4.1.4(1).}
\end{aligned}
$$

Thus, $\lim\limits_{n \to \infty} (f+g)(x_n) = (f+g)(c)$. So $(f+g)$ is continuous at c by Theorem 4.2.1. $\square$

Theorem 4.2.1 can now be used to prove that a function is continuous at a point. It can also be used to show that a function is not continuous at a point.

Corollary 4.2.2. Let $f\colon D \to \mathbb{R}$ and $c \in D$. If there is a sequence $\langle x_n \rangle$ of points in D such that $\lim\limits_{n \to \infty} x_n = c$ and $\lim\limits_{n \to \infty} f(x_n) \neq f(c)$, then f is not continuous at c.

Exercises 4.2

1. Define the function $f\colon \mathbb{R} \to \mathbb{R}$ by

$$f(x) = \begin{cases} 1, & \text{if } x \in \mathbb{Q}; \\ 0, & \text{if } x \notin \mathbb{Q}. \end{cases}$$

   Prove that f is not continuous at any point $c \in \mathbb{R}$.

2. Define the function $f\colon \mathbb{R} \to \mathbb{R}$ by

$$f(x) = \begin{cases} x, & \text{if } x \in \mathbb{Q}; \\ 0, & \text{if } x \notin \mathbb{Q}. \end{cases}$$

   Prove that f is continuous at 0.

3. Let $f\colon \mathbb{R} \to \mathbb{R}$ be continuous and $c \in \mathbb{R}$. Suppose that $f(x) = c$ for all $x \in \mathbb{Q}$. Prove that $f(x) = c$ for all $x \in \mathbb{R}$.

4. Let $f\colon (0,1) \to \mathbb{R}$ be the continuous function defined by $f(x) = \frac{1}{x}$. Give an example of a convergent sequence $\langle x_n \rangle$ where $x_n \in (0,1)$ for all $n \geq 1$, so that $\langle f(x_n) \rangle$ diverges.

5. Let $f\colon D \to \mathbb{R}$ and $g\colon D \to \mathbb{R}$ be continuous at $c \in D$. Use Theorem 4.2.1 and Theorem 3.2.3 to prove that $(fg)\colon D \to \mathbb{R}$ is continuous at c.

6. Let $f\colon D \to \mathbb{R}$ and $g\colon E \to \mathbb{R}$ be functions such that $f[D] \subseteq E$. Suppose that f is continuous at $c \in D$ and g is continuous at $f(c)$. Use Theorem 4.2.1 to prove that $(g \circ f)\colon D \to \mathbb{R}$ is continuous at c.

7. Let $f\colon D \to E$ be one-to-one and continuous where $E \subseteq \mathbb{R}$. Prove that if $c \in D$ is an accumulation point of D, then $f(c)$ is an accumulation point of E.

8. Let $f\colon D \to E$ be a bijection such that f^{-1} is continuous, where $E \subseteq \mathbb{R}$. Let $\langle x_n \rangle$ be a sequence of points in D and let $c \in D$. Suppose that $\langle f(x_n) \rangle$ converges to $f(c)$. Show that $\langle x_n \rangle$ converges to c.

Exercise Notes: For Exercises 1–3: Both $\mathbb{Q}$ and the set of irrational numbers $\mathbb{R} \setminus \mathbb{Q}$ are dense in $\mathbb{R}$. Apply Lemma 3.1.18 and Theorem 4.2.1.

4.3   LIMITS OF FUNCTIONS

In analysis, the limit of a function is a fundamental concept concerning the behavior of a function near a particular point c. Informally, a function has a limit L at the point c if $f(x)$ is "very close" to L whenever x is "very close" to c. Before we present the formal definition, we recall Definition 3.5.2 of Chapter 3.

Definition. Let $S \subseteq \mathbb{R}$. A point $x \in \mathbb{R}$ is said to be an *accumulation point of S* if every neighborhood of x contains an infinite number of points from S.

Definition 4.3.1. Let $f\colon D \to \mathbb{R}$, $L \in \mathbb{R}$, and c be an accumulation point of D. Then L **is the limit of** f **at** c, denoted by $\lim_{x \to c} f(x) = L$, if for every $\varepsilon > 0$ there exists a $\delta > 0$ such that for all $x \in D$, if $0 < |x - c| < \delta$, then $|f(x) - L| < \varepsilon$.

The definition of the limit at a point c is very similar to the definition of continuity at c, however, there are three key differences. First, the definition of the limit requires c to be an accumulation point. This is not required in the definition of continuity. Second, in the limit definition, the inequality $0 < |x - c| < \delta$ is used instead of the inequality $|x - c| < \delta$. Thus, in Definition 4.3.1, the point c is not assumed to be in the domain of the function, whereas in the definition of continuity, c must be in the domain. Finally, in the above definition, the limit value L is used rather than $f(c)$. The term $f(c)$ is crucial for the definition of continuity, but it is not relevant in the above limit definition.

Definition 4.3.1 can be used in a proof by means of the following proof strategy.

Proof Strategy 4.3.2. Let $f\colon D \to \mathbb{R}$ and let c be an accumulation point of D. To prove that $\lim_{x \to c} f(x) = L$, use the proof diagram:

Let $\varepsilon > 0$ be an arbitrary real number.
Let $\delta =$ (the positive value you found).
Let $x \in D$ be an arbitrary.
Assume $0 < |x - c| < \delta$.
Prove $|f(x) - L| < \varepsilon$.

To apply proof strategy 4.3.2 to a function f, defined by a formula, first let $\varepsilon > 0$. Then find a $\delta > 0$ such that when $0 < |x - c| < \delta$, one can prove that $|f(x) - L| < \varepsilon$. To find the desired δ, use algebra and properties of inequality on the expression $|f(x) - L|$ to "extract out" a larger value that contains $|x - c|$ and no other occurrences of x.

Theorem. Let $f \colon \mathbb{R} \to \mathbb{R}$ be defined by $f(x) = x^2 + 2x + 4$. Then $\lim_{x \to 3} f(x) = 19$.

Proof Analysis. Let $\varepsilon > 0$. From the expression $|f(x) - 19|$ we need to "extract out" $|x - 3|$ and no other occurrences of x. Observe that

$$|f(x) - 19| = \left| x^2 + 2x - 15 \right| = |x + 5| \, |x - 3|$$

where we have $|x - 3|$ and the term $|x + 5|$ which contains x. Let us assume that $|x - 3| < 1$. Then

$$|x + 5| = |x - 3 + 3 + 5| \leq |x - 3| + 8 < 1 + 8 = 9.$$

Thus, $|f(x) - 19| < 9\,|x - 3|$. So, if we have that $|x - 3| < 1$ and $9\,|x - 3| < \varepsilon$, we can conclude that $|f(x) - 19| < \varepsilon$. Solving the inequality $9\,|x - 3| < \varepsilon$ for $|x - 3|$, we obtain $|x - 3| < \frac{\varepsilon}{9}$. Hence, we will let $\delta = \min\{1, \frac{\varepsilon}{9}\}$. We can now compose the following logically correct proof.

Proof. Let $\varepsilon > 0$ and let $\delta = \min\{1, \frac{\varepsilon}{9}\}$. Assume that $0 < |x - 3| < \delta$. So in particular, $|x - 3| < 1$. It follows that $|x + 5| < 9$. Therefore,

$$|f(x) - 19| = \left| x^2 + 2x - 15 \right| = |x + 5| \, |x - 3| < 9\,|x - 3| < 9\delta \leq 9\frac{\varepsilon}{9} = \varepsilon. \qquad \square$$

In the proof of Theorem 4.3.4, below, we will be assuming the negation of the definition of a limit. Our next remark clarifies the meaning of this negation.

Remark. Let $f \colon D \to \mathbb{R}$, $L \in \mathbb{R}$, and let c be an accumulation point of D. Then $\lim_{x \to c} f(x) \neq L$, if and only if there is an $\varepsilon > 0$ such that for all $\delta > 0$ there is an $x \in D$ such that $0 < |x - c| < \delta$ and $|f(x) - L| \geq \varepsilon$.

As noted earlier, there is a close connection between the definition of continuity of a function f at a point c and the definition of the limit of f at c. This intimacy leads to our next result which provides an alternate way to interpret continuity at c, when c is in the domain of f and c is also an accumulation point of the domain.

Lemma 4.3.3. Let $f \colon D \to \mathbb{R}$ and let $c \in D$ is an accumulation point of D. Then f is continuous at c if and only if $\lim_{x \to c} f(x) = f(c)$.

Proof. This is easy to see by comparing the definition of the continuity of f at c, Definition 4.1.1, and the definition of the limit of f at c, namely Definition 4.3.1. $\quad \square$

We next present an important link between the limit of functions and the limit of sequences. As we have proven many results about sequences, this link will allow us to derive theorems on the limits of functions by appealing to previously established results on sequences.

Theorem 4.3.4 (Sequential Criterion for Functional Limits). Let $f\colon D \to \mathbb{R}$, $L \in \mathbb{R}$, and c be an accumulation point of D. Then $\lim\limits_{x \to c} f(x) = L$ if and only if for every sequence $\langle s_n \rangle$ of points in $D \setminus \{c\}$, if $\lim\limits_{n \to \infty} s_n = c$, then $\lim\limits_{n \to \infty} f(s_n) = L$.

Proof. Let $f\colon D \to \mathbb{R}$ and let c be an accumulation point of D. We shall prove the above stated equivalence.

($\Rightarrow$). Assume that $\lim\limits_{x \to c} f(x) = L$. Let $\langle s_n \rangle$ be a sequence of points in $D \setminus \{c\}$ such that $\lim\limits_{n \to \infty} s_n = c$. Let $\varepsilon > 0$. Since $\lim\limits_{x \to c} f(x) = L$, there is a $\delta > 0$ such that

$$\text{for all } x \in D, \text{ if } 0 < |x - c| < \delta, \text{ then } |f(x) - L| < \varepsilon. \tag{4.12}$$

As $\lim\limits_{n \to \infty} s_n = c$, there is an $N \in \mathbb{N}$ such that $0 < |s_n - c| < \delta$ for all $n > N$. Thus, (4.12) implies that $|f(s_n) - L| < \varepsilon$ for all $n > N$. Hence, $\lim\limits_{n \to \infty} f(s_n) = L$.

($\Leftarrow$). Assume that for every sequence $\langle s_n \rangle$ of points in $D \setminus \{c\}$,

$$\text{if } \lim\limits_{n \to \infty} s_n = c, \text{ then } \lim\limits_{n \to \infty} f(s_n) = L. \tag{4.13}$$

To prove that $\lim\limits_{x \to c} f(x) = L$, assume to the contrary that L is not a limit of f at c. Thus, there is an $\varepsilon > 0$ such that

$$\text{for all } \delta > 0, \text{ there is an } x \in D \text{ such that } 0 < |x - c| < \delta \text{ and } |f(x) - L| \geq \varepsilon.$$

So for each $n \in \mathbb{N}$, there is an $s_n \in D$ such that $0 < |s_n - c| < \frac{1}{n}$ and $|f(s_n) - L| \geq \varepsilon$. Therefore, $\langle s_n \rangle$ is such that (i) $s_n \in D$, (ii) $0 < |s_n - c| < \frac{1}{n}$ and (▲) $|f(s_n) - L| \geq \varepsilon$, for all $n \in \mathbb{N}$. It follows, from (i) and (ii), that

$$\langle s_n \rangle \text{ is a sequence of points in } D \setminus \{c\} \text{ such that } \lim\limits_{n \to \infty} s_n = c. \tag{4.14}$$

However, (4.13) and (4.14) imply that $\lim\limits_{n \to \infty} f(s_n) = L$. This contradicts (▲). □

The following corollary offers a method for showing that a functional limit does not exist. This method avoids having to work with the negation of the limit definition.

Corollary 4.3.5. Let $f\colon D \to \mathbb{R}$ and let c be an accumulation point of D. If there is a sequence $\langle s_n \rangle$ that converges to c such that $\langle f(s_n) \rangle$ does not converge, then the limit of f at c does not exist.

The proof of Theorem 4.3.4 can easily be modified (see the proof of Theorem 3.5.4) to derive our next corollary which will be applied in Chapter 5.

Corollary 4.3.6. Let $f\colon D \to \mathbb{R}$, $L \in \mathbb{R}$, and c be an accumulation point of D. Then $\lim\limits_{x \to c} f(x) = L$ if and only if for every sequence $\langle s_n \rangle$ of *distinct* points in $D \setminus \{c\}$, if $\lim\limits_{n \to \infty} s_n = c$, then $\lim\limits_{n \to \infty} f(s_n) = L$.

In the following theorem, we record the basic algebraic properties of functional limits. For example, we will show that if f and g have limits at c, then so does the sum function $(f + g)$. In the proof of this theorem, we use Theorem 4.3.4 and some relevant results about sequences.

Theorem 4.3.7. Suppose that $\lim_{x \to c} f(x) = L$ and $\lim_{x \to c} g(x) = M$, where $f \colon D \to \mathbb{R}$ and $g \colon D \to \mathbb{R}$ are functions and c is an accumulation point of D. Then

(1) $\lim_{x \to c}(f + g)(x) = L + M$.

(2) $\lim_{x \to c}(fg)(x) = LM$.

(3) $\lim_{x \to c}(kf)(x) = kL$, where $k \in \mathbb{R}$ is any constant.

(4) If $g(x) \neq 0$ for all $x \in D$ and $M \neq 0$, then $\lim_{x \to c} \left(\frac{f}{g} \right)(x) = \frac{L}{M}$.

Proof. Let $f \colon D \to \mathbb{R}$ and $g \colon D \to \mathbb{R}$ be functions and let c be an accumulation point of D. Assume $\lim_{x \to c} f(x) = L$ and $\lim_{x \to c} g(x) = M$. We shall only prove (1) and leave (2)–(4) for Exercise 7. To prove that $\lim_{x \to c}(f + g)(x) = L + M$, let $\langle s_n \rangle$ be an arbitrary sequence of points in $D \setminus \{c\}$ that converges to c. Since $\lim_{x \to c} f(x) = L$ and $\lim_{x \to c} g(x) = M$, Theorem 4.3.4 implies that

$$\lim_{n \to \infty} f(s_n) = L \quad \text{and} \quad \lim_{n \to \infty} g(s_n) = M. \tag{4.15}$$

Thus,

$$
\begin{aligned}
\lim_{n \to \infty}(f + g)(s_n) &= \lim_{n \to \infty}[f(s_n) + g(s_n)] && \text{by Definition 4.1.4(1)} \\
&= \lim_{n \to \infty} f(s_n) + \lim_{n \to \infty} g(s_n) && \text{by Theorem 3.2.2} \\
&= L + M && \text{by (4.15).}
\end{aligned}
$$

Theorem 4.3.4 now implies $\lim_{x \to c}(f + g)(x) = L + M$. □

We now present a squeeze theorem for functions. This theorem is analogous to Theorem 3.2.8, the squeeze theorem for sequences (see Exercise 8).

Theorem 4.3.8 (Squeeze Theorem). Let f, g, h be functions from D to $\mathbb{R}$ such that

$$f(x) \leq g(x) \leq h(x), \text{ for all } x \in D.$$

Let c be an accumulation point of D. If $\lim_{x \to c} f(x) = L = \lim_{x \to c} h(x)$, then $\lim_{x \to c} g(x) = L$.

Theorems 4.2.1 and 4.3.4 imply the following result (see Exercise 9) on the limit of a composition.

Theorem 4.3.9. Let $g \colon J \to I$ and $h \colon I \to \mathbb{R}$, where I and J are intervals. Suppose that g is continuous at $c \in J$. If $\lim_{y \to g(c)} h(y) = L$, then $\lim_{x \to c} h(g(x)) = L$.

One-Sided Limits

Let $f \colon I \to \mathbb{R}$ be a function where I is an interval. Let c be an accumulation point of I. Thus, Definition 4.3.1 precisely defines the limit notation $\lim_{x \to c} f(x) = L$. However, there are times when we need to emphasize how x approaches c in the limit process. As c is an accumulation point of the interval I, it follows that c is either an interior point or an endpoint of I. We now introduce the notion of a one-sided limit.

Definition 4.3.10. Let $f\colon I \to \mathbb{R}$ be a function where I is an interval, and let c be an accumulation point of I.

(1) *Right-Hand Limit*: Suppose that c is an interior point or a left endpoint of I. We write $\lim\limits_{x \to c^+} f(x) = L$ if and only if for every $\varepsilon > 0$, there is a $\delta > 0$ such that for all $x \in I$, if $0 < |x - c| < \delta$ and $c < x$, then $|f(x) - L| < \varepsilon$.

(2) *Left-Hand Limit*: Suppose that c is an interior point or a right endpoint of I. We write $\lim\limits_{x \to c^-} f(x) = L$ if and only if for every $\varepsilon > 0$, there is a $\delta > 0$ such that for all $x \in I$, if $0 < |x - c| < \delta$ and $x < c$, then $|f(x) - L| < \varepsilon$.

Remark 4.3.11. Let c be an interior point or an endpoint of an interval I. Now let $f\colon I \to \mathbb{R}$ be a function. Corollary 4.3.6 implies the following:

(1) $\lim\limits_{x \to c^+} f(x) = L$ if and only if for every sequence $\langle s_n \rangle$ of distinct points in I where $c < s_n$ for all $n \in \mathbb{N}$, if $\lim\limits_{n \to \infty} s_n = c$, then $\lim\limits_{n \to \infty} f(s_n) = L$.

(2) $\lim\limits_{x \to c^-} f(x) = L$ if and only if for every sequence $\langle s_n \rangle$ of distinct points in I where $s_n < c$ for all $n \in \mathbb{N}$, if $\lim\limits_{n \to \infty} s_n = c$, then $\lim\limits_{n \to \infty} f(s_n) = L$.

Let $f\colon I \to \mathbb{R}$ be a function on an interval I and let c be a left endpoint of I. In this case, Definition 4.3.1 and Definition 4.3.10(1) are equivalent, that is, $\lim\limits_{x \to c} f(x) = L$ if and only if $\lim\limits_{x \to c^+} f(x) = L$. The analogous equivalence also holds when c is a right endpoint of I. However, this is not quite the case when c is an interior point.

The next theorem shows that in order for the limit of a function to exist at an interior point, both the limit from the right and the limit from the left must exist and produce the same value.

Theorem 4.3.12. Let $f\colon I \to \mathbb{R}$, $L \in \mathbb{R}$, and let c be an interior point of I, an interval. Then $\lim\limits_{x \to c} f(x) = L$ if and only if $\lim\limits_{x \to c^-} f(x) = \lim\limits_{x \to c^+} f(x) = L$.

Infinite Limits

A function f that is not defined at a point c may have values $f(x)$ that become very large as x gets very close to c.

Definition 4.3.13. Let $f\colon D \to \mathbb{R}$ and let c be an accumulation point of D. We shall write $\lim\limits_{x \to c} f(x) = \infty$ if and only if for every $M > 0$ there exists a $\delta > 0$ such that for all $x \in D$, if $0 < |x - c| < \delta$, then $f(x) > M$.

Notice that $\lim\limits_{x \to 0} 1/x^2 = \infty$. However, $\lim\limits_{x \to 0^+} 1/x = \infty$ and $\lim\limits_{x \to 0^-} 1/x = -\infty$ (see Exercise 23(b)(c)). It follows that $\lim\limits_{x \to 0} 1/x$ is meaningless in the sense of Definition 4.3.13. The definition of $\lim\limits_{x \to c} f(x) = -\infty$ is similar to Definition 4.3.13, and will be left to the reader to formulate (see Exercise 23).

Limits at Infinity

On the other hand, as x gets very large, the values $f(x)$ may become very close to a real number, or the values $f(x)$ may become very large.

Definition 4.3.14. Let $f: (a, \infty) \to \mathbb{R}$ and $L \in \mathbb{R}$. Then $\lim_{x \to \infty} f(x) = L$ if and only if for every $\varepsilon < 0$ there exists a $M > a$ such that for all $x \in (a, \infty)$, if $x > M$, then $|f(x) - L| < \varepsilon$.

Definition 4.3.15. Let $f: (a, \infty) \to \mathbb{R}$. Then $\lim_{x \to \infty} f(x) = \infty$ if and only if for every $K > 0$ there exists a $M > a$ such that for all $x \in (a, \infty)$, if $x > M$, then $f(x) > K$.

Exercises 4.3

1. Use Definition 4.3.1 to prove each of the limits.

    (a) $\lim_{x \to 5} x^2 - 3x + 1 = 11$

    (b) $\lim_{x \to -2} x^2 + 2x + 7 = 7$.

2. Let $f: D \to \mathbb{R}$ be a function and let $c \in D$. Suppose that c is an accumulation point of D. Prove that f is continuous at c if and only if $\lim_{x \to c} f(x) = f(c)$.

3. Let $f: \mathbb{R} \to \mathbb{R}$ be defined by

$$f(x) = \begin{cases} 1, & \text{if } x \in \mathbb{Q}; \\ -1, & \text{if } x \notin \mathbb{Q}. \end{cases}$$

    Show that for every $c \in \mathbb{R}$ the limit of f at c does not exist.

4. Let $f: D \to \mathbb{R}$ and let c be an accumulation point of D. Prove that if $\lim_{x \to c} f(x) = L$ where L is a real number, then there is a $\delta > 0$ and an $M > 0$ such that $|f(x)| \le M$ whenever $0 < |x - c| < \delta$ and $x \in D$.

5. Let $f: D \to \mathbb{R}$ and let c be an accumulation point of D where $\lim_{x \to c} f(x) = L \ne 0$ and L is a real number. Prove that there is an $\varepsilon > 0$ and a $\delta > 0$ such that for all $x \in D$, if $0 < |x - c| < \delta$, then $|f(x)| \ge \varepsilon$.

6. Let $f: D \to \mathbb{R}$ and let c be an accumulation point of D where $\lim_{x \to c} f(x) = L > 0$ and L is a real number. Prove that there is a $\delta > 0$ such that $f(x) > 0$ when $0 < |x - c| < \delta$ and $x \in D$.

*7. Complete the proof of Theorem 4.3.7, that is, prove items (2)–(4) of this theorem.

*8. Prove Theorem 4.3.8.

*9. Prove Theorem 4.3.9.

10. Let $f: D \to \mathbb{R}$, c be an accumulation point of D, and $\lim_{x \to c} f(x) = L \in \mathbb{R}$.

    (a) Suppose that $f(x) \ge 0$ for all $x \in D$. Prove that $L \ge 0$.

    (b) Suppose that $f(x) \le 0$ for all $x \in D$. Prove that $L \le 0$.

*11. Let $f\colon I \to \mathbb{R}$, were I is an interval, and let $a \in I$ and $b \in I$ be such that $a < b$. Suppose that $\lim_{x \to a} f(x) < 0 < \lim_{x \to b} f(x)$. Prove that there exists $c, d \in I$ such that $a < c < d < b$ and $f(c) < 0 < f(d)$.

12. Let $f\colon I \to \mathbb{R}$ be increasing on an interval I. Let $c \in I$ be an interior point. Let $S = \{f(x) : x \in I \text{ and } x < c\}$. Prove that $\beta = \sup(S)$ exists and $\lim_{x \to c^-} f(x) = \beta$.

13. Let $f\colon I \to \mathbb{R}$ be increasing on an interval I. Let $c \in I$ be an interior point. Let $T = \{f(x) : x \in I \text{ and } c < x\}$. Prove that $\alpha = \inf(T)$ exists and $\lim_{x \to c^+} f(x) = \alpha$.

14. Let $f\colon I \to \mathbb{R}$, were I is an interval. Let c be an interior point of I. Suppose that f is increasing on I. Prove that $\lim_{x \to c^-} f(x) \leq f(c) \leq \lim_{x \to c^+} f(x)$.

15. Prove Corollary 4.3.6.

16. Prove Theorem 4.3.12.

*17. Let $f\colon D \to \mathbb{R}$ and let c be an accumulation point of D. Prove that $\lim_{x \to c} f(x) = \infty$ if and only if for every sequence $\langle s_n \rangle$ of points in $D \setminus \{c\}$, if $\lim_{n \to \infty} s_n = c$, then $\lim_{n \to \infty} f(s_n) = \infty$.

18. Let $f\colon D \to \mathbb{R}$, c be an accumulation point of D, and $\langle s_n \rangle$ be a sequence of points in D that converges to c. Prove that if $\lim_{x \to c} f(x) = \infty$, then for every $\varepsilon > 0$, there is an $N \in \mathbb{N}$ such that $\frac{1}{f(s_n)} < \varepsilon$ for all $n > N$.

*19. Let $f\colon (a, \infty) \to \mathbb{R}$ and $L \in \mathbb{R}$. Prove that $\lim_{x \to \infty} f(x) = L$ if and only if for every sequence $\langle s_n \rangle$ of *distinct* points in (a, ∞), if $\lim_{n \to \infty} s_n = \infty$, then $\lim_{n \to \infty} f(s_n) = L$.

20. Let $f\colon (a, \infty) \to \mathbb{R}$. Prove that $\lim_{x \to \infty} f(x) = \infty$ if and only if for every sequence $\langle s_n \rangle$ of points in (a, ∞), if $\lim_{n \to \infty} s_n = \infty$, then $\lim_{n \to \infty} f(s_n) = \infty$.

21. Let $h\colon \mathbb{R} \to \mathbb{R}$ be a function and let $a \in \mathbb{R}$. Suppose that $\lim_{x \to a^+} h(x) = L$.

    (a) Prove that if h is odd, then $\lim_{x \to -a^-} h(x) = -L$.

    (b) Prove that if h is even, then $\lim_{x \to -a^-} h(x) = L$.

22. Let $f\colon D \to \mathbb{R}$ and $g\colon D \to \mathbb{R}$, and let c be an accumulation point of D. Suppose that there is an $M > 0$ such that $|g(x)| \leq M$ for all $x \in D$, and $\lim_{x \to c} f(x) = 0$. Prove that $\lim_{x \to c}(fg)(x) = 0$.

*23. Provide definitions for the following limit concepts:

    (a) $\lim_{x \to c} f(x) = -\infty$.   (c) $\lim_{x \to c^-} f(x) = -\infty$.   (e) $\lim_{x \to \infty} f(x) = \infty$.

    (b) $\lim_{x \to c^+} f(x) = \infty$.   (d) $\lim_{x \to -\infty} f(x) = L$.   (f) $\lim_{x \to -\infty} f(x) = \infty$.

24. Let $f\colon (-\infty, b) \to \mathbb{R}$. So, the composition $f(-x)$ is defined on $(-b, \infty)$. Suppose that $\lim_{x \to -\infty} f(x) = L$. Show that $\lim_{x \to \infty} f(-x) = L$.

Exercise Notes: For Exercise 5, let $\varepsilon = \frac{|L|}{2}$. For Exercise 6, let $\varepsilon = L$. For Exercise 8, use Theorem 4.3.4 and Theorem 3.2.8. For Exercises 10, use Theorem 4.3.4 and Theorem 3.2.14. For Exercise 11, review the proof of Lemma 4.1.14. For Exercise 21, apply Remark 4.3.11. In particular, for (a), let $\langle s_n \rangle$ be a sequence of points in $\mathbb{R}$ such that $s_n < -a$ for all $n \in \mathbb{N}$. Suppose that $\lim_{n \to \infty} s_n = -a$. Show that $\lim_{n \to \infty} h(s_n) = -L$.

4.4 CONSEQUENCES OF CONTINUITY

In this section we establish several important properties that continuous functions possess. First we review some relevant features that have been shown to hold for sequences and continuous functions.

Concept Questions: Let $f \colon [a,b] \to \mathbb{R}$ be continuous and let $\langle x_n \rangle$ be a sequence of points such that $x_n \in [a,b]$ for all $n \geq 1$.

1. What can you conclude about the sequence $\langle x_n \rangle$ from Theorem 3.5.1? Answer: $\langle x_n \rangle$ has a convergent subsequence.
2. If the sequence $\langle f(x_n) \rangle$ converges, what can you conclude from Theorem 3.1.24? Answer: $\langle f(x_n) \rangle$ is bounded.
3. If $\langle x_n \rangle$ has a subsequence $\langle x_{n_k} \rangle$ that converges to c. What can you infer about c from Corollary 3.2.15? Answer: $c \in [a,b]$. What can you infer from Theorem 4.2.1 about the sequence $\langle f(x_{n_k}) \rangle$? Answer: $\lim_{k \to \infty} f(x_{n_k}) = f(c)$.
4. If $a < c < b$ and $f(c) > 0$, then what can you infer from Lemma 4.1.14(a)?
5. If $a < c < b$ and $f(c) < 0$, then what can you infer from Lemma 4.1.14(b)?

Recall that a function $f \colon [a,b] \to \mathbb{R}$ is bounded if and only if there is a real number $M > 0$ such that $|f(x)| \leq M$ for all $x \in [a,b]$ (see Remark 2.3.15).

Theorem 4.4.1. If $f \colon [a,b] \to \mathbb{R}$ is continuous, then f is bounded.

Proof. Let $f \colon [a,b] \to \mathbb{R}$ be continuous. Suppose that f is unbounded. Thus, for all $n \in \mathbb{N}$, there exists an $x_n \in [a,b]$ such that $|f(x_n)| > n$. So the sequence $\langle f(x_n) \rangle$ is unbounded and (▲) any subsequence of $\langle f(x_n) \rangle$ is unbounded. Consider the sequence $\langle x_n \rangle$ of points in $[a,b]$. Theorem 3.5.1 and Corollary 3.2.15 imply that $\langle x_n \rangle$ has a subsequence $\langle x_{n_k} \rangle$ that converges to a point $c \in [a,b]$. Theorem 4.2.1 implies that $\lim_{k \to \infty} f(x_{n_k}) = f(c)$. By Theorem 3.1.24 $\langle f(x_{n_k}) \rangle$ is bounded, contradicting (▲). □

4.4.1 The Extreme Value Theorem

The extreme value theorem states that if a function is continuous on a closed interval, then the function attains maximum and minimum values. This theorem is usually stated without proof in most calculus books. In this section, we shall prove this important result.

Definition 4.4.2. Let $f \colon D \to \mathbb{R}$ be a function. We say that

- *f* **attains a maximum value on** D if there is $c \in D$ such that $f(x) \leq f(c)$ for all $x \in D$, that is, $\sup(f[D]) = f(c)$ for some $c \in D$;

- *f* **attains a minimum value on** D if there is $d \in D$ such that $f(d) \leq f(x)$ for all $x \in D$, that is, $\inf(f[D]) = f(d)$ for some $d \in D$.

We will next show that a continuous function, whose domain is a closed and bounded interval, attains both a maximum and minimum value.

Theorem 4.4.3 (Extreme Value Theorem). If $f\colon [a,b] \to \mathbb{R}$ is continuous, then f attains maximum and minimum values on $[a,b]$.

Proof. Let $f\colon [a,b] \to \mathbb{R}$ be continuous. By Theorem 4.4.1, the set $f([a,b])$ is bounded. Let $M = \sup(f([a,b]))$. For each $n \in \mathbb{N}$, let $x_n \in [a,b]$ be such that $M - \frac{1}{n} < f(x_n)$. Thus, $M - f(x_n) < \frac{1}{n}$ for all $n \geq 1$. Since $f(x) \leq M$ for all x in $[a,b]$, we see that $|f(x_n) - M| < \frac{1}{n}$ for every $n \geq 1$. Thus, (▲) $\lim_{n\to\infty} f(x_n) = M$ (see Theorem 3.1.11).

By Theorem 3.5.1 and Corollary 3.2.15, the sequence $\langle x_n \rangle$ has a subsequence $\langle x_{n_k} \rangle$ that converges to a point $c \in [a,b]$. Since f is continuous, Theorem 4.2.1 implies that

$$\lim_{k\to\infty} f(x_{n_k}) = f(c). \tag{4.16}$$

Moreover, $\langle f(x_{n_k}) \rangle$ is a subsequence of $\langle f(x_n) \rangle$. Theorem 3.3.4 and (▲) thus imply that $\lim_{k\to\infty} f(x_{n_k}) = M$. So, from (4.16), we have $f(c) = M$ and therefore, f attains a maximum value on $[a,b]$. A similar argument, using $m = \inf(f([a,b]))$, shows that f attains a minimum value on $[a,b]$. ☐

The conclusion of Theorem 4.4.3 may not hold if the closed interval $[a,b]$ is replace by an open interval. For example, let $f\colon (0,1) \to \mathbb{R}$ be the continuous function defined by $f(x) = x$. The function f does not attain a maximum value on $(0,1)$.

4.4.2 The Intermediate Value Theorem

The Intermediate Value Theorem proclaims that if a continuous function, on an interval, attains two distinct values, then it also attains each value strictly between these two values. Before proving this theorem, we will first establish a key lemma which has a simple geometric interpretation. Let $f\colon I \to \mathbb{R}$ be continuous where I is an interval. Let $a,b \in I$ be such that $a < b$. Now suppose that $f(a) < 0 < f(b)$. Since f is continuous, the graph of f moves from the point $(a, f(a))$ to the point $(b, f(b))$ without any breaks. So the graph must cross the x-axis at a point c between the points a and b. So $f(c) = 0$ (see Figure 4.2). In spite of its apparent simplicity, the proof of this lemma requires the completeness axiom.

Lemma 4.4.4 (Intermediate Root Lemma). Let $f\colon I \to \mathbb{R}$ be continuous where I is an interval. Let $a,b \in I$ be such that $a < b$. If $f(a) < 0 < f(b)$, then there is a $c \in (a,b)$ such that $f(c) = 0$.

Proof. Let $f\colon I \to \mathbb{R}$ be continuous, $a,b \in I$ be such that $a < b$, and $f(a) < 0 < f(b)$. Let $S = \{x \in [a,b] : f(x) < 0\}$. As $f(a) < 0$, the set S is nonempty. Since $a \leq x \leq b$ for all $x \in S$, we conclude that S is bounded, $c = \sup(S)$ exists, and $c \in [a,b]$.

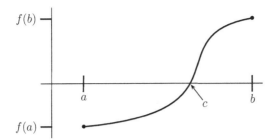

Figure 4.2: Illustration for Lemma 4.4.4.

Claim. $f(c) = 0$.

Proof of Claim. Suppose, for a contradiction, that either $f(c) > 0$ or $f(c) < 0$.

CASE 1: $f(c) > 0$. Lemma 4.1.14(a) implies that there is a $\delta > 0$ such that

$$f(x) > 0 \text{ for all } x \in [a, b] \text{ satisfying } |x - c| < \delta. \tag{4.17}$$

Since $c = \sup(S)$ and $c - \delta < c$, there is an $x \in S$ such that $c - \delta < x$ and $x \le c$. Thus, $x \in [a, b]$ and $|x - c| < \delta$. So $f(x) > 0$ by (4.17). Since $x \in S$, we also have that $f(x) < 0$. This contradiction implies that $f(c) \not> 0$.

CASE 2: $f(c) < 0$. Lemma 4.1.14(b) implies that there is a $\delta > 0$ such that

$$f(x) < 0 \text{ for all } x \in [a, b] \text{ satisfying } |x - c| < \delta. \tag{4.18}$$

As $c \le b$, $f(c) < 0$, and $0 < f(b)$, we have that $c < b$. Thus, $c < \min\{b, c + \delta\}$. Let x be such that $c < x < \min\{b, c + \delta\}$. Thus, $x \in [a, b]$ and $|x - c| < \delta$. Hence, (4.18) implies that $f(x) < 0$. So $x \in S$. Since $c < x$, this contradicts the fact that c is an upper bound for S. Thus, $f(c) \not< 0$.

Therefore, $f(c) = 0$ and this completes the proof of the claim. □

By the Claim, $f(c) = 0$ where $c \in [a, b]$. Since $f(a) < 0 < f(b)$, we must have that $a < c < b$ and thus, $c \in (a, b)$. The proof of the lemma is now complete. □

For a simple application of Lemma 4.4.4, let $f \colon [2, 3] \to \mathbb{R}$ be the continuous function defined by $f(x) = x^5 - 32x - 3$. Since $f(2) < 0 < f(3)$. There is a $c \in (2, 3)$ such that $f(c) = 0$.

The following definition resembles the statement of the Lemma 4.4.4, except that we allow $f(a)$ and $f(b)$ to be any two distinct values–not just one negative and the other positive (see Figure 4.3).

Definition 4.4.5. Let I be interval. A function $f \colon I \to \mathbb{R}$ has the **intermediate value property on** I if the following condition holds: Whenever $a, b \in I$ are such that $a < b$ and k is any real number strictly between $f(a)$ and $f(b)$, there is a $c \in (a, b)$ such that $f(c) = k$.

The following theorem generalizes Lemma 4.4.4. This generalization will be applied in this section and in later sections of the text.

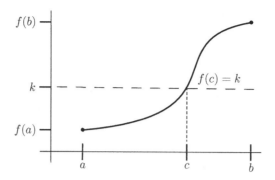

Figure 4.3: Illustration for Theorem 4.4.6.

Theorem 4.4.6 (Intermediate Value Theorem). Let I be an interval. If $f\colon I \to \mathbb{R}$ is continuous, then f has the intermediate value property on I.

Proof. Let $f\colon I \to \mathbb{R}$ be continuous and let $a, b \in I$ be such that $a < b$. Let $k \in \mathbb{R}$ be strictly between $f(a)$ and $f(b)$. If $f(a) < k < f(b)$, then let $g\colon I \to \mathbb{R}$ be defined by $g(x) = f(x) - k$. Because f is continuous, it follows that g is continuous. Since $g(a) = f(a) - k < 0 < f(b) - k = g(b)$, Lemma 4.4.4 implies that $g(c) = 0$ for some $c \in (a, b)$. Hence, $g(c) = f(c) - k = 0$ and therefore, $f(c) = k$. If $f(a) > k > f(b)$, then a similar argument applies using the function $g(x) = k - f(x)$. □

The Intermediate Value Theorem is an important theoretical tool that is used to prove a number of results in calculus and in real analysis. For example, it can be used to show that every positive real number has a positive nth root. To see this, let $k > 0$ and let $n \in \mathbb{N}$. Let $f\colon [0, k+1] \to \mathbb{R}$ be the continuous function defined by $f(x) = x^n$. As $f(0) < k < f(k+1)$, Theorem 4.4.6 implies that there is a $c \in (0, k+1)$ such that $f(c) = c^n = k$. Thus, c is the nth root of k.

One may suspect that the converse of Theorem 4.4.6 holds; that is, if a function has the intermediate property, then it must be continuous. However, this is false. Let $f\colon [0, 1] \to \mathbb{R}$ be defined by $f(x) = \sin(\frac{1}{x})$ for $x \neq 0$ and $f(0) = 0$. Then f has the intermediate value property on $[0, 1]$, but it is not continuous at 0 (see Exercise 13).

The Intermediate Value Theorem implies that the continuous image of an interval is an interval.

Theorem 4.4.7. Let $I \subseteq \mathbb{R}$ be an interval. Suppose that $f\colon I \to \mathbb{R}$ is continuous and nonconstant. Then $f[I]$ is an interval.

Proof. This follows from Theorem 4.4.6 and Definition 1.1.1 on page 2. □

Our next theorem shows that the continuous image of a closed interval is a closed interval.

Theorem 4.4.8. Let $f\colon [a, b] \to \mathbb{R}$ be continuous and nonconstant. Then there exists real numbers $m < M$ such that $f([a, b]) = [m, M]$.

Proof. Let $f\colon [a,b] \to \mathbb{R}$ be continuous and nonconstant. By Theorem 4.4.3 and Definition 4.4.2, there are $c, d \in [a, b]$ such that

$$f(c) = M = \sup(f([a,b])) \text{ and } f(d) = m = \inf(f([a,b])).$$

So $m, M \in f([a,b])$. Since f is a nonconstant function, $m < M$. As $m \le f(x) \le M$ for all $x \in [a,b]$, it follows that $f([a,b]) \subseteq [m, M]$. Moreover, Theorem 4.4.7 implies that $f([a,b])$ is an interval. As $m, M \in f([a,b])$, it follows that $[m, M] \subseteq f([a,b])$. Hence, $f([a,b]) = [m, M]$. □

We end this section with another application of Lemma 4.4.4.

Theorem 4.4.9 (Fixed Point Theorem). Let $f\colon [a,b] \to [a,b]$ be continuous. Then there is a point c in $[a,b]$ such that $f(c) = c$.

Proof. If $f(a) = a$ or $f(b) = b$, then the result holds. Suppose that $f(a) \ne a$ and $f(b) \ne b$. Since $f\colon [a,b] \to [a,b]$, it follows that $a < f(a)$ and $f(b) < b$. Now consider the continuous function $g\colon [a,b] \to \mathbb{R}$ defined by $g(x) = x - f(x)$. Since $a < f(a)$ and $f(b) < b$, we see that $g(a) < 0 < g(b)$. Lemma 4.4.4 implies that there is a point c in $[a,b]$ such that $g(c) = 0$. Thus, $f(c) = c$. □

Exercises 4.4

1. Prove Theorem 4.4.7.

2. Let $f\colon [a,b] \to \mathbb{R}$ be continuous. Suppose that $\inf(f([a,b])) \le k \le \sup(f([a,b]))$. Show that there exists an $x \in [a,b]$ such that $f(x) = k$.

3. Give an example of a continuous function on an open interval that is not bounded.

4. Let $\mathcal{I} = \{x \in \mathbb{Q} : 1 \le x \le 3\}$. $\mathcal{I}$ is called a *closed* $\mathbb{Q}$-*interval* (see page 51). Define $f\colon \mathcal{I} \to \mathbb{Q}$ by $f(x) = x^2 - 2$. Observe that $f(1) < 0 < f(3)$. Show that there is no $c \in \mathcal{I}$ such that $f(c) = 0$.

5. Let $f\colon I \to \mathbb{R}$ where I is an interval. Suppose that for all $a, b \in I$, if $a < b$, then $f([a,b])$ is an interval. Prove that f has the intermediate value property on I.

6. Given that $f(x) = \sin(x)$ is a continuous function, show that there is an open interval whose image under f is not an open interval.

7. Consider the function $f\colon [0,2] \to \mathbb{R}$ defined by $f(x) = x^4 - 3x^2 + 1$. Show that there are distinct points $x, y \in [0,2]$ such that $f(x) = 0$ and $f(y) = 0$.

8. Complete the proof of Theorem 4.4.3; that is, prove that f attains a minimum value on $[a,b]$.

9. Let $f\colon [0,1] \to \mathbb{R}$ be a continuous function. Suppose that $f(0) = f(1)$. Prove that there is a c in $[0, \frac{1}{2}]$ such that $f(c) = f(c + \frac{1}{2})$.

10. Let $f\colon [a,b] \to \mathbb{R}$ be a continuous function. Suppose that $f(x) \in \mathbb{Q}$, for all x in $[a,b]$. Prove that f is constant on $[a,b]$.

11. Let $f\colon [a,b] \to \mathbb{R}$ and $g\colon [a,b] \to \mathbb{R}$ be continuous. Suppose that $f(a) < g(a)$ and $f(b) > g(b)$. Prove that there is a point c in (a,b) such that $f(c) = g(c)$.

12. Show that there exists an $x \in (0, \pi/2)$ such that $\cos(x) = x$.

*13. Let $f\colon [0,1] \to \mathbb{R}$ be defined by

$$f(x) = \begin{cases} \sin(\frac{1}{x}), & \text{if } x \neq 0; \\ 0, & \text{if } x = 0. \end{cases}$$

Consider the sequence $\left\langle \frac{1}{2n\pi + \frac{\pi}{2}} \right\rangle$. Show that f is not continuous at 0.

Exercise Notes: For Exercise 8, since $m = \inf(f([a,b]))$, we see that for all $n \in \mathbb{N}$, there is an $x_n \in [a,b]$ such that $f(x_n) < m + \frac{1}{n}$. For Exercise 9, consider the function $h\colon [0, \frac{1}{2}] \to \mathbb{R}$ defined by $h(x) = f(x) - f(x + \frac{1}{2})$. Explain why h is continuous.

4.5 UNIFORM CONTINUITY

Recalling Definition 4.1.1, a function $f\colon D \to \mathbb{R}$ is continuous on D when for each $c \in D$ and for every $\varepsilon > 0$, there exists a $\delta > 0$ such that for all $x \in D$, if $|x - c| < \delta$, then $|f(x) - f(c)| < \varepsilon$. Given two points $c, d \in D$ and an $\varepsilon > 0$, the δ that you find to establish the continuity of f at c may not work to establish the continuity of f at d. That is, one may have to find a $\delta' \neq \delta$ which will show that f is continuous at d. When you can find a δ that works for all the points in D simultaneously, we shall say the function f is uniformly continuous.[2]

Definition 4.5.1. A function $f\colon D \to \mathbb{R}$ is **uniformly continuous on** D when the following holds: For every $\varepsilon > 0$ there exists a $\delta > 0$ such that for all $x, c \in D$, if $|x - c| < \delta$, then $|f(x) - f(c)| < \varepsilon$.

Proof Strategy 4.5.2. Let $f\colon D \to \mathbb{R}$. To prove that f is uniformly continuous on D, use the proof diagram:

> Let $\varepsilon > 0$ be an arbitrary real number.
> Let $\delta =$ (the positive value you found).
> Let $x \in D$ and $c \in D$ be arbitrary.
> Assume $|x - c| < \delta$.
> Prove $|f(x) - f(c)| < \varepsilon$.

Example 4.5.3. Let $f\colon \mathbb{R} \to \mathbb{R}$ be defined by $f(x) = 2x$. Prove that f is uniformly continuous on $\mathbb{R}$.

Solution. To prove that f is uniformly continuous on $\mathbb{R}$, we need to first do some "scratch work." Let $\varepsilon > 0$. We must find a δ that verifies the continuity of f at all points $c \in \mathbb{R}$ simultaneously, that is, we must find a δ such that *for all $c, x \in \mathbb{R}$, if*

---

[2]The word **uniform** means: *not changing in form or character; remaining the same in all cases and at all times.*

$|x - c| < \delta$, then $|f(x) - f(c)| < \varepsilon$. Recall the strategy to prove continuity of f at c. Working with $|f(x) - f(c)|$ we try to "extract out" $|x - c|$. Let us try this strategy for the function $f(x) = 2x$.

$$|f(x) - f(c)| = |2x - 2c| = 2|x - c|.$$

By solving the inequality $2|x - c| < \varepsilon$ for $|x - c|$, we obtain $|x - c| < \frac{\varepsilon}{2}$. Using $\delta = \frac{\varepsilon}{2}$, we can prove that f is continuous at c. *Notice that the δ we found does not depend on c.* Hence, we can also use this δ to prove that f is uniformly continuous on $\mathbb{R}$.

Proof. Let $f: \mathbb{R} \to \mathbb{R}$ be defined by $f(x) = 2x$. Let $\varepsilon > 0$. Take $\delta = \frac{\varepsilon}{2}$. Let $c, x \in \mathbb{R}$ and assume that $|x - c| < \delta$. Thus, $|f(x) - f(c)| = 2|x - c| < 2\delta = 2\frac{\varepsilon}{2} = \varepsilon$. Therefore, f is uniformly continuous on $\mathbb{R}$. □

Let $f: D \to \mathbb{R}$ and let $c \in D$. Suppose that you can prove that f is continuous at c using Definition 4.1.1. In addition, suppose that the δ you obtain depends on c. As a consequence, this δ will work for c; but it may not work for *all* points in D. In this case, the function f may not be uniformly continuous on D.

To prove that a function is not uniformly continuous can be a bit challenging because one has to establish the negation of Definition 4.5.1. We provide such a proof in our next example. To prepare us for this proof, we state the negation of uniform continuity in the following comment.

Comment. Let $f: D \to \mathbb{R}$. Then f is **not** uniformly continuous on D if and only if there exists an $\varepsilon > 0$ such that for all $\delta > 0$, there are $x, c \in D$ such that $|x - c| < \delta$ and $|f(x) - f(c)| \geq \varepsilon$.

Example 4.5.4. Let $f: \mathbb{R} \to \mathbb{R}$ be defined by $f(x) = x^2$. Prove that f is not uniformly continuous on $\mathbb{R}$.

Proof Analysis. We will apply the above comment. We need to first identify ε. Let us try $\varepsilon = 1$. Now, let $\delta > 0$ be arbitrary. We need to find $c, x \in \mathbb{R}$ such that $|x - c| < \delta$ and $|f(x) - f(c)| = |x + c||x - c| \geq \varepsilon = 1$. To make sure that $|x - c| < \delta$, let us take $(\star)$ $x = c + \frac{\delta}{2}$ with $c > 0$. Hence, $|x - c| = \frac{\delta}{2}$. To make $|x + c||x - c| \geq 1$ hold, we need (after substituting) $|x + c|\frac{\delta}{2} \geq 1$; that is, we need $|x + c| \geq \frac{2}{\delta}$. Since $x, c > 0$ we have that $|x + c| \geq |c|$. So, any c that satisfies $|c| \geq \frac{2}{\delta}$ will do. So, we let $c = \frac{2}{\delta}$ and $x = \frac{2}{\delta} + \frac{\delta}{2}$ (see $(\star)$). We can now prove that f is **not** uniformly continuous on $\mathbb{R}$.

Proof. Let $f: \mathbb{R} \to \mathbb{R}$ be defined by $f(x) = x^2$. Let $\varepsilon = 1$. Let $\delta > 0$. Let $c = \frac{2}{\delta}$ and $x = \frac{2}{\delta} + \frac{\delta}{2}$. Clearly, $|x - c| = \frac{\delta}{2} < \delta$ and

$$|f(x) - f(c)| = |x + c||x - c| = \left(\frac{4}{\delta} + \frac{\delta}{2}\right)\frac{\delta}{2} = 2 + \frac{\delta^2}{4} > 1.$$

Thus, $|x - c| < \delta$ and $|f(x) - f(c)| \geq 1$. Therefore, f is not uniformly continuous. □

We see from Example 4.5.4 that there are continuous functions that are not uniformly continuous on certain intervals. Nevertheless, there are conditions that will ensure that a continuous function is uniformly continuous. The following theorem presents one such condition: if the domain is a closed and bounded interval.

Theorem 4.5.5. Let $a, b \in \mathbb{R}$ where $a < b$. If $f : [a, b] \to \mathbb{R}$ is continuous, then f is uniformly continuous on $[a, b]$.

Proof. Let $f : [a, b] \to \mathbb{R}$ be continuous. Suppose, for a contradiction, that f is not uniformly continuous on $[a, b]$. Thus, there exists an $\varepsilon > 0$ such that for all $\delta > 0$, there are $x, c \in [a, b]$ such that $|x - c| < \delta$ and $|f(x) - f(c)| \geq \varepsilon$. Hence, for every natural number $n \geq 1$, there are points x_n, c_n in $[a, b]$ such that

$$|x_n - c_n| < \frac{1}{n} \text{ and } |f(x_n) - f(c_n)| \geq \varepsilon. \tag{4.19}$$

Since the sequence $\langle x_n \rangle$ consists of points in $[a, b]$, Theorem 3.5.1 and Corollary 3.2.15 imply that $\langle x_n \rangle$ has a subsequence $\langle x_{n_k} \rangle$ such that $\lim_{k \to \infty} x_{n_k} = c$ where $c \in [a, b]$.

Consider the corresponding subsequence $\langle c_{n_k} \rangle$ of $\langle c_n \rangle$. By Lemma 3.3.3 and (4.19), we see that $|x_{n_k} - c_{n_k}| < \frac{1}{n_k} \leq \frac{1}{k}$ for all $k \geq 1$. So $\lim_{k \to \infty} c_{n_k} = c$ by Lemma 3.2.11. As f is continuous, Theorem 4.2.1 implies that $\lim_{k \to \infty} f(x_{n_k}) = f(c)$ and $\lim_{k \to \infty} f(c_{n_k}) = f(c)$. Thus, $|f(x_{n_k}) - f(c)| < \frac{\varepsilon}{2}$ and $|f(c_{n_k}) - f(c)| < \frac{\varepsilon}{2}$ for some $k \geq 1$. Therefore,

$$|f(x_{n_k}) - f(c_{n_k})| = |f(x_{n_k}) - f(c) + f(c) - f(c_{n_k})|$$
$$\leq |f(x_{n_k}) - f(c)| + |f(c_{n_k}) - f(c)| < \varepsilon.$$

So $|f(x_{n_k}) - f(c_{n_k})| < \varepsilon$. However, (4.19) implies that $|f(x_{n_k}) - f(c_{n_k})| \geq \varepsilon$. This contradiction completes the proof. □

Let $f : D \to \mathbb{R}$ be a continuous function. Theorem 4.2.1 implies that for any sequence $\langle x_n \rangle$ of points in D, if $\langle x_n \rangle$ converges to a point $c \in D$, then the sequence $\langle f(x_n) \rangle$ converges to $f(c)$. However, if $c \notin D$, then the sequence $\langle f(x_n) \rangle$ may not converge. For example, let $D = (0, 2)$ and define $f : (0, 2) \to \mathbb{R}$ by $f(x) = \frac{1}{x}$. The function f is continuous. Now consider the sequence of points in $(0, 2)$ given by $\langle \frac{1}{n} \rangle$ for all $n \in \mathbb{N}$. The sequence $\langle \frac{1}{n} \rangle$ converges to 0 while the sequence $\langle f(\frac{1}{n}) \rangle = \langle n \rangle$ does not converge. The next theorem shows that this cannot occur for a function that is uniformly continuous.

Theorem 4.5.6. Suppose that $f : D \to \mathbb{R}$ is uniformly continuous. Let $\langle x_n \rangle$ be any sequence where $x_n \in D$ for all $n \geq 1$. If $\langle x_n \rangle$ converges, then $\langle f(x_n) \rangle$ converges.

Proof. Suppose that $f : D \to \mathbb{R}$ is uniformly continuous. Let $\langle x_n \rangle$ be a convergent sequence of points in D. By Theorem 3.6.5, $\langle x_n \rangle$ is a Cauchy sequence. We will now prove that $\langle f(x_n) \rangle$ is also a Cauchy sequence. Let $\varepsilon > 0$. Since f is uniformly continuous, there is a $\delta > 0$ such that

$$\text{for all } x, c \in D, \text{ if } |x - c| < \delta, \text{ then } |f(x) - f(c)| < \varepsilon. \tag{4.20}$$

As $\langle x_n \rangle$ is a Cauchy sequence, there is an N such that $|x_n - x_m| < \delta$ for all $n, m > N$. Thus, (4.20) implies that $|f(x_n) - f(x_m)| < \varepsilon$ for all $n, m > N$. Therefore, $\langle f(x_n) \rangle$ is a Cauchy sequence. Theorem 3.6.5 now implies that $\langle f(x_n) \rangle$ converges. □

Definition 4.5.7. A function $f : D \to \mathbb{R}$ **preserves convergent sequences** if for all convergent sequences $\langle x_n \rangle$ with points in D, we have that $\langle f(x_n) \rangle$ converges.

Thus, by Theorem 4.5.6, each uniformly continuous function preserves convergent sequences. The converse also holds when the domain of the function is bounded. Before we record this in the form of a theorem, we present a supportive lemma.

Lemma 4.5.8. Suppose that $f\colon D \to \mathbb{R}$ preserves convergent sequences and that $\langle x_n \rangle$ and $\langle c_n \rangle$ are convergent sequences of points in D. If $\lim_{n\to\infty} x_n = \lim_{n\to\infty} c_n$, then $\lim_{n\to\infty} f(x_n) = \lim_{n\to\infty} f(c_n)$.

Proof. Suppose that $f\colon D \to \mathbb{R}$ preserves convergent sequences and assume that $\lim_{n\to\infty} x_n = \ell = \lim_{n\to\infty} c_n$, where $\langle x_n \rangle$ and $\langle c_n \rangle$ are two sequences of points from D. By Proposition 3.3.9, the sequence $\langle x_1, c_1, x_2, c_2, x_3, c_3, \dots \rangle$ converges. Since f preserves convergent sequences, we conclude that sequence

$$\langle f(x_1), f(c_1), f(x_2), f(c_2), f(x_3), f(c_3), \dots \rangle$$

converges to a number L. Thus, $\lim_{n\to\infty} f(x_n) = L = \lim_{n\to\infty} f(c_n)$, by Theorem 3.3.4. □

Theorem 4.5.9. Suppose that $f\colon D \to \mathbb{R}$ preserves convergent sequences. If D is bounded, then f is uniformly continuous.

Proof. See Exercise 8. □

Exercises 4.5

1. Let m and c be real numbers where $m \neq 0$. Consider the function $f\colon \mathbb{R} \to \mathbb{R}$ defined by $f(x) = mx + c$. Prove that f is uniformly continuous.

2. Find a function $f\colon D \to \mathbb{R}$ such that f is uniformly continuous, but f^2 is not uniformly continuous. [Thus, the product of uniformly continuous functions is not necessarily uniformly continuous.]

3. Prove that $f\colon \mathbb{R} \to \mathbb{R}$ defined by $f(x) = x^3$ is not uniformly continuous on $\mathbb{R}$.

4. Let $f\colon D \to \mathbb{R}$ and $M > 0$ be such that $|f(x) - f(y)| \leq M\,|x - y|$ for all $x, y \in D$. Prove that f is uniformly continuous.

5. Suppose that $f\colon D \to \mathbb{R}$ is uniformly continuous and let $k \in \mathbb{R}$ be nonzero. Prove that kf is uniformly continuous.

6. Suppose that $f\colon D \to \mathbb{R}$ and $g\colon D \to \mathbb{R}$ are uniformly continuous. Prove that $(f + g)\colon D \to \mathbb{R}$ is uniformly continuous.

7. Suppose that $f\colon D \to \mathbb{R}$ is uniformly continuous. Let $g\colon D \to \mathbb{R}$ be defined by $g(x) = |f(x)|$, for all $x \in D$. Prove that g is uniformly continuous.

*8. Prove Theorem 4.5.9.

9. Let $f\colon D \to \mathbb{R}$ be uniformly continuous and $\langle x_n \rangle$ be a sequence of points in D. Prove that if $\langle x_n \rangle$ is a Cauchy sequence, then $\langle f(x_n) \rangle$ is Cauchy sequence.

10. Prove that if $f\colon D \to \mathbb{R}$ preserves convergent sequences, then f is continuous.

11. Let $f\colon D \to \mathbb{R}$ and $g\colon E \to \mathbb{R}$ be functions such that $f[D] \subseteq E$. Prove that if f and g are uniformly continuous, then $(g \circ f)\colon D \to \mathbb{R}$ is uniformly continuous.

12. Let $f\colon D \to \mathbb{R}$ and $g\colon D \to \mathbb{R}$ be uniformly continuous, where D is bounded. Using Theorems 4.5.6 and 4.5.9, prove that $(fg)\colon D \to \mathbb{R}$ is uniformly continuous.

13. Let $f\colon D \to \mathbb{R}$ be uniformly continuous. Suppose that $\gamma > 0$ such that $f(x) \geq \gamma$ for all $x \in D$. Prove that $(\frac{1}{f})\colon D \to \mathbb{R}$ is uniformly continuous.

14. Suppose that $f\colon D \to \mathbb{R}$ is uniformly continuous and D is bounded. Prove that f is bounded, using Theorem 4.5.6.

15. Let $f\colon D \to \mathbb{R}$ be uniformly continuous and let $c \notin D$ be an accumulation point of D.

    (a) Let $\langle x_n \rangle$ be any sequence of points in D that converges to c. Show that $\langle f(x_n) \rangle$ converges.
    (b) Let $\langle z_n \rangle$ and $\langle y_n \rangle$ be sequences of points in D that both converge to c. Show that $\lim_{n \to \infty} f(z_n) = \lim_{n \to \infty} f(y_n)$.
    (c) Conclude that $\lim_{x \to c} f(x)$ exists.

Exercise Notes: For Exercise 1, review Example 4.5.3. For Exercise 3, review Example 4.5.4 and note that $|x^3 - c^3| = |x^2 + xc + c^2| \, |x - c|$. For Exercise 6, review the proof of Theorem 4.1.6. For Exercise 8, modify the proof of Theorem 4.5.5 by appealing to Lemma 4.5.8. For Exercise 10, apply Lemma 4.5.8 and Theorem 4.2.1. For Exercise 11, modify the proof of Theorem 4.1.13.

CHAPTER 5

Differentiation

An important foundational concept in calculus with respect to a function f is the quotient of two differences $\frac{f(x)-f(c)}{x-c}$, called the **difference quotient** of f between the real numbers x and c. Of course, if $x = c$, then this difference quotient yields the meaningless ratio $\frac{0}{0}$. So, in a difference quotient, we assume that the real numbers x and c are distinct. As you may recall from calculus, the difference quotient is used to find the slope of a secant line to the graph of f, where the secant is the line joining the distinct points $(c, f(c))$ and $(x, f(x))$. This slope also identifies the average rate of change of f between c and x. By taking the limit of the difference quotient as x approaches c, we arrive at a mathematical precise definition of the slope of the tangent line to f at the point c (see Figure 5.1).

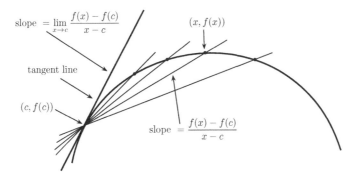

Figure 5.1: Limit of difference quotients.

In this chapter, we will focus on functions whose domain is an interval. For such functions, we shall present a theoretical treatment of the derivative and explore some of its properties. We also identify and establish three significant results that concern the derivative: The Mean Value Theorem, L'Hôpital's Rule, and Taylor's Theorem.

5.1 THE DERIVATIVE

In this section, our attention will be on the definition of the derivative. We will present some formulas and procedures for finding derivatives. The key tool for establishing these procedures involves the algebraic manipulation of difference quotients.

Definition 5.1.1. Let $f: I \to \mathbb{R}$ be a function where I is an interval. Let $c \in I$. We shall say that f **is differentiable at** c to mean that the limit

$$\lim_{x \to c} \frac{f(x) - f(c)}{x - c}$$

exists and is a real number denoted by $f'(c)$. So, when f is differentiable at c, we shall write

$$f'(c) = \lim_{x \to c} \frac{f(x) - f(c)}{x - c}.$$

The value $f'(c)$ is the **derivative** of f at c. Let $S \subseteq I$. Then f is **differentiable on** S when f is differentiable at all points in S, and the function $f': S \to \mathbb{R}$ is called the **derivative** of f on S. Finally, f is **differentiable** if f is differentiable at every point in I and, in this case, $f': I \to \mathbb{R}$ is said to be the **derivative** of f.

The limit operation $\lim_{x \to c} G(x)$ (see Definition 4.3.1) only requires that c be an accumulation point of the domain of G. So in Definition 5.1.1, c can be an endpoint of the interval I, as c is an accumulation point of I.

Example. Let $f: \mathbb{R} \to \mathbb{R}$ be defined by $f(x) = x^2 + 1$. Then for any $c \in \mathbb{R}$ we have

$$\begin{aligned}
f'(c) &= \lim_{x \to c} \frac{f(x) - f(c)}{x - c} & \text{by definition of } f'(c) \\
&= \lim_{x \to c} \frac{x^2 + 1 - (c^2 + 1)}{x - c} & \text{by definition } f \\
&= \lim_{x \to c} \frac{(x + c)(x - c)}{x - c} & \text{by algebra} \\
&= \lim_{x \to c} (x + c) = 2c & \text{by algebra and Theorem 4.3.7.}
\end{aligned}$$

Therefore, $f'(c) = 2c$.

Example. Let $f: \mathbb{R} \to \mathbb{R}$ be defined by $f(x) = |x|$. For $c = 0$, $\lim_{x \to 0^+} \frac{|x| - |0|}{x - 0} = 1$ and $\lim_{x \to 0^-} \frac{|x| - |0|}{x - 0} = -1$. By Theorem 4.3.12, f is not differentiable at 0. However, f is continuous at 0.

The above example shows that a function may be continuous at a point without being differentiable at the point. So continuity does not imply differentiability. On the other hand, our next theorem shows that if a function is differentiable at a point, then it is continuous at this point. For the remainder of the text, I denotes an interval.

Theorem 5.1.2. If $f: I \to \mathbb{R}$ is differentiable at $c \in I$, then f is continuous at c.

Proof. Let $f: I \to \mathbb{R}$ be differentiable at c where $c \in I$. To prove that f is continuous at c, we will show that $\lim_{x \to c} f(x) = f(c)$. Since f is differentiable at c, we have that

$$\lim_{x \to c} \frac{f(x) - f(c)}{x - c} = f'(c).$$

Notice that for $x \in I$ where $x \neq c$ we have

$$f(x) = (x - c)\frac{f(x) - f(c)}{x - c} + f(c).$$

Hence,

$$\lim_{x \to c} f(x) = \left(\lim_{x \to c}(x - c)\right)\left(\lim_{x \to c} \frac{f(x) - f(c)}{x - c}\right) + f(c) \tag{5.1}$$

by Theorem 4.3.7. Since $\lim_{x \to c}(x - c) = 0$, equation (5.1) implies that $\lim_{x \to c} f(x) = f(c)$. Hence, Lemma 4.3.3 implies that f is continuous at c. □

Corollary 5.1.3. If $f : I \to \mathbb{R}$ is differentiable on I, then f is continuous on I.

5.1.1 The Rules of Differentiation

In general, computing the derivative of a function using Definition 5.1.1 is difficult. Fortunately, there are rules and procedures for computing derivatives. Our next next theorem establishes these familiar rules of differentiation that we first learned in calculus. Moreover, the proofs of these rules use Definition 5.1.1 and the algebraic maneuvering of expressions that involve difference quotients. After the proof of this theorem, we will give a proof of the chain rule. Recall that I represents an interval.

Theorem 5.1.4. Let $f : I \to \mathbb{R}$ and $g : I \to \mathbb{R}$ be differentiable at $c \in I$. Then

(1) (Constant Multiple Rule) For $k \in \mathbb{R}$, the function kf is differentiable at c and

$$(kf)'(c) = k \cdot f'(c).$$

(2) (Sum Rule) The function $f + g$ is differentiable at c and

$$(f + g)'(c) = f'(c) + g'(c).$$

(3) (Product Rule) The function fg is differentiable at c and

$$(fg)'(c) = f(c)g'(c) + g(c)f'(c).$$

(4) (Quotient Rule) If $g(c) \neq 0$, then the function $\frac{f}{g}$ is differentiable at c and

$$\left(\frac{f}{g}\right)'(c) = \frac{g(c)f'(c) - f(c)g'(c)}{[g(c)]^2}.$$

Proof. Assume that (▲) $f : I \to \mathbb{R}$ and $g : I \to \mathbb{R}$ are differentiable at $c \in I$. For parts (1) and (2), see Exercise 3. We will just prove (3) and (4).

PROOF OF (3). We will first start with the difference quotient $\frac{(fg)(x) - (fg)(c)}{x - c}$ and algebraically "extract out" the difference quotients $\frac{f(x) - f(c)}{x - c}$ and $\frac{g(x) - g(c)}{x - c}$. For $x \in I$ where $x \neq c$, we have

$$\frac{(fg)(x) - (fg)(c)}{x - c} = \frac{f(x)g(x) - f(c)g(c)}{x - c} \qquad \text{by Def. 4.1.4(3)}$$

$$= \frac{[f(x)g(x) - f(x)g(c)] + [f(x)g(c) - f(c)g(c)]}{x - c} \qquad \text{by algebra}$$

$$= f(x)\frac{g(x) - g(c)}{x - c} + g(c)\frac{f(x) - f(c)}{x - c} \qquad \text{by algebra.}$$

Thus,
$$\frac{(fg)(x) - (fg)(c)}{x - c} = f(x)\frac{g(x) - g(c)}{x - c} + g(c)\frac{f(x) - f(c)}{x - c}.$$

Theorem 4.3.7 implies that

$$\lim_{x \to c} \frac{(fg)(x) - (fg)(c)}{x - c} = \left(\lim_{x \to c} f(x)\right)\left(\lim_{x \to c} \frac{g(x) - g(c)}{x - c}\right) + g(c)\left(\lim_{x \to c} \frac{f(x) - f(c)}{x - c}\right). \; ★$$

Since f is continuous at c, we have that $\lim_{x \to c} f(x) = f(c)$. In addition, by (▲), we have that $\lim_{x \to c} \frac{f(x) - f(c)}{x - c} = f'(c)$ and $\lim_{x \to c} \frac{g(x) - g(c)}{x - c} = g'(c)$. Thus, equation ★ yields

$$\lim_{x \to c} \frac{(fg)(x) - (fg)(c)}{x - c} = f(c)g'(c) + g(c)f'(c).$$

PROOF OF (4). Assume that $g(c) \neq 0$. Theorem 5.1.2 and Lemma 4.1.14 imply that there is an interval $J \subseteq I$ with $c \in J$ such that $g(x) \neq 0$ for all $x \in J$. Again, we will start with the difference quotient $\frac{(\frac{f}{g})(x) - (\frac{f}{g})(c)}{x - c}$ and algebraically "extract out" the difference quotients $\frac{f(x) - f(c)}{x - c}$ and $\frac{g(x) - g(c)}{x - c}$. For $x \in J$ where $x \neq c$, we have

$$\frac{\left(\frac{f}{g}\right)(x) - \left(\frac{f}{g}\right)(c)}{x - c} = \frac{\frac{f(x)}{g(x)} - \frac{f(c)}{g(c)}}{x - c}$$

$$= \frac{1}{(x - c)}\left(\frac{f(x)}{g(x)} - \frac{f(c)}{g(c)}\right)$$

$$= \frac{1}{(x - c)}\left(\frac{f(x)g(c) - f(c)g(x)}{g(x)g(c)}\right)$$

$$= \frac{1}{(x - c)}\left(\frac{[f(x)g(c) - f(c)g(c)] - [f(c)g(x) - f(c)g(c)]}{g(x)g(c)}\right)$$

$$= \frac{1}{g(x)g(c)}\left(\frac{g(c)[f(x) - f(c)] - f(c)[g(x) - g(c)]}{(x - c)}\right)$$

$$= \frac{1}{g(x)g(c)}\left(g(c)\frac{f(x) - f(c)}{(x - c)} - f(c)\frac{g(x) - g(c)}{(x - c)}\right).$$

Thus,
$$\frac{\left(\frac{f}{g}\right)(x) - \left(\frac{f}{g}\right)(c)}{x - c} = \frac{g(c)\frac{f(x) - f(c)}{(x - c)} - f(c)\frac{g(x) - g(c)}{(x - c)}}{g(x)g(c)}.$$

Theorem 4.3.7 now implies that

$$\lim_{x \to c} \frac{\left(\frac{f}{g}\right)(x) - \left(\frac{f}{g}\right)(c)}{x - c} = \frac{g(c)\left(\lim\limits_{x \to c} \frac{f(x)-f(c)}{x-c}\right) - f(c)\left(\lim\limits_{x \to c} \frac{g(x)-g(c)}{x-c}\right)}{g(c) \lim\limits_{x \to c} g(x)} \quad (5.2)$$

Since $\lim\limits_{x \to c} \frac{f(x)-f(c)}{x-c} = f'(c)$ and $\lim\limits_{x \to c} \frac{g(x)-g(c)}{x-c} = g'(c)$ as f and g are differentiable at c, equation (5.2), Theorem 5.1.2, and Lemma 4.3.3 yield the equation

$$\lim_{x \to c} \frac{\left(\frac{f}{g}\right)(x) - \left(\frac{f}{g}\right)(c)}{x - c} = \frac{g(c)f'(c) - f(c)g'(c)}{[g(c)]^2}. \qquad \square$$

5.1.2 The Chain Rule

The chain rule is a procedure for differentiating the composition $f \circ g$ of differentiable functions f and g. Let I and J be intervals. Suppose that $f \colon I \to \mathbb{R}$ and $g \colon J \to I$, where g is differentiable at $c \in J$ and f is differentiable at $g(c)$. Before we present the main theorem, let us assume that the function g is one-to-one. Then for all x and c in J, if $x \neq c$, then $g(x) - g(c) \neq 0$. Thus, Theorems 5.1.2 and 4.3.9 imply that

$$\lim_{x \to c} \frac{f(g(x)) - f(g(c))}{g(x) - g(c)} = \lim_{y \to g(c)} \frac{f(y) - f(g(c))}{y - g(c)} = f'(g(c)). \quad (5.3)$$

Hence, we have that

$$\lim_{x \to c} \frac{(f \circ g)(x) - (f \circ g)(c)}{x - c} = \lim_{x \to c} \frac{f(g(x)) - f(g(c))}{x - c}$$

$$= \lim_{x \to c} \left(\frac{f(g(x)) - f(g(c))}{g(x) - g(c)}\right)\left(\frac{g(x) - g(c)}{x - c}\right)$$

$$= \lim_{x \to c} \left(\frac{f(g(x)) - f(g(c))}{g(x) - g(c)}\right) \lim_{x \to c} \left(\frac{g(x) - g(c)}{x - c}\right)$$

$$= f'(g(c))g'(c) \qquad \text{by (5.3)}.$$

Thus, when g is one-to-one, the above proof shows that $(f \circ g)'(c) = f'(g(c))g'(c)$. If g is not one-to-one, then we may have $x \neq c$ and $g(x) - g(c) = 0$. Thus, the above argument would not hold, because we cannot divide by 0. The following proof, however, shows that the chain rule holds even if g is not one-to-one.

Theorem 5.1.5 (Chain Rule). Let I and J be intervals. Suppose that $f \colon I \to \mathbb{R}$ and $g \colon J \to I$, where g is differentiable at $c \in J$ and f is differentiable at $g(c) \in I$. Then $(f \circ g) \colon J \to \mathbb{R}$ is differentiable at c and

$$(f \circ g)'(c) = f'(g(c)) \cdot g'(c).$$

Proof. Since g is differentiable at c, we have that

$$\lim_{x \to c} \frac{g(x) - g(c)}{x - c} = g'(c) \qquad (5.4)$$

and because f is differentiable at $g(c)$, we have that

$$\lim_{y \to g(c)} \frac{f(y) - f(g(c))}{y - g(c)} = f'(g(c)). \tag{5.5}$$

Consider the function $h\colon J \to \mathbb{R}$ defined by

$$h(y) = \begin{cases} \frac{f(y) - f(g(c))}{y - g(c)}, & \text{if } y \neq g(c); \\ f'(g(c)), & \text{if } y = g(c). \end{cases}$$

Using equation (5.5) and the definition of h, one can see that $\lim\limits_{y \to g(c)} h(y) = h(g(c))$.[1]
Lemma 4.3.3 now implies that h is continuous at $g(c)$, and Theorem 5.1.2 implies that g is continuous at c. Hence, $h \circ g$ is continuous at c by Theorem 4.1.13 and so,

$$\lim_{x \to c}(h \circ g)(x) = h(g(c)) = f'(g(c)). \tag{5.6}$$

Note that

$$(h \circ g)(x) = h(g(x)) = \begin{cases} \frac{f(g(x)) - f(g(c))}{g(x) - g(c)}, & \text{if } g(x) \neq g(c); \\ f'(g(c)), & \text{if } g(x) = g(c). \end{cases}$$

Thus, for $x \in J$ and $x \neq c$, we have (see Exercise 5)

$$\frac{(f \circ g)(x) - (f \circ g)(c)}{x - c} = (h \circ g)(x)\left(\frac{g(x) - g(c)}{x - c}\right). \tag{5.7}$$

Hence,

$$\lim_{x \to c} \frac{(f \circ g)(x) - (f \circ g)(c)}{x - c} = \left(\lim_{x \to c}(h \circ g)(x)\right)\left(\lim_{x \to c} \frac{g(x) - g(c)}{x - c}\right)$$

by Theorem 4.3.7(2). Therefore, equations (5.6) and (5.4) now yield the equation

$$\lim_{x \to c} \frac{(f \circ g)(x) - (f \circ g)(c)}{x - c} = f'(g(c)) \cdot g'(c). \qquad \square$$

Exercises 5.1 _____

1. Let $f\colon \mathbb{R}^+ \to \mathbb{R}^+$ be defined by $f(x) = \sqrt{x}$. Using Definition 5.1.1, find the derivative $f'(c)$ when $c > 0$.

2. Let $f\colon \mathbb{R} \setminus \{0\} \to \mathbb{R}$ be defined by $f(x) = \frac{1}{x}$. Using Definition 5.1.1, find the derivative $f'(c)$ when $c \neq 0$.

*3. Prove parts (1) and (2) of Theorem 5.1.4.

4. Let $f\colon \mathbb{R} \to \mathbb{R}$ and $g\colon \mathbb{R} \to \mathbb{R}$ be differentiable functions. Suppose that $f(2) = 4$, $g(2) = 2$, $f'(2) = 5$, and $g'(2) = -3$. Find $(fg)'(2)$, $\left(\frac{f}{g}\right)'(2)$, and $(f \circ g)'(2)$.

---
[1]The notation $\lim\limits_{y \to g(c)}$ implies that $y \neq g(c)$.

*5. Show that (5.7), in the proof of Theorem 5.1.5, holds for all $x \in J$ when $x \neq c$.

6. Let I be an interval. Suppose that f and g are differentiable functions on I. Suppose that $f(x)g(x) = 1$ for all $x \in I$. Show that $\frac{f'(x)}{f(x)} + \frac{g'(x)}{g(x)} = 0$ for all $x \in I$.

7. Let I be an interval and f be a differentiable function on I such that $f(x) \leq f(y)$ whenever $x \leq y$ are in I. Prove that $f'(c) \geq 0$ for all $c \in I$.

8. Let $f \colon \mathbb{R} \to \mathbb{R}$ be differentiable on $\mathbb{R}$.

   (a) Prove that if f an even function, then f' is an odd function.

   (b) Prove that if f an odd function, then f' is an even function.

9. Let $f \colon \mathbb{R} \to \mathbb{R}$ be defined by

$$f(x) = \begin{cases} x^2 \sin(\frac{1}{x}), & \text{if } x \neq 0; \\ 0, & \text{if } x = 0. \end{cases}$$

   (a) Using the chain and product rules, show that f is differentiable at any $c \neq 0$. Find $f'(c)$.

   (b) Using Definition 5.1.1, show that f is differentiable at $c = 0$. Find $f'(0)$.

   (c) Show that f' is not continuous at $c = 0$.

10. Let $f \colon \mathbb{R} \to \mathbb{R}$ be defined by

$$f(x) = \begin{cases} x^2, & \text{if } x \geq 0; \\ 0, & \text{if } x < 0. \end{cases}$$

Evaluate $\lim\limits_{x \to 0^+} \frac{f(x)-f(0)}{x-0}$ and $\lim\limits_{x \to 0^-} \frac{f(x)-f(0)}{x-0}$. Is f is differentiable at 0? If so, then find $f'(0)$.

Exercise Notes: For Exercise 5, there are two cases to consider: $g(x) = g(c)$ and $g(x) \neq g(c)$. For Exercise 7, use Exercise 10 on page 117. For Exercise 9(b), see Exercise 22 on page 118. This exercise shows that the derivative of a differentiable function may not be continuous.

5.2   THE MEAN VALUE THEOREM

The Mean Value Theorem is one of the most important theorems in calculus, and it is a key tool in the proof of the Fundamental Theorem of Calculus (see Theorem 6.4.2). In this section, we establish the Mean Value Theorem and derive several of its consequences. We also identify and prove Cauchy's Mean Value Theorem, a generalization of the Mean Value Theorem.

Theorem 5.2.1. Let f be a differentiable function on the open interval (a, b). If f attains a maximum value or minimum value at a point $c \in (a, b)$, then $f'(c) = 0$.

Proof. Let f be a differentiable function on the open interval (a, b). We shall assume that f attains a maximum value at $c \in (a, b)$. [The proof for the case that f attains a minimum value is similar.] Since $f(c)$ is a maximum value, we have that $f(x) \leq f(c)$ for all $x \in (a, b)$. We shall prove that $f'(c) = 0$. Let $\langle x_n \rangle$ be a sequence of points in (a, b) that converges to c such that $c < x_n$ for all $n \in \mathbb{N}$. Theorem 4.3.4 implies that

$$\lim_{n \to \infty} \frac{f(x_n) - f(c)}{x_n - c} = f'(c).$$

Since $f(x_n) \leq f(c)$ and $c < x_n$ for all $n \in \mathbb{N}$, it follows that $\frac{f(x_n) - f(c)}{x_n - c} \leq 0$ for all $n \in \mathbb{N}$. Thus, Theorem 3.2.12 implies that $f'(c) \leq 0$.

Also, let $\langle y_n \rangle$ be a sequence of points in (a, b) that converges to c such that $y_n < c$ for all $n \in \mathbb{N}$. Theorem 4.3.4 implies that

$$\lim_{n \to \infty} \frac{f(y_n) - f(c)}{y_n - c} = f'(c).$$

Since $f(y_n) \leq f(c)$ and $c > y_n$ for all $n \in \mathbb{N}$, it follows that $\frac{f(y_n) - f(c)}{y_n - c} \geq 0$ for all $n \in \mathbb{N}$. Thus, $f'(c) \geq 0$. Since $f'(c) \geq 0$ and $f'(c) \leq 0$, we conclude that $f'(c) = 0$. □

As often occurs in mathematics, one first proves a special case of a more general theorem. Then the special case is used to prove the general theorem. The special case that we next prove is called Rolle's Theorem, and this result will be used to prove the Mean Value Theorem.

Theorem 5.2.2 (Rolle's Theorem). Let $f \colon [a, b] \to \mathbb{R}$ be a continuous function that is differentiable on the open interval (a, b). If $f(a) = f(b)$, then there is a point $c \in (a, b)$ such that $f'(c) = 0$.

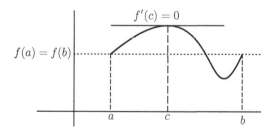

Figure 5.2: Illustration for Theorem 5.2.2.

Proof. Let $f \colon [a, b] \to \mathbb{R}$ be a continuous function that is differentiable on the open interval (a, b). Assume $f(a) = f(b)$. If f were a constant function, then $f'(c) = 0$ for all c in $[a, b]$. So, assume that f is not a constant function. Let $k = f(a) = f(b)$. As f is not a constant function, there is an x in (a, b) where $f(x) \neq k$. Since $f \colon [a, b] \to \mathbb{R}$ is continuous, Theorem 4.4.3 implies that f attains a maximum value or minimum value at a point $c \in (a, b)$. Theorem 5.2.1 asserts that $f'(c) = 0$. □

As portrayed in Figure 5.2, Rolle's Theorem 5.2.2 implies that the graph of any differentiable function which has the same value at two distinct points must have a tangent line that is parallel to the x-axis. Using Theorem 5.2.2, we will now state and prove the Mean Value Theorem,

Theorem 5.2.3 (Mean Value Theorem). Let $f\colon [a,b] \to \mathbb{R}$ be a continuous function that is differentiable on the open interval (a,b). Then there is a point $c \in (a,b)$ such that $f'(c) = \frac{f(b)-f(a)}{b-a}$.

Proof. Let $f\colon [a,b] \to \mathbb{R}$ be a continuous function. Suppose that f is differentiable on the open interval (a,b). Define the function $h\colon [a,b] \to \mathbb{R}$ by

$$h(x) = f(x) - \left(\frac{f(b)-f(a)}{b-a}(x-a) + f(a)\right). \tag{5.8}$$

Note that h is a continuous on $[a,b]$ and differentiable on (a,b). Moreover, $h(a) = 0$ and $h(b) = 0$. Theorem 5.2.2 implies that there is a $c \in (a,b)$ such that $h'(c) = 0$. Equation (5.8) implies that $h'(c) = f'(c) - \frac{f(b)-f(a)}{b-a} = 0$. Thus, $f'(c) = \frac{f(b)-f(a)}{b-a}$. $\square$

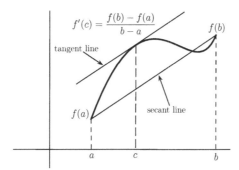

Figure 5.3: Illustration for Theorem 5.2.3.

Figure 5.3 offers a geometric interpretation of Theorem 5.2.3. The value $\frac{f(b)-f(a)}{b-a}$ is the slope of the secant line between the points $(a, f(a))$ and $(b, f(b))$. The theorem states that there is a $c \in (a,b)$ such that the slope of the associated tangent line, $f'(c)$, is equal to the slope of this secant line.

The Mean Value Theorem can be used to derive inequalities. For example, let $0 \le a < b$. We will prove that $b^3 - a^3 < 3b^2(b-a)$. Let $f\colon [a,b] \to \mathbb{R}$ be defined by $f(x) = x^3$. Then f is continuous on $[a,b]$ and differentiable on (a,b). By Theorem 5.2.3, there is a $c \in (a,b)$ such that $b^3 - a^3 = 3c^2(b-a)$. Since $c < b$, we conclude that $b^3 - a^3 < 3b^2(b-a)$.

We now show how the Mean Value Theorem implies some familiar relationships that hold between a function and its derivative.

Theorem 5.2.4. Let $f\colon [a,b] \to \mathbb{R}$ be a continuous function. Suppose that f is differentiable on the open interval (a,b). If $f'(x) = 0$ for all $x \in (a,b)$, then f is a constant function on $[a,b]$.

Proof. Let $f\colon [a,b] \to \mathbb{R}$ be a continuous function. Suppose that f is differentiable on the open interval (a,b). Assume that

$$f'(x) = 0 \text{ for all } x \in (a,b). \tag{5.9}$$

Suppose, for a contradiction, that f is not constant of $[a,b]$. Thus, there are y, z in $[a,b]$ such that $y < z$ and $f(y) \neq f(z)$. So,

$$\frac{f(z) - f(y)}{z - y} \neq 0. \tag{5.10}$$

Consider the closed interval $[y, z]$. Since f is continuous on $[y, z]$ and differentiable on (y, z), there is a $c \in (y, z)$ such that $f'(c) = \frac{f(z) - f(y)}{z - y}$ by Theorem 5.2.3. Hence, (5.10) implies that $f'(c) \neq 0$. However, (5.9) implies that $f'(c) = 0$, a contradiction. $\square$

Corollary 5.2.5. Let $f\colon [a,b] \to \mathbb{R}$ and $g\colon [a,b] \to \mathbb{R}$ be a continuous functions that are differentiable on the open interval (a,b). If $f'(x) = g'(x)$ for all $x \in (a,b)$, then there is a constant $C \in \mathbb{R}$ such that $f(x) = g(x) + C$ for all $x \in [a,b]$.

Proof. Let $f\colon [a,b] \to \mathbb{R}$ and $g\colon [a,b] \to \mathbb{R}$ be a continuous functions. Suppose that f and g are differentiable on the open interval (a,b). Assume that

$$f'(x) = g'(x) \text{ for all } x \in (a,b). \tag{5.11}$$

Let $h\colon [a,b] \to \mathbb{R}$ be defined by $h(x) = f(x) - g(x)$. Since $h'(x) = f'(x) - g'(x)$, (5.11) implies that $h'(x) = 0$ for all $x \in [a,b]$. Thus, by Theorem 5.2.4, there is a constant C such that $h(x) = C$ for all $x \in [a,b]$. Hence, $f(x) = g(x) + C$ for all $x \in [a,b]$. $\square$

In the proof of Theorem 5.2.8 below, we use the Mean Value Theorem to establish a connection between the derivative of a function and the function's behavior. First, we need to identify these particular behaviors. Recall that "iff" is an abbreviation for "if and only if."

Definition 5.2.6. Let $f\colon I \to \mathbb{R}$ be a function where I is an interval.

- f is **strictly increasing** on I iff for all $y, z \in I$, if $y < z$, then $f(y) < f(z)$.
- f is **increasing** on I iff for all $y, z \in I$, if $y < z$, then $f(y) \leq f(z)$.
- f is **strictly decreasing** on I iff for all $y, z \in I$, if $y < z$, then $f(y) > f(z)$.
- f is **decreasing** on I iff for all $y, z \in I$, if $y < z$, then $f(y) \geq f(z)$.

Theorem 5.2.7. Let I be an interval. If $f\colon I \to \mathbb{R}$ is strictly increasing or strictly decreasing, then f is one-to-one.

Theorem 5.2.8. Let $f\colon I \to \mathbb{R}$ be a function where I is an interval. Suppose that f is differentiable on I. Then the following hold:

(a) If $f'(x) > 0$ for all $x \in I$, then f is strictly increasing on I.
(b) If $f'(x) < 0$ for all $x \in I$, then f is strictly decreasing on I.

Proof. Let $f\colon I \to \mathbb{R}$ be differentiable on an interval I. We prove (a) and leave (b) as an Exercise 12 (the proof is very similar). Assume that (▲) $f'(x) > 0$ for all $x \in I$. Let $y, z \in I$ be such that $y < z$. We must show that $f(y) < f(z)$. By Corollary 5.1.3 and the Mean Value Theorem 5.2.3, there is a $c \in (y, z)$ such that

$$f(z) - f(y) = f'(c)(z - y).$$

By (▲) we have that $f'(c) > 0$. Since $z - y > 0$, we conclude that $f(y) < f(z)$. □

We now extend the Mean Value Theorem to one that involves two functions f and g. The proof of this extension relies on the introduction of a new function h.

Theorem 5.2.9 (Cauchy Mean Value Theorem). Let $f\colon [a, b] \to \mathbb{R}$ and $g\colon [a, b] \to \mathbb{R}$ be continuous on $[a, b]$ and differentiable on (a, b). Then there is $c \in (a, b)$ such that

$$[f(b) - f(a)]g'(c) = [g(b) - g(a)]f'(c). \tag{5.12}$$

If $g'(x) \neq 0$ for all $x \in (a, b)$, then the above equation can be expressed as

$$\frac{f(b) - f(a)}{g(b) - g(a)} = \frac{f'(c)}{g'(c)}. \tag{5.13}$$

Proof. Let $f\colon [a, b] \to \mathbb{R}$ and $g\colon [a, b] \to \mathbb{R}$ be as stated. Define $h\colon [a, b] \to \mathbb{R}$ by

$$h(x) = (f(b) - f(a))g(x) - (g(b) - g(a))f(x).$$

Hence, h is a continuous on $[a, b]$ and differentiable on (a, b). In addition, $h(a) = 0$ and $h(b) = 0$. By Theorem 5.2.2, there is a $c \in (a, b)$ such that $h'(c) = 0$. Thus,

$$h'(c) = (f(b) - f(a))g'(c) - (g(b) - g(a))f'(c) = 0$$

which implies (5.12). Finally, if $g'(x) \neq 0$ for all $x \in (a, b)$, then Theorem 5.2.2 implies that $g(b) \neq g(a)$. Therefore (5.12) implies (5.13). □

5.2.1 L'Hôpital's Rule

As you may recall from calculus, the "indeterminate forms" $\frac{0}{0}$ and $\frac{\infty}{\infty}$ can result from attempting to evaluate the limit of a ratio of functions. L'Hôpital's Rule offers a method for evaluating such limits (if they exist). We will prove, independently, four versions of L'Hôpital's Rule using Theorem 5.2.9 and Theorem 4.3.4. The first two rules address the indeterminate form $\frac{0}{0}$, while last two rules deal with the form $\frac{\infty}{\infty}$. L'Hôpital's rules are fairly easy to apply; however, their proofs are fairly involved.

Theorem 5.2.10 (L'Hôpital's Rule I). Let f and g be continuous on an interval I, and differentiable on I except possibly at $c \in I$. Suppose that $f(c) = g(c) = 0$ and $g'(x) \neq 0$ for all $x \in I \setminus \{c\}$. If $\lim\limits_{x \to c} \frac{f'(x)}{g'(x)} = L$ and $L \in \mathbb{R}$, then $\lim\limits_{x \to c} \frac{f(x)}{g(x)} = L$.

Proof. Let f and g be continuous on I and differentiable on $I \setminus \{c\}$ where $c \in I$ and I is an interval. Suppose that

$$f(c) = g(c) = 0. \tag{5.14}$$

Assume that $g'(x) \neq 0$ for all $x \in I \setminus \{c\}$ and $L \in \mathbb{R}$ is such that

$$\lim_{x \to c} \frac{f'(x)}{g'(x)} = L. \tag{5.15}$$

Let $\langle s_n \rangle$ be a sequence of points in $I \setminus \{c\}$ such that $\lim_{n \to \infty} s_n = c$. Theorem 5.2.9 implies that for each $n \in \mathbb{N}$, there is a point c_n strictly between s_n and c such that

$$\frac{f(c) - f(s_n)}{g(c) - g(s_n)} = \frac{f'(c_n)}{g'(c_n)}. \tag{5.16}$$

Thus, equations (5.14) and (5.16) imply that

$$\frac{f(s_n)}{g(s_n)} = \frac{f'(c_n)}{g'(c_n)}, \text{ for all } n \in \mathbb{N}. \tag{5.17}$$

Since $\lim_{n \to \infty} s_n = c$ and each c_n is between s_n and c, we have that $\lim_{n \to \infty} c_n = c$ by Corollary 3.1.13. Theorem 4.3.4 and equation (5.15) imply that $\lim_{n \to \infty} \frac{f'(c_n)}{g'(c_n)} = L$. From (5.17) we infer that $\lim_{n \to \infty} \frac{f(s_n)}{g(s_n)} = L$. Hence, $\lim_{x \to c} \frac{f(x)}{g(x)} = L$ by Theorem 4.3.4. □

For each limit $\lim_{x \to c}$ in Theorem 5.2.10, it is understood that x is in $I \setminus \{c\}$. We now present our second version of L'Hôpital's Rule, which involves a limit at infinity.

Theorem 5.2.11 (L'Hôpital's Rule II). Let f and g be differentiable on (a, ∞) such that $\lim_{x \to \infty} f(x) = \lim_{x \to \infty} g(x) = 0$ and $g'(x) \neq 0$ for all $x \in (a, \infty)$. If $\lim_{x \to \infty} \frac{f'(x)}{g'(x)} = L$ and $L \in \mathbb{R}$, then $\lim_{x \to \infty} \frac{f(x)}{g(x)} = L$.

Proof. Let f and g be differentiable on (a, ∞) and satisfy the above conditions. Since $g'(x) \neq 0$ for all $x \in (a, \infty)$, Rolle's Theorem implies that there is a $D_1 \geq a$ such that $g(x) \neq 0$ for all $x > D_1$. Assume that $\lim_{x \to \infty} \frac{f'(x)}{g'(x)} = L$. Let $\langle s_n \rangle$ be a sequence of distinct points in (a, ∞) such that $\lim_{n \to \infty} s_n = \infty$. Since $\lim_{x \to \infty} f(x) = \lim_{x \to \infty} g(x) = 0$, it follows that $\lim_{n \to \infty} f(s_n) = \lim_{n \to \infty} g(s_n) = 0$ by Exercise 19 on page 118. We now prove that $\lim_{n \to \infty} \frac{f(s_n)}{g(s_n)} = L$. Let $\varepsilon > 0$. Since $\lim_{x \to \infty} \frac{f'(x)}{g'(x)} = L$, there is a $D_2 \geq D_1$ such that

$$\left| \frac{f'(x)}{g'(x)} - L \right| < \frac{\varepsilon}{2} \text{ for all } x \in (a, \infty) \text{ such that } x > D_2. \tag{5.18}$$

Since $\lim_{n \to \infty} s_n = \infty$, there is an $N_1 \in \mathbb{N}$ such that $s_n > D_2$, for all $n > N_1$. Let $m > N_1$. For each $n > m$, there is a point c_n between s_n and s_m such that

$$\frac{f(s_m) - f(s_n)}{g(s_m) - g(s_n)} = \frac{f'(c_n)}{g'(c_n)}, \tag{5.19}$$

by Theorem 5.2.9. Since $c_n > D_2$ for all $n > N_1$, (5.18) and (5.19) imply that

$$\left| \frac{f(s_m) - f(s_n)}{g(s_m) - g(s_n)} - L \right| < \frac{\varepsilon}{2} \quad \text{for all } n > N_1. \tag{5.20}$$

As $\lim\limits_{n \to \infty} f(s_n) = \lim\limits_{n \to \infty} g(s_n) = 0$, it follows that

$$\lim_{n \to \infty} \frac{f(s_m) - f(s_n)}{g(s_m) - g(s_n)} - L = \frac{f(s_m)}{g(s_m)} - L.$$

Therefore, $\left| \frac{f(s_m)}{g(s_m)} - L \right| \le \frac{\varepsilon}{2}$ by (5.20) and Corollary 3.2.16. Thus, $\left| \frac{f(s_m)}{g(s_m)} - L \right| < \varepsilon$ for all $m > N_1$. Hence, $\lim\limits_{m \to \infty} \frac{f(s_m)}{g(s_m)} = L$. So $\lim\limits_{x \to \infty} \frac{f(x)}{g(x)} = L$ by Exercise 19 on page 118. □

The following version of L'Hôpital's Rule involves a infinite limit which is usually identified as having the form $\frac{\infty}{\infty}$; however, the proof only requires that the limit of the denominator be infinite.

Theorem 5.2.12 (L'Hôpital's Rule III)**.** Let f and g be differentiable on $I \setminus \{c\}$, where I is an interval and c is an interior or end point of I. Suppose that $\lim\limits_{x \to c} g(x) = \infty$ and $g'(x) \ne 0$ for all $x \in I \setminus \{c\}$. If $\lim\limits_{x \to c} \frac{f'(x)}{g'(x)} = L$ and $L \in \mathbb{R}$, then $\lim\limits_{x \to c} \frac{f(x)}{g(x)} = L$.

Proof. Let f, g, c, and I be as stated. Let c be an interior point and assume that $\lim\limits_{x \to c} \frac{f'(x)}{g'(x)} = L$. Let $\langle s_n \rangle$ be a sequence of distinct points in I that converges to c where $s_n > c$ for all $n \in \mathbb{N}$. Since $\lim\limits_{x \to c} g(x) = \infty$ and $\lim\limits_{n \to \infty} s_n = c$, it follows that $\lim\limits_{n \to \infty} g(s_n) = \infty$ by Exercise 17 on page 118. We now prove that $\lim\limits_{n \to \infty} \frac{f(s_n)}{g(s_n)} = L$. Let $\varepsilon > 0$. Since $\lim\limits_{x \to c} \frac{f'(x)}{g'(x)} = L$, there is a $\delta > 0$ such that

$$\left| \frac{f'(x)}{g'(x)} - L \right| < \frac{\varepsilon}{2} \quad \text{for all } x \in I \text{ such that } 0 < |x - c| < \delta. \tag{5.21}$$

Since $\lim\limits_{n \to \infty} g(s_n) = \infty$ and $\lim\limits_{n \to \infty} s_n = c$, there is an $N_1 \in \mathbb{N}$ such that $g(s_n) > 0$ and $|s_n - c| < \delta$, for all $n \ge N_1$. For each $n > N_1$, there is a point c_n between s_n and s_{N_1} such that

$$\frac{f(s_n) - f(s_{N_1})}{g(s_n) - g(s_{N_1})} = \frac{f'(c_n)}{g'(c_n)}, \tag{5.22}$$

by Theorem 5.2.9. Since $0 < |c_n - c| < \delta$ for all $n > N_1$, (5.21) and (5.22) imply that

$$\left| \frac{f(s_n) - f(s_{N_1})}{g(s_n) - g(s_{N_1})} - L \right| < \frac{\varepsilon}{2} \quad \text{for all } n > N_1. \tag{5.23}$$

As $\lim\limits_{n \to \infty} g(s_n) = \infty$, let $N_2 \in \mathbb{N}$ be such that $N_2 \ge N_1$ and

$$0 < \frac{g(s_n) - g(s_{N_1})}{g(s_n)} = 1 - \frac{g(s_{N_1})}{g(s_n)} < 1 \quad \text{for all } n > N_2.$$

Thus, from (5.23), we have

$$\frac{g(s_n) - g(s_{N_1})}{g(s_n)} \left| \frac{f(s_n) - f(s_{N_1})}{g(s_n) - g(s_{N_1})} - L \right| < \frac{g(s_n) - g(s_{N_1})}{g(s_n)} \frac{\varepsilon}{2} < \frac{\varepsilon}{2} \quad \text{for all } n > N_2.$$

So, by algebra,

$$\left| \frac{f(s_n) - f(s_{N_1})}{g(s_n)} - \left(1 - \frac{g(s_{N_1})}{g(s_n)} \right) L \right| < \frac{\varepsilon}{2} \quad \text{for all } n > N_2.$$

Hence, by more algebra,

$$\left| \left(\frac{f(s_n)}{g(s_n)} - L \right) - \left(\frac{f(s_{N_1}) - Lg(s_{N_1})}{g(s_n)} \right) \right| < \frac{\varepsilon}{2} \quad \text{for all } n > N_2.$$

Theorem 2.2.7(iii), the backward triangle inequality, thus implies that

$$\left| \frac{f(s_n)}{g(s_n)} - L \right| < \frac{\varepsilon}{2} + \left| \frac{f(s_{N_1}) - Lg(s_{N_1})}{g(s_n)} \right| \quad \text{for all } n > N_2. \tag{5.24}$$

Since $\lim_{n \to \infty} g(s_n) = \infty$, there is an $N_3 \in \mathbb{N}$ such that $N_3 \geq N_2$ and

$$\left| \frac{f(s_{N_1}) - Lg(s_{N_1})}{g(s_n)} \right| < \frac{\varepsilon}{2} \quad \text{for all } n > N_3. \tag{5.25}$$

Thus, (5.25) and (5.24) imply that $\left| \frac{f(s_n)}{g(s_n)} - L \right| < \varepsilon$, for all $n > N_3$. We conclude that $\lim_{n \to \infty} \frac{f(s_n)}{g(s_n)} = L$. Therefore, $\lim_{x \to c^+} \frac{f(x)}{g(x)} = L$ (see Remark 4.3.11). A similar argument shows that $\lim_{x \to c^-} \frac{f(x)}{g(x)} = L$. Therefore, $\lim_{x \to c} \frac{f(x)}{g(x)} = L$ by Theorem 4.3.12. The above argument shows that the theorem also holds when c is an endpoint of I. □

Remark 5.2.13. We note that if $\lim_{x \to c} g(x) = -\infty$, then $\lim_{x \to c} -g(x) = \infty$. This fact can be used to show that Theorem 5.2.12 implies that the theorem also holds if condition $\lim_{x \to c} g(x) = \infty$ is replaced with $\lim_{x \to c} g(x) = -\infty$.

Theorem 5.2.12 and it's proof do *not require* that $\lim_{x \to c} f(x) = \infty$, that is, the numerator is not required to have an infinite limit. We take advantage of this fact in our next example.

Example 5.2.14. Let e^x be the natural exponential function. Let $h \colon \mathbb{R} \to \mathbb{R}$ be defined by

$$h(x) = \begin{cases} \dfrac{e^{-1/x^2}}{x}, & \text{if } x \neq 0; \\ 0, & \text{if } x = 0. \end{cases}$$

We shall show that $\lim_{x \to 0} h(x) = 0$. Note that

$$h(x) = \frac{e^{-1/x^2}}{x} = \frac{1/x}{e^{1/x^2}}, \tag{5.26}$$

whenever $x \neq 0$. Letting $f(x) = \frac{1}{x}$, $g(x) = e^{1/x^2}$, $I = \mathbb{R}$, and $c = 0$, we see that the conditions of Theorem 5.2.12 are satisfied. Since

$$\lim_{x \to 0} \frac{f'(x)}{g'(x)} = \lim_{x \to 0} \frac{\frac{-1}{x^2}}{e^{1/x^2} \frac{-2}{x^3}} = \lim_{x \to 0} \frac{x}{2e^{1/x^2}} = 0,$$

Theorem 5.2.12 implies that $\lim\limits_{x \to 0} \frac{f(x)}{g(x)} = \frac{1/x}{e^{1/x^2}} = 0$. Hence, by (5.26), $\lim\limits_{x \to 0} h(x) = 0$.

The proof of our final version of L'Hôpital's Rule is similar to the above proof of Theorem 5.2.12 (see Exercise 18).

Theorem 5.2.15 (L'Hôpital's Rule IV). Let f and g be differentiable on (a, ∞). Suppose that $\lim\limits_{x \to \infty} g(x) = \infty$ and $g'(x) \neq 0$ for all $x \in (a, \infty)$. If $\lim\limits_{x \to \infty} \frac{f'(x)}{g'(x)} = L$ and $L \in \mathbb{R}$, then $\lim\limits_{x \to \infty} \frac{f(x)}{g(x)} = L$.

5.2.2 The Intermediate Value Theorem for Derivatives

By Theorem 4.4.6, continuous functions satisfy the intermediate value property. In this section we will prove that the derivative of a function also has this property, even if the derivative is not continuous. First we show that the derivative of a differentiable function, satisfies an "intermediate root property."

Lemma 5.2.16 (Intermediate Root Lemma). Let $f : I \to \mathbb{R}$ be differentiable on I, an interval. Let $a \in I$ and $b \in I$ be such that $a < b$. Suppose that $f'(a) < 0 < f'(b)$. Then there is an $i \in (a, b)$ such that $f'(i) = 0$.

Proof. Assume that $f'(a) < 0 < f'(b)$. Thus,

$$f'(a) = \lim_{x \to a} \frac{f(x) - f(a)}{x - a} < 0 < f'(b) = \lim_{x \to b} \frac{f(x) - f(b)}{x - b}.$$

By Exercise 11 on page 118, there are $c \in I$ and $d \in I$ such that (▲) $a < c < d < b$ and

$$\frac{f(c) - f(a)}{c - a} < 0 < \frac{f(d) - f(b)}{d - b}. \tag{5.27}$$

Hence, (5.27) and (▲) imply that (▼) $f(c) < f(a)$ and $f(d) < f(b)$. Since f is continuous, Theorem 5.2.1 implies that f attains a minimum value at a point $i \in [a, b]$, and (▼), (▲) imply that $i \in (a, b)$. Theorem 5.2.1 now implies that $f'(i) = 0$. □

The next result is due to Jean Gaston Darboux. It shows that any function that is the derivative of a function has the intermediate value property (see Definition 4.4.5).

Theorem 5.2.17 (Intermediate Value Theorem). If $f : I \to \mathbb{R}$ is differentiable on an interval I, then $f' : I \to \mathbb{R}$ has the intermediate value property on I.

Proof. Assume that $f'(a) < k < f'(b)$ where $a < b$ and $a, b \in I$. Let $g : I \to \mathbb{R}$ be defined by $g(x) = f(x) - k$. Clearly, g differentiable on I. Note that $g'(a) < 0 < g'(b)$. Lemma 5.2.16 implies that there is an $i \in (a, b)$ such that $g'(i) = f'(i) - k = 0$. So $f'(i) = k$. If $f'(a) > k > f('b)$, then a similar argument applies using the function $g(x) = k - f(x)$. □

5.2.3 Inverse Function Theorems

There are two results about inverse functions that we will address in this section. We first prove that the inverse of a continuous function is continuous, and then prove that the inverse of a differentiable function is differentiable. The first result will be used to prove the latter. We begin by identifying a shorter way of saying that a function is "either increasing or decreasing."

Definition 5.2.18. Let $f: I \to \mathbb{R}$ be a function where I is an interval. We say that f is **monotone** on I if f is either increasing or decreasing on I. Moreover, if f is either strictly increasing or strictly decreasing, then we say that f is **strictly monotone**.

The Inverse of a Continuous Function

Let f be a continuous injection defined on an interval. So the inverse function f^{-1} exists; but, is it continuous? Since f is a continuous injection, f must be strictly monotone (see Theorem 4.4.6 and Exercise 23). The next lemma will be used to show that, in fact, f^{-1} is continuous.

Lemma 5.2.19. Let $g: I \to \mathbb{R}$ be strictly monotone on the interval I. If $g[I]$ is an interval, then g is continuous.

Proof. Let $g: I \to \mathbb{R}$ be strictly increasing on the interval I and assume that $g[I]$ is an interval. [The proof is similar when g is strictly decreasing.] To prove that g is continuous, let $c \in I$ and $\varepsilon > 0$. There are three cases to consider.

CASE 1: $g(c)$ is an interior point of the interval $g[I]$. Let $\varepsilon^* > 0$ be such that $\varepsilon^* < \varepsilon$ and

$$[g(c) - \varepsilon^*, g(c) + \varepsilon^*] \subseteq g[I].$$

Let $a, b \in I$ such that (▲) $g(a) = g(c) - \varepsilon^*$ and $g(b) = g(c) + \varepsilon^*$. Since g is strictly increasing and $g(a) < g(c) < g(b)$, it follows that $a < c < b$ (see Exercise 6). Let $\delta = \min\{|a - c|, |b - c|\}$. Let $x \in I$ be such that $|x - c| < \delta$. Hence, $a < x < b$. Thus, as g is strictly increasing, $g(a) < g(x) < g(b)$. By (▲), we have that

$$g(c) - \varepsilon^* < g(x) < g(c) + \varepsilon^*.$$

Therefore, $|g(x) - g(c)| < \varepsilon^* < \varepsilon$.

CASE 2: $g(c)$ is the left endpoint of $g[I]$. Since g is strictly increasing, c must be the left endpoint of I. The proof of Case 1 adapts to prove that g is continuous at c.

CASE 3: $g(c)$ is the right endpoint of $g[I]$. As g is strictly increasing, c must be the right endpoint of I. The proof of Case 1 adapts to prove that g is continuous at c. □

Theorem 5.2.20. Let $f: I \to \mathbb{R}$ be strictly monotone and continuous on I, an interval. Then $J = f[I]$ is an interval, f is one-to-one, and $f^{-1}: J \to I$ is strictly monotone and continuous on J.

Proof. Let $f: I \to \mathbb{R}$ be strictly monotone and continuous on an interval I, and let $J = f[I]$. Theorem 5.2.7 implies that f is one-to-one, and Theorem 4.4.7 implies that J is an interval. Thus, $f^{-1}: J \to I$ is strictly monotone, by Exercise 9. As $f^{-1}[J] = I$ where J and I are intervals, Lemma 5.2.19 implies that f^{-1} is continuous. □

For an interval I, let I^* denote the set of points in I that are not endpoints of I.

Corollary 5.2.21. Let $f\colon I \to \mathbb{R}$ be continuous on the interval I. Suppose that $f'(x) \neq 0$ for all $x \in I^*$. Then $J = f[I]$ is an interval, f is one-to-one, and $f^{-1}\colon J \to I$ is strictly monotone and continuous on J.

Proof. Let $f\colon I \to \mathbb{R}$ be continuous on the interval I where $f'(x) \neq 0$ for all $x \in I^*$. Theorem 5.2.17 implies that either $f'(x) > 0$ for all $x \in I^*$, or $f'(x) < 0$ for all $x \in I^*$. Hence, by Theorem 5.2.8, f is strictly monotone on I^*. The Mean Value Theorem 5.2.3 now implies that f is strictly monotone on I. Theorem 5.2.20 thus implies that f is one-to-one, $f[I]$ is an interval, and f^{-1} is continuous on $f[I]$. ☐

The Inverse of a Differentiable Function

We can now establish a relationship between the derivative of a strictly monotone function and the derivative of its inverse.

Theorem 5.2.22. Let $f\colon I \to \mathbb{R}$ be differentiable on the interval I. Suppose that $f'(x) \neq 0$ for all $x \in I$. Then f is one-to-one, $f[I]$ is an interval, and $f^{-1}\colon f[I] \to I$ is differentiable on $f[I]$. Moreover, $(f^{-1})'(y) = \frac{1}{f'(f^{-1}(y))}$, for all $y \in f[I]$.

Proof. Let $f\colon I \to \mathbb{R}$ be such that that $f'(x) \neq 0$ for all $x \in I$. By Corollary 5.2.21, f is one-to-one, $f[I]$ is an interval, and f^{-1} is continuous on $f[I]$.

To show that f^{-1} is differentiable on $f[I]$, let $y \in f[I]$. Now let $c \in I$ be such that **(1)** $f(c) = y$. Hence, **(2)** $c = f^{-1}(y)$. We will show that

$$\lim_{z \to y} \frac{f^{-1}(z) - f^{-1}(y)}{z - y} = \frac{1}{f'(f^{-1}(y))}$$

by applying Theorem 4.3.4. Let $\langle y_n \rangle$ be a sequence of points in $f[I] \setminus \{y\}$ such that $\lim_{n \to \infty} y_n = y$. Thus, there is a sequence of points $\langle c_n \rangle$ in $I \setminus \{c\}$ such that $f(c_n) = y_n$ for all $n \in \mathbb{N}$. So **(3)** $c_n = f^{-1}(y_n)$ for all $n \in \mathbb{N}$. Since f^{-1} is continuous, we see that **(▲)** $\lim_{n \to \infty} f^{-1}(y_n) = f^{-1}(y)$. So **(▲)**, **(2)**, and **(3)** imply that $\lim_{n \to \infty} c_n = c$. Because f is differentiable, Theorem 4.3.4 implies that **(▼)** $\lim_{n \to \infty} \frac{f(c_n) - f(c)}{c_n - c} = f'(c)$. Therefore,

$$\lim_{n \to \infty} \frac{f^{-1}(y_n) - f^{-1}(y)}{y_n - y} = \lim_{n \to \infty} \frac{c_n - c}{f(c_n) - f(c)} \quad \text{by (1), (2), and (3)}$$

$$= \lim_{n \to \infty} \frac{1}{\frac{f(c_n) - f(c)}{c_n - c}} \quad \text{by algebra}$$

$$= \frac{1}{f'(c)} = \frac{1}{f'(f^{-1}(y))} \quad \text{by Theorem 3.2.6, (▼), and (2).}$$

Thus, f^{-1} is differentiable on $f[I]$, and $(f^{-1})'(y) = \frac{1}{f'(f^{-1}(y))}$ for all $y \in f[I]$. ☐

The next theorem, which follows from the proof of Theorem 5.2.22, focuses on a single point in the domain of an inverse function.

Theorem 5.2.23. Let $f: I \to f[I]$ be strictly monotone and continuous on the interval I. Then $f[I]$ is an interval and $f^{-1}: f[I] \to I$ is continuous. Let $d \in f[I]$ and $c \in I$ be such that $f(c) = d$. If $f'(c) \neq 0$, then $f^{-1}: f[I] \to I$ is differentiable at d and $(f^{-1})'(d) = \frac{1}{f'(f^{-1}(d))}$.

Exercises 5.2

1. Prove Theorem 5.2.7.

2. Let $f: [0,2] \to \mathbb{R}$ be continuous on $[0,2]$ and differentiable on $(0,2)$. Suppose that $f(0) = 0$ and $f(1) = f(2) = 1$.

   (a) Show that there is a $c \in (0,1)$ such that $f'(c) = 1$.
   (b) Show that there is a $c \in (1,2)$ such that $f'(c) = 0$.

3. Let $f(x) = \frac{1}{x}$ and let $[a, b]$ be an interval where $a > 0$. Find the point c guaranteed by the Mean Value Theorem. Since $a < c < b$, derive an interesting inequality.

4. Prove that the equation $x^7 + x^5 + x^3 + x + 1 = 0$ has exactly one real solution. Hint: For uniqueness, use Rolle's Theorem.

5. Let $f: (0,1] \to \mathbb{R}$ be differentiable on $(0,1]$ such that $|f'(x)| \leq 1$ for all $x \in (0,1]$. Define a sequence $\langle t_n \rangle$ by $t_n = f(\frac{1}{n})$ for all $n \geq 1$. Prove that $\langle t_n \rangle$ converges.

*6. Let $f: I \to \mathbb{R}$ be a function where I is an interval. Let f be strictly increasing on I and let $y, z \in I$. Suppose that $f(y) < f(z)$. Prove that $y < z$.

7. Let $f: [0, \infty) \to \mathbb{R}$ be continuous on $[0, \infty)$ and differentiable on $(0, \infty)$. Suppose $f(0) = 0$ and that f' is increasing on $(0, \infty)$. Define $g: (0, \infty) \to \mathbb{R}$ by $g(x) = \frac{f(x)}{x}$. Prove that g is increasing on $(0, \infty)$.

8. Let $f: I \to \mathbb{R}$ be differentiable on the interval I. Prove the following:

   (a) If f is increasing on I, then $f'(x) \geq 0$ for all $x \in I$.
   (b) If f is decreasing on I, then $f'(x) \leq 0$ for all $x \in I$.

*9. Let $f: I \to \mathbb{R}$ be strictly monotone on the interval I. Theorem 5.2.7 implies that f is one-to-one. Now suppose that $f[I]$ is an interval. Prove that the inverse function $f^{-1}: f[I] \to I$ is strictly monotone.

10. Complete the proofs of Case 2 and Case 3 in the proof of Lemma 5.2.19.

11. Let $f: I \to \mathbb{R}$ be differentiable on an interval I. Suppose that f' is not constant. Show that f' takes on some irrational values.

*12. Prove Theorem 5.2.8(b).

13. Provide an example of a function that satisfies the conditions of Theorem 5.2.8 and shows that the converse of Theorem 5.2.8(a) does not hold.

14. Suppose that $f: \mathbb{R} \to \mathbb{R}$ and $g: \mathbb{R} \to \mathbb{R}$ are both differentiable and strictly increasing on $\mathbb{R}$. Prove that $(f \circ g): \mathbb{R} \to \mathbb{R}$ is also strictly increasing on $\mathbb{R}$.

15. Let $i \in I$, where I is an interval. Suppose that $G: I \to \mathbb{R}$ and $f: I \to \mathbb{R}$ are continuous. Prove that if $G'(x) = f(x)$ for all x in $I \setminus \{i\}$, then $G'(i) = f(i)$.

16. Evaluate $\lim\limits_{x\to 1}\frac{\ln(x)}{x-1}$, $\lim\limits_{x\to 0^+}\frac{\sin(x)}{\sqrt{x}}$, and $\lim\limits_{x\to 0^+}\frac{\cot(x)}{\ln(x)}$.

*17. For every integer $n \geq 0$, define $h_n \colon \mathbb{R} \to \mathbb{R}$ by

$$h_n(x) = \begin{cases} \dfrac{e^{-1/x^2}}{x^n}, & \text{if } x \neq 0; \\ 0, & \text{if } x = 0. \end{cases}$$

It follows that $\lim\limits_{x\to 0} h_0(x) = 0$, and Example 5.2.14 shows that $\lim\limits_{x\to 0} h_1(x) = 0$. Prove, by an induction, that $\lim\limits_{x\to 0} h_n(x) = 0$ for all $n \geq 1$.

*18. Prove Theorem 5.2.15. (See Exercise 19 on page 118.)

19. Let $f \colon \mathbb{R} \to \mathbb{R}$ is differentiable on $\mathbb{R}$. Let $M > 0$ be such that $|f'(x)| \leq M$ for all $x \in \mathbb{R}$. Prove that f is uniformly continuous.

20. Let f and g be differentiable on $(-\infty, b)$. Suppose that

(1) $\lim\limits_{x\to -\infty} g(x) = \infty$,

(2) $g'(x) \neq 0$ for all $x \in (-\infty, b))$.

Show that Theorem 5.2.15 implies that if $\lim\limits_{x\to -\infty} \frac{f'(x)}{g'(x)} = L$ and $L \in \mathbb{R}$, then $\lim\limits_{x\to -\infty} \frac{f(x)}{g(x)} = L$.

21. Let $f \colon I \to \mathbb{R}$ be continuous on the interval I. Let $j \in I$. Suppose that $f'(x) > 0$ for all $x \in I \setminus \{j\}$ and $f'(j) = 0$. Prove that f is strictly increasing on I. Conclude that $f^{-1} \colon f[I] \to I$ exists and is continuous.

22. Let $f \colon I \to \mathbb{R}$ be one-to-one function with the intermediate value property on an interval I. Let $x, y, z \in I$ be such that $x < y < z$. Suppose that either $f(x) < f(y)$, $f(y) < f(z)$, or $f(x) < f(z)$. Prove that $f(x) < f(y) < f(z)$.

*23. Suppose that $f \colon I \to \mathbb{R}$ is one-to-one and has the intermediate value property on an interval I. Using Exercise 22, prove that f is strictly monotone.

Exercise Notes: For Exercise 5, use Theorem 5.2.3 and Exercise 4 on page 90 to show the sequence is a Cauchy sequence. For Exercise 7, prove that $g'(x) \geq 0$ for all $x \in (0, \infty)$. First evaluate $g'(x)$ by using the quotient rule and then show that there is a c such that $0 < c < x$ and $f(x) = f'(c)x$. For Exercise 8, use Exercise 10 on page 117. For Exercise 15, use Theorem 4.3.4 and Theorem 5.2.3. If i is an endpoint of I, then the derivative $G'(i)$ is a one-sided limit of the difference quotient. For Exercise 17, use Example 5.2.14 as a model. For Exercise 18, modify the proof of Theorem 5.2.12. For Exercise 20, see Exercise 24 on page 118. For Exercise 23, assume that f is not strictly decreasing, and then prove that f is strictly increasing.

5.3 TAYLOR'S THEOREM

We next discuss another generalization of the Mean Value Theorem which involves higher order derivatives. This generalization can allow one to approximate certain functions by polynomial functions.

Higher Order Derivatives

Let $f\colon I \to \mathbb{R}$ be differentiable on an interval I. Then, for all $c \in I$, we have that

$$f'(c) = \lim_{x \to c} \frac{f(x) - f(c)}{x - c}.$$

Thus, the derivative of f is itself a function from I to $\mathbb{R}$, that is, $f'\colon I \to \mathbb{R}$. If $c \in I$ and f' is differentiable at c, then the derivative of f' at c is denoted by $f''(c)$; that is,

$$f''(c) = \lim_{x \to c} \frac{f'(x) - f'(c)}{x - c}.$$

The value $f''(c)$ is called the **second derivative** of f at c. In this case, f is said to be **twice differentiable** at c. The value $f''(c)$ is also denoted by $f^{(2)}(c)$. In many cases, this process can be repeated. We will say that a function $f\colon I \to \mathbb{R}$ is n-**times differentiable** on I, if f can be differentiated n-times at each point $c \in I$, where $n \geq 1$ is a natural number. In this case, we denote the nth derivative of f at c by $f^{(n)}(c)$. Moreover, we let $f^{(0)} = f$.

A Mean Value Theorem for Higher Derivatives

Let I be an open interval and let $f\colon I \to \mathbb{R}$ be differentiable on I. The Mean Value Theorem 5.2.3 implies that for any two distinct points $a \in I$ and $b \in I$, there exists a point c between a and b such that $f'(c) = \frac{f(b) - f(a)}{b - a}$, that is,

$$f(b) = f(a) + f'(c)(b - a).$$

We now present a Mean Value Theorem for the second derivative of a function.

Theorem 5.3.1. Let $f\colon I \to \mathbb{R}$ be twice differentiable on an open interval I. Then for any two points $a \in I$ and $b \in I$, there is a point c between a and b such that

$$f(b) = f(a) + f'(a)(b - a) + \frac{f''(c)}{2}(b - a)^2. \tag{5.28}$$

Proof. If $a = b$, then (5.28) holds for $c = a$. So, assume that $a \neq b$. Let M be the unique real number that satisfies the equation

$$f(b) = f(a) + f'(a)(b - a) + M(b - a)^2. \tag{5.29}$$

Define $F\colon I \to \mathbb{R}$ by

$$F(t) = f(t) + f'(t)(b - t) + M(b - t)^2. \tag{5.30}$$

It follows that F is differentiable on I. So, by Corollary 5.1.3, F is continuous on I. In particular, F is differentiable on the open interval with endpoints a and b, and F is continuous on the closed interval with endpoints a and b. Note that

$$F(a) = f(a) + f'(a)(b - a) + M(b - a)^2.$$

Thus, by (5.29), $F(a) = f(b)$. Moreover, by equation (5.30), we see that $F(b) = f(b)$. Hence, $F(a) = F(b)$. Theorem 5.2.2 implies that there is a c between a and b such that (▲) $F'(c) = 0$. Equation (5.30) implies that

$$F'(c) = f'(c) + f''(c)(b - c) - f'(c) - 2M(b - c) = f''(c)(b - c) - 2M(b - c).$$

Hence, from (▲), we conclude that $F'(c) = f''(c)(b-c) - 2M(b-c) = 0$. Solving the latter equation for M, we obtain $M = \frac{f''(c)}{2}$. Therefore, equation (5.29) implies that

$$f(b) = f(a) + f'(a)(b - a) + \frac{f''(c)}{2}(b - a)^2. \qquad \square$$

The proof of Theorem 5.3.1 can be extended to prove the following theorem (see Exercise 4).

Theorem 5.3.2 (Taylor's Theorem). Let $f\colon I \to \mathbb{R}$ be $(n + 1)$-times differentiable on an open interval I. For any two points $a \in I$ and $b \in I$, there is a point c between a and b such that

$$f(b) = f(a) + f'(a)(b - a) + \frac{f''(a)}{2!}(b - a)^2 + \cdots$$
$$+ \frac{f^{(n)}(a)}{n!}(b - a)^n + \frac{f^{(n+1)}(c)}{(n + 1)!}(b - a)^{n+1}. \tag{5.31}$$

Equation (5.31), in Theorem 5.3.2, can be written as $f(b) = \sum_{k=0}^{n} \frac{f^{(k)}(a)}{k!}(b - a)^k$. Taylor's Theorem can be used to approximate f near a point a by a polynomial function. Let $f\colon I \to \mathbb{R}$ be $(n+1)$-times differentiable on an open interval I containing a, and for all $x \in I$ let

$$P_n(x) = \sum_{k=0}^{n} \frac{f^{(k)}(a)}{k!}(x - a)^k \text{ and } R_n(x) = f(x) - P_n(x).$$

P_n is called the nth **Taylor polynomial** of f centered at a, and R_n is said to be the **remainder term**, which measures the difference between f and P_n. For any $b \in I$, Theorem 5.3.2 states that there is a point c between a and b such that

$$R_n(b) = \frac{f^{(n+1)}(c)}{(n + 1)!}(b - a)^{n+1}.$$

This result offers another way to measure $R_n(b) = f(b) - P_n(b)$, the difference between $f(b)$ and $P_n(b)$.

Example. To illustrate the usefulness of Taylor's theorem in approximations, let $f(x) = e^x$ for all $x \in \mathbb{R}$. Recall from calculus, that $f^{(n)}(x) = e^x$ for all $n \in \mathbb{N}$. We are interested in approximating e^x near $a = 0$. Since $f(0) = 1$ and $f^{(n)}(0) = 1$ for all $n \in \mathbb{N}$, we see that nth Taylor polynomial of f centered at 0 is

$$P_n(x) = \sum_{k=0}^{n} \frac{f^{(k)}(0)}{k!} x^k = 1 + x + \frac{x^2}{2!} + \frac{x^3}{3!} + \cdots + \frac{x^n}{n!}.$$

Let $n = 5$ and let $b \in [-1, 1]$. Since $R_5(b) = \frac{f^{(6)}(c)}{6!}b^6 = \frac{e^c}{6!}b^6$ for some c between 0 and b, we conclude that

$$R_5(b) = \frac{e^c}{6!}b^6 \leq \frac{e^c}{6!}1^6 = \frac{e}{6!} < .0038.$$

So, for $b \in [-1, 1]$, $P_5(b) = 1 + b + \frac{b^2}{2!} + \frac{b^3}{3!} + \frac{b^4}{4!} + \frac{b^5}{5!}$ differs from e^b by less than .0038.

Exercises 5.3

1. Let $f \colon \mathbb{R} \to \mathbb{R}$ be twice differentiable on $\mathbb{R}$ and let $c \in \mathbb{R}$. Show that
$$f''(c) = \lim_{x \to c} \frac{f(x) + f(2c - x) - 2f(c)}{(x - c)^2}.$$

2. Prove Theorem 5.3.2 for $n = 2$.

3. Prove the following by mathematical induction on $n \geq 1$: Let $f \colon I \to \mathbb{R}$ be $(n+1)$-times differentiable on an open interval I. Define $F_n \colon I \to \mathbb{R}$ by
$$F_n(t) = f(t) + f'(t)(b - t) + \frac{f''(t)}{2!}(b - t)^2 + \cdots + \frac{f^{(n)}(t)}{n!}(b - t)^n. \quad (5.32)$$
Then $F_n'(t) = \frac{f^{(n+1)}(t)}{n!}(b - t)^n$. [Hint: $F_{n+1}(t) = F_n(t) + \frac{f^{(n+1)}(t)}{(n+1)!}(b - t)^{n+1}$.]

*4. Using the result of Exercise 3, prove Theorem 5.3.2.

5. Let $f \colon I \to \mathbb{R}$ be $(n+1)$-times differentiable on an open interval I. Suppose that $f^{(n+1)}(z) = 0$ for every $z \in I$. Let $a \in I$ and $b \in I$. Show that
$$f(b) = f(a) + f'(a)(b - a) + \frac{f''(a)}{2!}(b - a)^2 + \cdots + \frac{f^{(n)}(a)}{n!}(b - a)^n.$$

6. Let $f \colon I \to \mathbb{R}$ be $(n+1)$-times differentiable on an open interval I. Let $M > 0$ and $d > 0$. Suppose that $\left|f^{(n+1)}(c)\right| \leq M$ for every $c \in I$. Let $a \in I$ and $b \in I$ be such that $|b - a| \leq d$. Show that
$$\left|f(b) - \sum_{k=0}^{n} \frac{f^{(k)}(a)}{k!}(b - a)^k\right| \leq \frac{Md^{n+1}}{(n + 1)!}.$$

7. Let $f \colon [k, \ell] \to \mathbb{R}$ be twice differentiable on the open interval (k, ℓ). Suppose that f and f' are continuous on $[k, \ell]$. Let $a \in [k, \ell]$ and $b \in [k, \ell]$. Show that the proof of Theorem 5.3.1 can be slightly modified to demonstrate that there is a point c between a and b such that
$$f(b) = f(a) + f'(a)(b - a) + \frac{f''(c)}{2}(b - a)^2.$$

8. Let $f(x) = \sin(x)$. Find $P_4(x)$, the 4th Taylor polynomial of f centered at 0. For $b \in [-1, 1]$, estimate how close is $P_4(b)$ to $\sin(b)$?

Exercise Notes: For Exercise 1, see Theorem 5.2.10. For Exercise 4, note that $F_1(t) = f(t) + f'(t)(b - t)$. Modify the proof of Theorem 5.3.1 (see equations (5.32) and (5.30)).

CHAPTER 6

Riemann Integration

One of the oldest problems in mathematics is called the area problem:

How to find the area of a plane figure whose boundary has one or more curved edges?

Archimedes (287–212 BC) developed an ingenious method for finding the area of a circle and a few other figures with curved edges. Unfortunately, Archimedes method could not be widely applied. The general problem of evaluating the area of a region with a curved boundary plagued mathematicians for nearly two thousand years. Then, in the late 17th century, Isaac Newton and Gottfried Leibniz independently developed a method that can be used to solve the area problem for a wide variety of regions bounded by curves.

Let us consider one version of the area problem: Let $f \colon [a, b] \to \mathbb{R}$ be continuous and positive. Let R be the region under the curved graph of f and above the interval $[a, b]$ on the x-axis. Newton and Leibniz showed that one can evaluate the area of R, by finding an antiderivative of f. This discovery is generally identified as the beginning of calculus and allowed mathematicians to solve a score of applied problems. Such success may have caused mathematicians to overlook an important foundational question: What does the "area of R" really mean? Eventually, in the middle of the 19th century, Bernhard Riemann gave a mathematical precise definition of the area under a function, now called the Riemann integral.

In this chapter we present a rigorous development of the theory of Riemann integration. The reader should be familiar with the concept of integration from calculus. However, in this chapter, we focus on the proofs of many of the theorems that are applied in a calculus course. We follow a development due to Jean Gaston Darboux, which, when compared to Riemann's original idea, is easier to understand.

6.1   THE RIEMANN INTEGRAL

The definition of the Riemann integral is technically more involved than many of the other concepts discussed in this book. Before presenting this definition, we first develop the necessary terminology, notation, and preliminary lemmas. This is done in the next two sections.

6.1.1 Partitions and Darboux Sums

The notion of a partition of a closed interval is needed in order to define the Riemann integral. A partition allows one to split an interval into a finite number of subintervals.

Definition 6.1.1 (Partition of an Interval). Let $[a, b]$ be an interval.

- A **partition** P of $[a, b]$ is a finite set $P = \{x_0, x_1, x_2, \ldots, x_n\}$ of points such that $a = x_0 < x_1 < x_2 < \cdots < x_n = b$.

- If P and P^* are partitions of $[a, b]$ and $P \subseteq P^*$, then P^* is a **refinement** of P.

The points in a partition P are always listed in strictly increasing order where the first point is the left endpoint of the interval and the last point is the right endpoint of the interval. A refinement of a partition P is just a partition that is obtained by adding more points, from the interval, to P.

Remark. Given two partitions P and Q of $[a, b]$. By listing the points of $P \cup Q$ in increasing order, we see that $P \cup Q$ is also a partition of $[a, b]$. Thus, $P \cup Q$ is a refinement of both P and Q.

Example. Let $a = 0$ and $b = 2$. Then $P = \{0, \frac{1}{4}, \frac{1}{3}, 2\}$ and $Q = \{0, \frac{1}{4}, \frac{1}{2}, 1, 2\}$ are two partitions of $[0, 2]$. So, $P \cup Q = \{0, \frac{1}{4}, \frac{1}{3}, \frac{1}{2}, 1, 2\}$ is also a partition of $[0, 2]$ and is a refinement of both P and Q.

Definition 6.1.2 (Summation Notation). Let $\langle a_1, a_2, \ldots a_n \rangle$ be a finite sequence of real numbers. Then $\sum_{i=1}^{n} a_i = a_1 + a_2 + \cdots + a_n$. Whenever $1 \leq k \leq n$, $\sum_{i \neq k} a_i$ represents the sum of that a_i's where $i \neq k$. Moreover, if $A \subseteq \{1, 2, \ldots, n\}$ is nonempty, then $\sum_{i \in A} a_i$ denotes the sum of the a_i's where $i \in A$. If $A = \varnothing$, then $\sum_{i \in A} a_i = 0$.

Let $f \colon [a, b] \to \mathbb{R}$ be a positive bounded function. The next definition identifies Darboux idea on how to approximate the area between the graph of f and the x-axis. This is done by summing the area of rectangles that are circumscribed above that graph of f (see Figure 6.1) and by summing the area of rectangles that are inscribed below that graph of f (see Figure 6.2).

Definition 6.1.3 (Darboux Sums). Let $f \colon [a, b] \to \mathbb{R}$ be bounded. Given a partition $P = \{x_0, x_1, x_2, \ldots, x_n\}$ of $[a, b]$, for each $i = 1, 2, \ldots, n$, let

$$M_i(f) = \sup\{f(x) : x \in [x_{i-1}, x_i]\},$$
$$m_i(f) = \inf\{f(x) : x \in [x_{i-1}, x_i]\},$$
$$\Delta x_i = x_i - x_{i-1}.$$

When there is no ambiguity, we shall write $M_i = M_i(f)$ and $m_i = m_i(f)$. In addition, we define the **upper sum** of f with respect to P by $U(f, P) = \sum_{i=1}^{n} M_i \Delta x_i$ and define the **lower sum** of f with respect to P by $L(f, P) = \sum_{i=1}^{n} m_i \Delta x_i$.

Let $f\colon [a,b] \to \mathbb{R}$ be a positive bounded function and let P be a partition as in Definition 6.1.3. To obtain the upper sum $U(f,P)$, for each $i = 1, 2, \ldots, n$ one obtains a rectangle with base Δx_i, height M_i, and area $M_i \Delta x_i$. Then $U(f,P)$ is the sum of these rectangular areas. Similarly, each lower sum $L(f,P)$ is equal to the sum of the areas of rectangles with base Δx_i and height m_i. Since $m_i \leq M_i$ for each $i = 1, 2, \ldots, n$, we see that $L(f,P) \leq U(f,P)$.

Remark 6.1.4. In Definition 6.1.3, letting $f([x_{i-1}, x_i]) = \{f(x) : x \in [x_{i-1}, x_i]\}$, we can write $M_i(f) = \sup[f([x_{i-1}, x_i])]$ and $m_i(f) = \inf[f([x_{i-1}, x_i])]$.

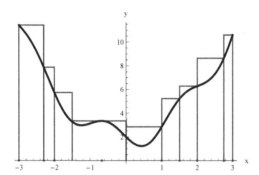

Figure 6.1: Upper sum of f with respect to P equals $U(f,P) = 35.9796$.

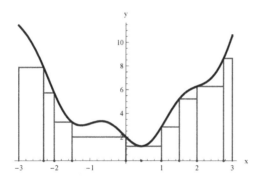

Figure 6.2: Lower sum of f with respect to P equals $L(f,P) = 23.9974$.

Example. Define $f\colon [-3,3] \to \mathbb{R}$ by $f(x) = x^2 - \sin(3x) + 2$. Consider the partition $P = \{-3, -2.3, -2, -1.5, 0, 1, 1.5, 2, 2.75, 3\}$ of $[-3,3]$. The upper sum $U(f,P)$ and the lower sum $L(f,P)$ are illustrated in Figures 6.1 and 6.2, respectively. Observe that $\Delta x_1 = -2.3 - (-3) = .7$ and $\Delta x_6 = 1.5 - 1 = .5$.

6.1.2 Basic Results Regarding Darboux Sums

Lemma 6.1.5. Let $f\colon [a,b] \to \mathbb{R}$ be a bounded function. Let $P = \{x_0, x_1, x_2, \ldots, x_n\}$ be a partition of $[a,b]$. Then for all $i = 1, 2, \ldots, n$, we have $m_i \leq M_i$.

Proof. Clearly, $m_i = \inf[f([x_{i-1}, x_i])] \leq \sup[f([x_{i-1}, x_i])] = M_i$, for $i = 1, \ldots, n$. □

It should be clear that for any particular partition, the lower sum will be less than or equal to the upper sum; however, we will now formally verify this.

Lemma 6.1.6. Let $f\colon [a,b] \to \mathbb{R}$ be a bounded function. Let $P = \{x_0, x_1, x_2, \ldots, x_n\}$ be a partition of $[a,b]$. Then $L(f,P) \le U(f,P)$.

Proof. Let $f\colon [a,b] \to \mathbb{R}$ be a bounded function. Let $P = \{x_0, x_1, x_2, \ldots, x_n\}$ be a partition of $[a,b]$. By Lemma 6.1.5, $m_i \le M_i$ for all $i = 1, 2, \ldots, n$. Hence,

$$L(f,P) = \sum_{i=1}^{n} m_i \Delta x_i \le \sum_{i=1}^{n} M_i \Delta x_i = U(f,P). \qquad \square$$

In the following two lemmas, we confirm two algebraic identities that will be applied later.

Lemma 6.1.7. Let $f\colon [a,b] \to \mathbb{R}$ be a bounded function. Let $P = \{x_0, x_1, x_2, \ldots, x_n\}$ be a partition of $[a,b]$. Then $U(f,P) - L(f,P) = \sum_{i=1}^{n} (M_i - m_i)\Delta x_i$.

Proof. By algebra,

$$U(f,P) - L(f,P) = \sum_{i=1}^{n} M_i \Delta x_i - \sum_{i=1}^{n} m_i \Delta x_i = \sum_{i=1}^{n}(M_i - m_i)\Delta x_i. \qquad \square$$

Lemma 6.1.8. $\sum_{i=1}^{n} \Delta x_i = b - a$ for any partition $P = \{x_0, x_1, x_2, \ldots, x_n\}$ of $[a,b]$.

Proof. Since $x_0 = a$ and $x_n = b$, we have $\sum_{i=1}^{n} \Delta x_i = \sum_{i=1}^{n}(x_i - x_{i-1}) = x_n - x_0 = b - a$ (see Exercise 9 on page 27). $\qquad \square$

The Refinement Theorem

Given a partition, we can evaluate the upper and lower sums of a function. We will now show that a refinement of this partition will comparatively decrease the upper sum and increase the lower sum; that is, by refining a partition "the upper sum will go down and the lower sum will go up."

Theorem 6.1.9 (Refinement Theorem). Let $f\colon [a,b] \to \mathbb{R}$ be a bounded function. If P is a partition of $[a,b]$ and P^* is a refinement of P, then

$$L(f,P) \le L(f,P^*) \le U(f,P^*) \le U(f,P). \qquad (6.1)$$

Proof. Let $f\colon [a,b] \to \mathbb{R}$ be a bounded function. Let P and P^* be partitions of $[a,b]$ where P^* is a refinement of P. Let $P = \{x_0, x_1, x_2, \ldots, x_n\}$. So, the upper sum of f with respect to P is

$$U(f,P) = \sum_{i=1}^{n} M_i \Delta x_i. \qquad (6.2)$$

Lemma 6.1.6 implies that $L(f,P^*) \le U(f,P^*)$, the "middle" inequality of (6.1). We now prove that $U(f, P^*) \leq U(f, P)$. First, suppose that the partition P^* contains just one more point than P. So $P^* = P \cup \{x^*\}$ and $x_{k-1} < x^* < x_k$ for some natural number k where $1 \leq k \leq n$. In equation (6.2), let us "isolate the interval $[x_{k-1}, x_k]$" in the sum (6.2) for $U(f, P)$ by separating the term $M_k \Delta x_k$ from the sum $\sum\limits_{i=1}^{n} M_i \Delta x_i$ and rewriting this sum as

$$U(f, P) = \left(\sum_{i \neq k} M_i \Delta x_i \right) + M_k \Delta x_k. \tag{6.3}$$

Recall that $M_k \Delta x_k = M_k(x_k - x_{k-1})$ where $M_k = \sup[f([x_{k-1}, x_k])]$. By "isolating the two intervals $[x_{k-1}, x^*]$ and $[x^*, x_k]$" in the sum for $U(f, P^*)$, we see that

$$U(f, P^*) = \left(\sum_{i \neq k} M_i \Delta x_i \right) + s_1(x^* - x_{k-1}) + s_2(x_k - x^*), \tag{6.4}$$

where $s_1 = \sup[f([x_{k-1}, x^*])]$ and $s_2 = \sup[f([x^*, x_k])]$. Since $s_1 \leq M_k$ and $s_2 \leq M_k$, we have that

$$s_1(x^* - x_{k-1}) + s_2(x_k - x^*) \leq M_k(x^* - x_{k-1}) + M_k(x_k - x^*)$$
$$= M_k(x_k - x_{k-1}) = M_k \Delta x_k. \tag{6.5}$$

Hence, (6.3), (6.4), and (6.5) imply that $U(f, P^*) \leq U(f, P)$ when P^* contains one additional point. Now suppose that P^* contains $j > 1$ many more points than P. So $P^* = P \cup \{x_1^*, x_2^*, \ldots, x_j^*\}$. Let $P_i^* = P \cup \{x_1^*, x_2^*, \ldots, x_i^*\}$ whenever $1 \leq i \leq j$. Thus, $P_1^* = P^* \cup \{x_1^*\}$, $P_2^* = P^* \cup \{x_1^*, x_2^*\}$, $\ldots$, and $P_j^* = P^*$. For each $i < j$, the partition P_{i+1}^* contains just one more point than P_i^* and thus, the above argument shows that $U(f, P_{i+1}^*) \leq U(f, P_i^*)$. We conclude that

$$U(f, P^*) \leq U(f, P_{j-1}^*) \leq \cdots \leq U(f, P_{i+1}^*) \leq U(f, P_i^*) \leq \cdots \leq U(f, P_1^*) \leq U(f, P).$$

Therefore, $U(f, P^*) \leq U(f, P)$. The proof that $L(f, P) \leq L(f, P^*)$ is similar. □

We now show that *every* lower sum is less than or equal to *every* upper sum, of the function.

Lemma 6.1.10. Let $f : [a, b] \to \mathbb{R}$ be a bounded function. If P and Q are partitions of $[a, b]$, then $L(f, P) \leq U(f, Q)$.

Proof. Let $f : [a, b] \to \mathbb{R}$ be a bounded function and let P and Q be partitions of $[a, b]$. Since $P^* = P \cup Q$ is a refinement of P, Theorem 6.1.9 implies that

$$L(f, P) \leq U(f, P^*). \tag{6.6}$$

Since $P^* = P \cup Q$ is a refinement of Q, Theorem 6.1.9 implies that

$$U(f, P^*) \leq U(f, Q). \tag{6.7}$$

Hence, (6.6) and (6.7) imply that $L(f, P) \leq U(f, Q)$. □

The Boundedness Lemma

Let $f\colon [a,b] \to \mathbb{R}$ be bounded. Is the set of all Darboux sums of f bounded? That is, are there real numbers α and β such that $\alpha \leq L(f,P) \leq U(f,P) \leq \beta$ for *all* partitions P of $[a,b]$? The following lemmas answers this question in the affirmative.

Lemma 6.1.11. Let $f\colon [a,b] \to \mathbb{R}$ be a bounded function and let m and M be such that $m \leq f(x) \leq M$ for all $x \in [a,b]$. Let P be a partition of $[a,b]$. Then

$$m(b-a) \leq L(f,P) \leq U(f,P) \leq M(b-a).$$

Proof. Suppose that $m \leq f(x) \leq M$ for all $x \in [a,b]$, and let $P = \{x_0, x_1, x_2, \ldots, x_n\}$ be a partition of $[a,b]$. So, by Lemma 6.1.6, $L(f,P) \leq U(f,P)$. We now prove that $U(f,P) \leq M(b-a)$. Note that $M_i \leq M$ for all $i = 1, 2, \ldots, n$. Hence,

$$U(f,P) = \sum_{i=1}^{n} M_i \Delta x_i \leq \sum_{i=1}^{n} M \Delta x_i = M \sum_{i=1}^{n} \Delta x_i = M(b-a).$$

The proof that $m(b-a) \leq L(f,P)$ is similar. $\qquad\square$

So if $f\colon [a,b] \to \mathbb{R}$ is bounded, then the sets of real numbers

$$\{U(f,P) : P \text{ is a partition of } [a,b]\} \text{ and } \{L(f,P) : P \text{ is a partition of } [a,b]\} \quad (6.8)$$

are also bounded.

6.1.3   The Definition of the Riemann Integral

Since Lemma 6.1.11 implies that the sets in (6.8) are bounded, the completeness axiom ensures that the following upper and lower integrals of f exist.

Definition 6.1.12. Let $f\colon [a,b] \to \mathbb{R}$ be a bounded function.

- The **upper integral** of f on $[a,b]$, denoted by $U(f)$, is defined by

$$U(f) = \inf\{U(f,P) : P \text{ is a partition of } [a,b]\}.$$

- The **lower integral** of f on $[a,b]$, denoted by $L(f)$, is defined by

$$L(f) = \sup\{L(f,P) : P \text{ is a partition of } [a,b]\}.$$

Thus, the upper integral is equal to the infimum of the upper sums and the lower integral is equal to the supremum of the lower sums.

Comment. Let $f\colon [a,b] \to \mathbb{R}$ be a bounded function. Given any partition P of $[a,b]$, we always have that $L(f,P) \leq L(f)$ and $U(f) \leq U(f,P)$.

Let $f\colon [a,b] \to \mathbb{R}$ be positive and bounded. Suppose we know that A is the area of the region under the graph of f and above the interval $[a,b]$. As in Figure 6.1, one can view each upper sum $U(f,P)$ as an overestimate of A, that is, $A \leq U(f,P)$. Thus, A is a lower bound for the set $\{U(f,P) : P$ is a partition of $[a,b]\}$. Hence, $A \leq U(f)$. Similarly, $L(f) \leq A$. Thus, $L(f) \leq A \leq U(f)$. Therefore, if $L(f) = U(f)$, then $L(f) = A = U(f)$. So if $L(f) = U(f)$, then the area under f can be defined to be this common value.

Definition 6.1.13. Let $f\colon [a,b] \to \mathbb{R}$ be a bounded function. If $L(f) = U(f)$, we say that f is **Riemann integrable** on $[a,b]$ and we shall denote this common value by $\int_a^b f$ or by $\int_a^b f(x)\,dx$.

We often say that a function f is *integrable* on $[a,b]$, rather than stating that f is Riemann integrable. The term $\int_a^b f$ is called the *integral* of f over $[a,b]$. We now present three basic lemmas that establish three useful inequalities.

Lemma 6.1.14. Let $f\colon [a,b] \to \mathbb{R}$ be a Riemann integrable function. If P is any partition of $[a,b]$, then $L(f,P) \leq \int_a^b f \leq U(f,P)$.

Proof. Let $f\colon [a,b] \to \mathbb{R}$ be Riemann integrable. So (▲) $\int_a^b f = L(f) = U(f)$. Let P be a partition of $[a,b]$. As $L(f,P) \leq L(f)$ and $U(f) \leq U(f,P)$, (▲) implies that $L(f,P) \leq \int_a^b f \leq U(f,P)$. □

Lemma 6.1.15. Let $f\colon [a,b] \to \mathbb{R}$ be Riemann integrable. Suppose $m, M \in \mathbb{R}$ are such that $m \leq f(x) \leq M$ for all $x \in [a,b]$. Then $m(b-a) \leq \int_a^b f \leq M(b-a)$.

Proof. Let $f\colon [a,b] \to \mathbb{R}$ be Riemann integrable. Let $m \in \mathbb{R}$ and $M \in \mathbb{R}$ be such that $m \leq f(x) \leq M$ for all $x \in [a,b]$. Lemma 6.1.11 implies that

$$m(b-a) \leq L(f,P) \leq U(f,P) \leq M(b-a)$$

for any partition P of $[a,b]$. Therefore, Lemma 6.1.14 now implies that

$$m(b-a) \leq \int_a^b f \leq M(b-a). \qquad \square$$

Lemma 6.1.16. Let $f\colon [a,b] \to \mathbb{R}$ be a Riemann integrable function. Suppose that $M \geq 0$ is such that $|f(x)| \leq M$ for all $x \in [a,b]$. Then $\left| \int_a^b f \right| \leq M(b-a)$.

Proof. Let $f\colon [a,b] \to \mathbb{R}$ be Riemann integrable such that $|f(x)| \leq M$ for all $x \in [a,b]$, where $M \geq 0$. Thus, $-M \leq f(x) \leq M$ for all $x \in [a,b]$. Lemma 6.1.15 implies that

$$-M(b-a) \leq \int_a^b f \leq M(b-a).$$

Consequently, $\left| \int_a^b f \right| \leq M(b-a)$ by Theorem 2.2.6(c). □

6.1.4   A Necessary and Sufficient Condition

To formally prove that a function is Riemann integrable, it can be difficult to apply Definition 6.1.13. Fortunately, there is a more accessible method for showing that a function is Riemann integral. In this section we shall establish an important necessary and sufficient condition (see Theorem 6.1.18) which we will use to show that many functions are Riemann integrable.

Theorem 6.1.17. Let $f\colon [a, b] \to \mathbb{R}$ be a bounded function. Then $L(f) \le U(f)$.

Proof. Let $f\colon [a, b] \to \mathbb{R}$ be bounded. Recall that

$$L(f) = \sup\{L(f, P) : P \text{ is a partition of } [a, b]\}$$
$$U(f) = \inf\{U(f, Q) : Q \text{ is a partition of } [a, b]\}.$$

Lemma 6.1.10 implies that $L(f, P) \le U(f, Q)$ for all partitions P and Q of $[a, b]$. Theorem 2.3.17 implies that $L(f) \le U(f)$. $\qquad\square$

From the above theorem, we see that $L(f, P) \le L(f) \le U(f) \le U(f, P)$ for all partitions P. This indicates that a function is Riemann integrable if and only if one can find partitions whose upper and lower sums are arbitrarily close together.

Theorem 6.1.18 (Necessary and Sufficient Condition). Let $f\colon [a, b] \to \mathbb{R}$ be a bounded function. Then f is Riemann integrable if and only if for every $\varepsilon > 0$ there exists a partition P of $[a, b]$ such that $U(f, P) - L(f, P) < \varepsilon$.

Proof. Let $f\colon [a, b] \to \mathbb{R}$ be a bounded function.

($\Rightarrow$). Assume that f is Riemann integrable. So, $L(f) = U(f)$. Let $\varepsilon > 0$. Recall that

$$L(f) = \sup\{L(f, P) : P \text{ is a partition of } [a, b]\} \tag{6.9}$$
$$U(f) = \inf\{U(f, Q) : Q \text{ is a partition of } [a, b]\}. \tag{6.10}$$

By (6.9) there is a partition P' such that $L(f) - L(f, P') < \frac{\varepsilon}{2}$, and by (6.10) there is a partition Q such that $U(f, Q) - U(f) < \frac{\varepsilon}{2}$.[1] Let $P = P' \cup Q$. Because P is a refinement of P', Theorem 6.1.9 implies that $L(f, P') \le L(f, P) \le L(f)$ and since $L(f) - L(f, P') < \frac{\varepsilon}{2}$, we conclude that

$$L(f) - L(f, P) < \frac{\varepsilon}{2}. \tag{6.11}$$

Similarly, Theorem 6.1.9 implies that

$$U(f, P) - U(f) < \frac{\varepsilon}{2}. \tag{6.12}$$

Since $L(f) = U(f)$, by adding (6.11) and (6.12), we have that $U(f, P) - L(f, P) < \varepsilon$.

---

[1]See Exercises 7 and 8 on page 46.

($\Leftarrow$). Suppose that for every $\varepsilon > 0$ we have $U(f, P) - L(f, P) < \varepsilon$ for some partition P of $[a, b]$. Let $\varepsilon > 0$. Thus, (▲) $U(f, P) - L(f, P) < \varepsilon$ for some partition P. Since

$$L(f, P) \leq L(f) \leq U(f) \leq U(f, P),$$

we conclude[2] that $U(f) - L(f) \leq U(f, P) - L(f, P)$. So, by (▲), $0 \leq U(f) - L(f) < \varepsilon$. By Corollary 2.2.4, we infer that $U(f) = L(f)$. Thus, f Riemann integrable. $\qquad \square$

Definition 6.1.13 and Theorem 6.1.18 present us with two methods for proving that a function is Riemann integrable on a given interval.

Proof Strategy 6.1.19. Let f be a bounded function. To prove that f is Riemann integrable on the interval $[a, b]$, use one of the following two proof strategies:

(a) Apply Definition 6.1.13 directly and prove that $L(f) = U(f)$.

(b) Apply Theorem 6.1.18 by using the proof diagram:

Let $\varepsilon > 0$ be arbitrary.
Let P be the partition of $[a, b]$ that you found.
Prove $U(f, P) - L(f, P) < \varepsilon$.

Strategy 6.1.19(a) requires one to prove something about all partitions using Definitions 6.1.13 and 6.1.12, whereas Strategy 6.1.19(b) just involves finding one relevant partition. For this reason, Strategy 6.1.19(b) is typically easier to apply.

Example 6.1.20. Let $g \colon [0, 1] \to \mathbb{R}$ be defined by

$$g(x) = \begin{cases} 1, & \text{if } x < 1; \\ 0, & \text{if } x = 1. \end{cases}$$

We will use Proof Strategy 6.1.19(b) to show that g is Riemann integrable. Let $\varepsilon > 0$. Let $n \in \mathbb{N}$ be such that $\frac{1}{n} < \min\{\varepsilon, 1\}$ and let $P = \{0, \frac{n-1}{n}, 1\}$. We will show that $U(g, P) - L(g, P) < \varepsilon$. Since

$$U(g, P) = \sum_{i=1}^{2} M_i \Delta x_i = 1 \cdot \frac{n-1}{n} + 1 \left(1 - \frac{n-1}{n}\right) = 1$$

$$L(g, P) = \sum_{i=1}^{2} m_i \Delta x_i = 1 \cdot \frac{n-1}{n} + 0 \left(1 - \frac{n-1}{n}\right) = 1 - \frac{1}{n},$$

we see that $U(g, P) - L(g, P) = \frac{1}{n} < \varepsilon$. Thus, g is Riemann integrable.

Example 6.1.20 presents a function that is continuous everywhere except at an endpoint. It is then shown that this function is Riemann integrable. The argument used in this example, inspires the proof of our next theorem; but, before proving this theorem, we identify a relevant assumption strategy. Whenever we have a function that is Riemann integrable, Definition 6.1.13 and Theorem 6.1.18 allow us to reach two conclusions that can be applied in a proof.

---

[2]If $c \leq x \leq y \leq d$, then $y - x \leq d - c$.

Assumption Strategy 6.1.21. Let $f\colon [a,b] \to \mathbb{R}$ be bounded. If f is Riemann integrable on $[a,b]$, then both of the following assertions hold:

(a) $L(f) = U(f)$.

(b) For any $\varepsilon > 0$, there exists a partition P of $[a,b]$ such that $U(f,P) - L(f,P) < \varepsilon$.

Theorem 6.1.22. Let $f\colon [a,b] \to \mathbb{R}$ be bounded. If f is Riemann integrable on $[c,b]$ for every $c \in (a,b)$, then f is Riemann integrable on $[a,b]$. A similar result holds for the endpoint b.

Proof. Let $f\colon [a,b] \to \mathbb{R}$ be bounded and assume that (▲) f is Riemann integrable on $[c,b]$ for all $c \in (a,b)$. Let $\varepsilon > 0$. Let $S = \sup[f([a,b])]$ and let $T = \inf[f([a,b])]$. Let $n \in \mathbb{N}$ be such that $\frac{S-T}{n} < \frac{\varepsilon}{2}$ and $a + \frac{1}{n} < b$. By (▲) and Theorem 6.1.18, there exists a partition P_1 of $[a + \frac{1}{n}, b]$ such that $U(f,P_1) - L(f,P_1) < \frac{\varepsilon}{2}$. Now let $P = \{a\} \cup P_1$. Clearly, P is a partition of $[a,b]$. One can now show that $U(f,P) - L(f,P) < \varepsilon$. Thus, by Theorem 6.1.18, f is Riemann integrable on $[a,b]$. □

In Section 6.2.3, we will show that a function is integrable on $[a,b]$ if and only if it is integrable on $[a,c]$ and $[c,b]$ for all $c \in (a,b)$. Thus, Theorem 6.1.22, together with an induction argument, shows that a bounded function on $[a,b]$ having just a finite number of discontinuities is Riemann integrable.

Moreover, to show that a function f is not Riemann integrable on $[a,b]$, one can apply the negation of 6.1.21(b). To do so, one must identify an $\varepsilon > 0$ and show that $U(f,P) - L(f,P) \geq \varepsilon$ for every partition P of $[a,b]$.

Example 6.1.23. Let $h\colon [0,1] \to \mathbb{R}$ be defined by

$$h(x) = \begin{cases} 1, & \text{if } x \in \mathbb{Q}; \\ 0, & \text{if } x \notin \mathbb{Q}. \end{cases}$$

We show f is not Riemann integrable. Let $\varepsilon = \frac{1}{2}$ and let $P = \{x_0, x_1, x_2, \ldots, x_n\}$ be any partition of $[0,1]$. Since $\mathbb{Q}$ and $\mathbb{R} \setminus \mathbb{Q}$ are dense in $\mathbb{R}$, we see that $M_i = 1$ and $m_i = 0$ for each $i = 1, 2, \ldots, n$. Thus, $U(h,P) = \sum_{i=1}^{n} M_i \Delta x_i = \sum_{i=1}^{n} \Delta x_i = 1$ and $L(h,P) = \sum_{i=1}^{n} m_i \Delta x_i = 0$. So $U(h,P) - L(h,P) = 1 > \frac{1}{2}$. Thus, h is not integrable.

Example 6.1.23 shows that a function with an infinite number of discontinuities may not be Riemann integrable.

Remark. Georg Friedrich Bernhard Riemann (1826–1866) was famous for his work in non-Euclidean geometry, differential equations, and number theory. His results in physics and mathematics form the basis of Einstein's theory of general relativity.

Exercises 6.1

1. Let $f\colon [0,2] \to \mathbb{R}$ be defined by $f(x) = x^2 - x$ whose graph appears in Figure 6.3. Consider the partition $P = \{0, \frac{1}{2}, 1, \frac{3}{2}, 2\}$ of $[0,2]$. Find $U(f,P)$ and $L(f,P)$.

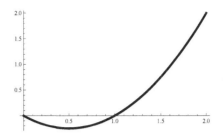

Figure 6.3: Graph of $f(x) = x^2 - x$ for Exercise 1.

2. Let c be a constant. Suppose that $f(x) = c$ for all x in $[a,b]$. Prove that f is Riemann integrable and that $\int_a^b f = c(b-a)$.

*3. Let $f: [a,b] \to \mathbb{R}$ and $g: [a,b] \to \mathbb{R}$ be bounded. Assume that $f(x) \le g(x)$ for all $x \in [a,b]$. Show that

   (a) $U(f,P) \le U(g,P)$ for every partition P of $[a,b]$.

   (b) $L(f,P) \le L(g,P)$ for every partition P of $[a,b]$.

   (c) $U(f) \le U(g)$ and $L(f) \le L(g)$.

*4. Let $f: [a,b] \to \mathbb{R}$ be bounded and let c be a constant. Define $g: [a,b] \to \mathbb{R}$ by $g(x) = f(x) + c$. Show that

   (a) $U(g,P) = U(f,P) + c(b-a)$ for every partition P of $[a,b]$.

   (b) $L(g,P) = L(f,P) + c(b-a)$ for every partition P of $[a,b]$.

   (c) $U(g) = U(f) + c(b-a)$ and $L(g) = L(f) + c(b-a)$.

5. Let $f: [0,1] \to \mathbb{R}$ be defined by

$$f(x) = \begin{cases} 1, & \text{if } 0 \le x < 1; \\ 2, & \text{if } x = 1. \end{cases}$$

   Show that f is Riemann integrable and find $\int_0^1 f$.

6. Complete that proof of Theorem 6.1.22 by showing that $U(f,P) - L(f,P) < \varepsilon$.

7. Let $f: [0,1] \to \mathbb{R}$ be defined by

$$f(x) = \begin{cases} 1, & \text{if } x \in \mathbb{Q}; \\ -1, & \text{if } x \notin \mathbb{Q}. \end{cases}$$

   Show that f is not Riemann integrable.

8. Let $f: [a,b] \to \mathbb{R}$ be bounded and let $\mathcal{K} \in \mathbb{R}$. Prove that if for every $\varepsilon > 0$ there is a partition P of $[a,b]$ such that $U(f,P) - L(f,P) < \varepsilon$ and $L(f,P) \le \mathcal{K} \le U(f,P)$, then f is Riemann integrable and $\int_a^b f = \mathcal{K}$.

9. Let $f: [a,b] \to \mathbb{R}$ be bounded. Suppose that there is a sequence $\langle P_n \rangle$ of partitions of $[a,b]$ such that

$$\lim_{n \to \infty} (U(f,P_n) - L(f,P_n)) = 0.$$

(a) Prove that f is Riemann integrable.

(b) Prove that $\lim_{n\to\infty} L(f, P_n) = \int_a^b f$ and $\lim_{n\to\infty} U(f, P_n) = \int_a^b f$.

10. Let $f\colon [0,1] \to [0,1]$ be defined by

$$f(x) = \begin{cases} 0, & \text{if } x \in \mathbb{R} \setminus \mathbb{Q}; \\ \frac{1}{n}, & \text{if } x \in \mathbb{Q}^+ \text{ and } x = \frac{m}{n} \text{ is in reduced form;} \\ 1, & \text{if } x = 0; \end{cases}$$

for each $x \in [0,1]$, where m and n are natural numbers. One can show that f is Riemann integrable as follows: Let $\varepsilon > 0$ be such that $\varepsilon < 1$.

(a) Let $S = \{\frac{m}{n} \in [0,1] : \frac{m}{n} \in \mathbb{Q}^+ \text{ (in reduced form) and } \frac{1}{n} > \frac{\varepsilon}{2}\}$. Show that $S \neq \varnothing$ and S is finite.

(b) Let $N = \#S \geq 1$, where $\#S$ denotes the number of elements in the finite set S. Let $P = \{x_0, x_1, \ldots, x_n\}$ be a partition of $[0,1]$ such that $P \cap S = \{x_n\} = \{1\}$ and $\Delta x_i < \frac{\varepsilon}{2N}$ for all $i = 1, 2, \ldots, n$. Break up the set $\{1, 2, \ldots, n\}$ into the following two disjoint parts:

$$A = \{i : [x_i, x_{i-1}] \cap S \neq \varnothing\} \text{ and } B = \{i : [x_i, x_{i-1}] \cap S = \varnothing\}.$$

(1) Show that $\sum_{i \in A} M_i \Delta x_i < \frac{\varepsilon}{2}$. Note: $1 \leq \#A \leq N$, and $M_i \leq 1$ for all $i \in A$.

(2) Show that $\sum_{i \in B} M_i \Delta x_i < \frac{\varepsilon}{2}$. Note: $\#B < n$, and $M_i \leq \frac{\varepsilon}{2}$ for all $i \in B$.

Conclude that $U(f, P) - L(f, P) < \varepsilon$.

Exercise Notes: For Exercise 3(a)(b), use Exercise 20 on page 47. For Exercise 4, use Exercise 13 on page 46. For Exercises 5 and 7, use proof strategy 6.1.19(b). For Exercise 9, use Exercise 11 on page 74.

6.2 PROPERTIES OF THE RIEMANN INTEGRAL

In the previous section, the focus was on the definition of the Riemann integral and on a useful necessary and sufficient condition for its existence. This section will be directed toward deriving the important properties of the Riemann integral.

6.2.1 Linearity Properties

In this section, we will identify and prove several algebraic properties of the integral. Exercise 11 on page 46 and Definition 6.1.3 imply the following result, which will be used to prove the constant multiple rule for the Riemann integral.

Lemma 6.2.1. Suppose that $f\colon [a, b] \to \mathbb{R}$ is bounded and let $k \in \mathbb{R}$. Let $P = \{x_0, x_1, x_2, \ldots, x_n\}$ be an arbitrary partition of $[a, b]$. If $k \geq 0$, then

1. $M_i(kf) = kM_i(f)$ and $m_i(kf) = km_i(f)$, for all $i = 1, \ldots, n$.

2. $U(kf, P) = kU(f, P)$ and $L(kf, P) = kL(f, P)$.

If $k < 0$, then

1. $M_i(kf) = km_i(f)$ and $m_i(kf) = kM_i(f)$, for all $i = 1, \ldots, n$.
2. $U(kf, P) = kL(f, P)$ and $L(kf, P) = kU(f, P)$.

Lemma 6.2.2 (Constant Multiple Rule). Suppose that $f \colon [a, b] \to \mathbb{R}$ is Riemann integrable and let $k \in \mathbb{R}$. Then kf is Riemann integrable and $\int_a^b kf = k \int_a^b f$.

Proof. Let $f \colon [a, b] \to \mathbb{R}$ be Riemann integrable and let $k \in \mathbb{R}$. We consider the cases (1) $k \geq 0$ and (2) $k < 0$ separately.

CASE 1: $k \geq 0$. First let P be a partition of $[a, b]$. Lemma 6.2.1 implies that

$$U(kf, P) = kU(f, P) \text{ and } L(kf, P) = kL(f, P). \tag{6.13}$$

Recalling Definition 6.1.12, we see from (6.13) that

$$U(kf) = \inf\{kU(f, P) : P \text{ is a partition of } [a, b]\}$$
$$L(kf) = \sup\{kL(f, P) : P \text{ is a partition of } [a, b]\}.$$

Theorem 2.3.13 implies that (i) $U(kf) = kU(f)$ and (ii) $L(kf) = kL(f)$. Since f is Riemann integrable, we know that $U(f) = L(f)$. Therefore, (i) and (ii) imply that $U(kf) = L(kf)$ and hence, kf is Riemann integrable. In addition, because $U(kf) = kU(f)$, we have that $\int_a^b kf = k \int_a^b f$.

CASE 2: $k < 0$. The proof is analogous to the proof of Case 1 and is Exercise 2. □

The following lemma will be used to prove the sum rule for integration.

Lemma 6.2.3. Let $f \colon [a, b] \to \mathbb{R}$ and $g \colon [a, b] \to \mathbb{R}$ be bounded. For any partition P of $[a, b]$, we have that

$$L(f, P) + L(g, P) \leq L(f + g, P) \leq U(f + g, P) \leq U(f, P) + U(g, P).$$

Proof. Let $f \colon [a, b] \to \mathbb{R}$ and $g \colon [a, b] \to \mathbb{R}$ be bounded and let $P = \{x_0, x_1, \ldots, x_n\}$ be a partition of $[a, b]$. For each $i = 1, \ldots, n$, letting $D_i = [x_{i-1}, x_i]$, Definition 6.1.3 and Theorem 2.3.16 imply that

$$m_i(f) + m_i(g) = \inf(f[D_i]) + \inf(g[D_i]) \leq \inf((f + g)[D_i]) = m_i(f + g)$$
$$M_i(f + g) = \sup((f + g)[D_i]) \leq \sup(f[D_i]) + \sup(g[D_i]) = M_i(f) + M_i(g).$$

Thus, $m_i(f) + m_i(g) \leq m_i(f + g)$ and $M_i(f + g) \leq M_i(f) + M_i(g)$ for all $i = 1, \ldots, n$. Hence, by Definition 6.1.3, we have

$$L(f, P) + L(g, P) \leq L(f + g, P) \text{ and } U(f + g, P) \leq U(f, P) + U(g, P).$$

Since $L(f + g, P) \leq U(f + g, P)$, we conclude that

$$L(f, P) + L(g, P) \leq L(f + g, P) \leq U(f + g, P) \leq U(f, P) + U(g, P). \qquad \square$$

Lemma 6.2.4 (Sum Rule). If $f\colon [a,b] \to \mathbb{R}$ and $g\colon [a,b] \to \mathbb{R}$ are both Riemann integrable, then $f + g$ is Riemann integrable and $\int_a^b (f+g) = \int_a^b f + \int_a^b g$.

Proof. Let $f\colon [a,b] \to \mathbb{R}$ and $g\colon [a,b] \to \mathbb{R}$ be Riemann integrable. We prove the theorem in two steps.

STEP 1: We will prove that $f + g$ is Riemann integrable by applying Theorem 6.1.18. Let $\varepsilon > 0$. Since f and g are Riemann integrable, Theorem 6.1.18 implies that there are partitions P_1 and P_2 such that

$$U(f, P_1) - L(f, P_1) < \frac{\varepsilon}{2} \tag{6.14}$$

$$U(g, P_2) - L(g, P_2) < \frac{\varepsilon}{2}. \tag{6.15}$$

Let $P^* = P_1 \cup P_2$. Since P^* is a refinement of P_1 and P_2, Theorem 6.1.9 implies that

$$L(f, P_1) \le L(f, P^*) \le U(f, P^*) \le U(f, P_1)$$
$$L(g, P_2) \le L(g, P^*) \le U(g, P^*) \le U(g, P_2)$$

and thus, (6.14) and (6.15) respectively imply[3] that

$$U(f, P^*) - L(f, P^*) < \frac{\varepsilon}{2} \tag{6.16}$$

$$U(g, P^*) - L(g, P^*) < \frac{\varepsilon}{2}. \tag{6.17}$$

Hence, (6.16) and (6.17) imply that

$$U(f, P^*) + U(g, P^*) - (L(f, P^*) + L(g, P^*)) < \varepsilon. \tag{6.18}$$

By Lemma 6.2.3, we have that

$$L(f, P^*) + L(g, P^*) \le L(f + g, P^*) \le U(f + g, P^*) \le U(f, P^*) + U(g, P^*). \tag{6.19}$$

Thus, from (6.19) and (6.18), we conclude that $U(f + g, P^*) - L(f + g, P^*) < \varepsilon$. Therefore, Theorem 6.1.18 implies that $f + g$ is Riemann integrable.

STEP 2: Now we shall prove that $\int_a^b (f + g) = \int_a^b f + \int_a^b g$. Let $\varepsilon > 0$. Using the same argument as in Step 1, there is a partition P^* satisfying (6.18) and (6.19). By Lemma 6.1.14 we also have that

$$L(f, P^*) \le \int_a^b f \le U(f, P^*) \tag{6.20}$$

$$L(g, P^*) \le \int_a^b g \le U(g, P^*) \tag{6.21}$$

$$L(f + g, P^*) \le \int_a^b (f + g) \le U(f + g, P^*). \tag{6.22}$$

---

[3]If $c \le x \le y \le d$ and $d - c < \frac{\varepsilon}{2}$, then $y - x < \frac{\varepsilon}{2}$.

By adding corresponding sides, (6.20) and (6.21) imply the next inequality (6.23). Also, (6.22) and (6.19) imply the subsequent inequality (6.24)

$$L(f, P^*) + L(g, P^*) \le \int_a^b f + \int_a^b g \le U(f, P^*) + U(g, P^*) \qquad (6.23)$$

$$L(f, P^*) + L(g, P^*) \le \int_a^b (f + g) \le U(f, P^*) + U(g, P^*). \qquad (6.24)$$

Inequalities (6.18), (6.23), and (6.24) now imply[4] that

$$\left| \int_a^b f + \int_a^b g - \int_a^b (f + g) \right| < \varepsilon.$$

Corollary 2.2.4 now implies that $\int_a^b (f + g) = \int_a^b f + \int_a^b g$. □

Lemma 6.2.2 and Lemma 6.2.4 establish two important algebraic properties of the Riemann integral. We now formally unite these two results into a single theorem.

Theorem 6.2.5 (Linearity Theorem). Suppose that $f \colon [a, b] \to \mathbb{R}$ and $g \colon [a, b] \to \mathbb{R}$ are Riemann integrable and let $k \in \mathbb{R}$. Then

(a) kf is Riemann integrable and $\int_a^b kf = k \int_a^b f$.

(b) $f + g$ is Riemann integrable and $\int_a^b (f + g) = \int_a^b f + \int_a^b g$.

Proof. Lemma 6.2.2 implies (a) and Lemma 6.2.4 implies (b). □

Corollary 6.2.6 (Difference Rule). Suppose that $f \colon [a, b] \to \mathbb{R}$ and $g \colon [a, b] \to \mathbb{R}$ are Riemann integrable. Then $f - g$ is Riemann integrable and $\int_a^b (f - g) = \int_a^b f - \int_a^b g$.

Proof. By Theorem 6.2.5(a), $-g$ is Riemann integrable. Since $f - g = f + (-g)$, we have that $f - g$ is integrable and $\int_a^b (f - g) = \int_a^b f - \int_a^b g$, by Theorem 6.2.5. □

6.2.2 Order Properties

We will show that Riemann integration preserves an order relation on functions.

Theorem 6.2.7. Suppose that $f \colon [a, b] \to \mathbb{R}$ is Riemann integrable and that $f(x) \ge 0$ for all $x \in [a, b]$. Then $\int_a^b f \ge 0$.

Proof. Let $f \colon [a, b] \to \mathbb{R}$ be Riemann integrable where $f(x) \ge 0$ for all $x \in [a, b]$. It follows that $0 \le L(P, f)$ for every partition P of $[a, b]$. Thus, $0 \le L(f)$. Since f is Riemann integrable, we have that $\int_a^b f = L(f)$. Therefore, $\int_a^b f \ge 0$. □

Corollary 6.2.8. Let $f \colon [a, b] \to \mathbb{R}$ and $g \colon [a, b] \to \mathbb{R}$ be Riemann integrable where $f(x) \le g(x)$ for all $x \in [a, b]$. Then $\int_a^b f \le \int_a^b g$.

Proof. Suppose that $f \colon [a, b] \to \mathbb{R}$ and $g \colon [a, b] \to \mathbb{R}$ are Riemann integrable and that $f(x) \le g(x)$ for all $x \in [a, b]$. Define the function $h \colon [a, b] \to \mathbb{R}$ by $h(x) = g(x) - f(x)$. Clearly, $h(x) \ge 0$ for all $x \in [a, b]$. Corollary 6.2.6 implies that h is Riemann integrable and Theorem 6.2.7 implies that $0 \le \int_a^b h$. Thus, $0 \le \int_a^b (f - g)$ and Corollary 6.2.6 implies that $\int_a^b f \le \int_a^b g$. □

---

[4]If $c \le x, y \le d$ and $d - c < \varepsilon$, then $|x - y| < \varepsilon$.

6.2.3 Integration over Subintervals

We will show that if a function is integrable on a closed interval $[a, b]$, then it is also integrable on any closed subinterval of $[a, b]$. This will allow us to split an integral into two parts (see Theorem 6.2.11).

Theorem 6.2.9. Let $f: [a, b] \to \mathbb{R}$ be Riemann integrable on $[a, b]$. Let c be such that $a < c < b$. Then f is integrable on $[a, c]$ and on $[c, b]$.

Proof. Suppose that f is Riemann integrable on $[a, b]$ and $a < c < b$. Let $\varepsilon > 0$. Since f is Riemann integrable on $[a, b]$, there is a partition Q of $[a, b]$ such that

$$U(f, Q) - L(f, Q) < \varepsilon.$$

By Theorem 6.1.9, we can assume that $c \in Q$. Let $P = Q \cap [a, c]$ and $P' = Q \cap [c, b]$. Thus, P is a partition of $[a, c]$, P' is a partition of $[c, b]$, and $Q = P \cup P'$. Hence,

$$U(f, Q) = U(f, P) + U(f, P')$$
$$L(f, Q) = L(f, P) + L(f, P'),$$

and thus,

$$U(f, Q) - L(f, Q) = [U(f, P) - L(f, P)] + [U(f, P') - L(f, P')] < \varepsilon.$$

We conclude that $U(f, P) - L(f, P) < \varepsilon$ and $U(f, P') - L(f, P') < \varepsilon$. Theorem 6.1.18 thus implies that f is Riemann integrable on $[a, c]$ and on $[c, b]$. □

Corollary 6.2.10. Let $f: [a, b] \to \mathbb{R}$ be Riemann integrable on $[a, b]$. Then for all c, d where $a \le c < d \le b$, we have that f is integrable on $[c, d]$.

Proof. See Exercise 3 (apply Theorem 6.2.9 twice, at most). □

We now prove the converse of Theorem 6.2.9

Theorem 6.2.11. Suppose that $f: [a, b] \to \mathbb{R}$ is Riemann integrable on $[a, c]$ and on $[c, b]$, where $a < c < b$. Then f is Riemann integrable on $[a, b]$ and $\int_a^b f = \int_a^c f + \int_c^b f$.

Proof. Let $f: [a, b] \to \mathbb{R}$ be Riemann integrable on $[a, c]$ and on $[c, b]$, where $a < c < b$. The proof will be done in two steps. In Step 1, we prove that f is Riemann integrable on $[a, b]$, and in Step 2 we prove $\int_a^b f = \int_a^c f + \int_c^b f$.

STEP 1: Let $\varepsilon > 0$. Since f is Riemann integrable on $[a, c]$ and on $[c, b]$, Theorem 6.1.18 implies that there are partitions P_1 of $[a, c]$ and P_2 of $[c, b]$ such that

$$U(f, P_1) - L(f, P_1) < \frac{\varepsilon}{2} \tag{6.25}$$

$$U(f, P_2) - L(f, P_2) < \frac{\varepsilon}{2}. \tag{6.26}$$

Let $P = P_1 \cup P_2$, which is a partition of $[a, b]$. Since P_1 and P_2 both have c as an end point, we conclude that

$$U(f, P) = U(f, P_1) + U(f, P_2) \qquad (6.27)$$
$$L(f, P) = L(f, P_1) + L(f, P_2), \qquad (6.28)$$

and thus,

$$U(f, P) - L(f, P) = [U(f, P_1) - L(f, P_1)] + [U(f, P_2) - L(f, P_2)] < \frac{\varepsilon}{2} + \frac{\varepsilon}{2} = \varepsilon.$$

So $U(f, P) - L(f, P) < \varepsilon$. Thus, f is Riemann integrable on $[a, b]$ by Theorem 6.1.18.

STEP 2: Now we prove that $\int_a^b f = \int_a^c f + \int_c^b f$. Let $\varepsilon > 0$. Arguing as in Step 1, there are partitions P_1, P_2, and $P = P_1 \cup P_2$ that satisfy (6.27), (6.28), and the inequality $U(f, P) - L(f, P) < \varepsilon$. Furthermore, Lemma 6.1.14 implies that

$$L(f, P_1) \le \int_a^c f \le U(f, P_1) \qquad (6.29)$$

$$L(f, P_2) \le \int_c^b f \le U(f, P_2) \qquad (6.30)$$

$$L(f, P) \le \int_a^b f \le U(f, P). \qquad (6.31)$$

After adding (6.29) and (6.30), equations (6.28) and (6.27) imply that

$$L(f, P) = L(f, P_1) + L(f, P_2) \le \int_a^c f + \int_c^b f \le U(f, P_1) + U(f, P_2) = U(f, P). \quad (6.32)$$

Since $U(f, P) - L(f, P) < \varepsilon$, inequalities (6.31) and (6.32) imply[5] that

$$\left| \left(\int_a^c f + \int_c^b f \right) - \int_a^b f \right| < \varepsilon.$$

Since $\varepsilon > 0$ was arbitrarily chosen, we must have $\int_a^b f = \int_a^c f + \int_c^b f$. □

The Riemann integral is defined for a function whose domain is identified to be a closed interval $[a, b]$, where $a < b$. We now introduce some standard conventions.

Definition 6.2.12. Let $f : [a, b] \to \mathbb{R}$ be Riemann integrable. Define $\int_a^a f = 0$ and $\int_b^a f = -\int_a^b f$.

Definition 6.2.12 and Theorem 6.2.11 now imply the following corollary.

Corollary 6.2.13. Let f be integrable on every closed subinterval of an interval I. Let a, b, c be in I. Then

$$\int_a^b f = \int_a^c f + \int_c^b f. \qquad (6.33)$$

---

[5]If $d \le x, y \le e$ and $e - d < \varepsilon$, then $|x - y| < \varepsilon$.

Proof. Let f be integrable on any closed subinterval of an interval I. Let a, b, c be in I. If $a = b$, $b = c$, or $a = c$, then (6.33) is easy to verify, using Definition 6.2.12. Assume that the points a, b, c are all distinct. These three points determine a subinterval of I; namely, the interval $J = [\min\{a, b, c\}, \max\{a, b, c\}]$, where a, b, c are in J. By assumption, f is integrable on the closed interval J. To prove (6.33), we must consider all the possible orderings of a, b, c. However, we will just consider one such ordering, as the proof is similar for all of the other orderings of a, b, c. So, suppose that $b < a < c$. Then, by Theorem 6.2.11, we have that (▲) $\int_b^c f = \int_b^a f + \int_a^c f$. Since $b < a < c$, Definition 6.2.12 and (▲) imply that $-\int_c^b f = -\int_a^b f + \int_a^c f$. Thus, by algebra, we have that $\int_a^b f = \int_a^c f + \int_c^b f$. □

6.2.4 The Composition Theorem

Theorem 4.1.13 implies that the composition of continuous functions is continuous. Moreover, Theorem 5.1.5 (the chain rule) shows that the composition of differentiable functions is differentiable. Is the composition of integrable functions integrable? The answer is, unfortunately, "not always." There are examples of integrable functions whose composition is not integrable (see Exercise 10). However, our next theorem identifies a condition under which the composition of Riemann integrable functions will be Riemann integrable.

Theorem 6.2.14 (Composition Theorem). If $f \colon [a, b] \to [c, d]$ is Riemann integrable and $g \colon [c, d] \to \mathbb{R}$ is continuous, then $(g \circ f) \colon [a, b] \to \mathbb{R}$ is Riemann integrable.

Proof. See Appendix A on page 235. □

The Composition Theorem is a very useful result. For example, it implies the following corollary which states that the square of a Riemann integrable function is also Riemann integrable.

Corollary 6.2.15. If $f \colon [a, b] \to \mathbb{R}$ is integrable, then f^2 is integrable on $[a, b]$.

Proof. Assume f is Riemann integrable. Thus, f is bounded. So, there exists $c, d \in \mathbb{R}$ such that $f([a, b]) \subseteq [c, d]$. Let $g \colon [c, d] \to \mathbb{R}$ be defined by $g(x) = x^2$. Since g is continuous (Exercise 5 on page 108), Theorem 6.2.14 implies that $g \circ f$ is integrable on $[a, b]$. Clearly, $g \circ f = f^2$ and thus, f^2 is Riemann integrable. □

Another important consequence of the Composition Theorem is that the product of Riemann integrable functions is Riemann integrable.

Corollary 6.2.16. If $f \colon [a, b] \to \mathbb{R}$ and $g \colon [a, b] \to \mathbb{R}$ are integrable, then fg is integrable on $[a, b]$.

Proof. Using algebra, we have $fg = \frac{1}{4}[(f + g)^2 - (f - g)^2]$. As f and g are Riemann integrable, Theorem 6.2.5, Corollary 6.2.6, and Corollary 6.2.15 imply that fg is Riemann integrable. □

The Composition Theorem also implies that the absolute value of a Riemann integrable function is Riemann integrable.

Corollary 6.2.17. If $f: [a, b] \to \mathbb{R}$ is integrable, then $|f|$ is integrable on $[a, b]$ and $\left| \int_a^b f \right| \le \int_a^b |f|$.

Proof. Let f be integrable. So f is bounded. Let $c, d \in \mathbb{R}$ be such that $f([a, b]) \subseteq [c, d]$. Let $g: [c, d] \to \mathbb{R}$ be defined by $g(x) = |x|$. As g is continuous (Exercise 4 on page 108), Theorem 6.2.14 implies that $g \circ f$ is integrable on $[a, b]$. Clearly, $g \circ f = |f|$ and so, $|f|$ is integrable. Since $-|f(x)| \le f(x) \le |f(x)|$ for all $x \in [a, b]$, Corollary 6.2.8 implies that $-\int_a^b |f| \le \int_a^b f \le \int_a^b |f|$. Therefore, $\left| \int_a^b f \right| \le \int_a^b |f|$. □

Exercises 6.2

1. Prove Lemma 6.2.1.

*2. Complete the proof of Lemma 6.2.2, by proving the lemma under the case $k < 0$.

*3. Prove Corollary 6.2.10.

4. Find a function f so that $|f|$ is integrable on an interval $[a, b]$ but f is not.

5. Let $f: [a, b] \to \mathbb{R}$ be Riemann integrable. Let $k > 0$ be such that $f(x) \ge k$ for all $x \in [a, b]$. Prove that $\frac{1}{f}$ is Riemann integrable on $[a, b]$.

6. Let $f: [a, b] \to \mathbb{R}$ be Riemann integrable. Suppose that $f(x) \ge 0$ for all $x \in [a, b]$. Prove that $\sqrt{f}$ is Riemann integrable on $[a, b]$.

7. Let $f: [a, b] \to \mathbb{R}$, $h: [a, b] \to \mathbb{R}$, $g: [a, b] \to \mathbb{R}$ be such that $f(x) \le h(x) \le g(x)$ for all $x \in [a, b]$. Suppose that f and g are integrable on $[a, b]$ and $\int_a^b f = \int_a^b g$. Prove that h is integrable and $\int_a^b h = \int_a^b f$.

8. Let f and g be Riemann integrable on $[c, d]$. Prove that

$$\left| \int_c^d fg \right| \le \left[\left(\int_c^d f^2 \right) \left(\int_c^d g^2 \right) \right]^{1/2}.$$

9. Let f and g be Riemann integrable on $[c, d]$. Prove that

$$\left[\int_c^d (f + g)^2 \right]^{1/2} \le \left(\int_c^d f^2 \right)^{1/2} + \left(\int_c^d g^2 \right)^{1/2}.$$

*10. Let $f: [0, 1] \to [0, 1]$ be the integrable function defined in Exercise 10 on page 162. The function $g: [0, 1] \to \mathbb{R}$ defined in Example 6.1.20 on page 159 is integrable. Show that $g \circ f = h$, where h is defined in Example 6.1.23 on page 160. Conclude that $g \circ f$ is not Riemann integrable on $[0, 1]$.

*11. Let $g: [a, b] \to \mathbb{R}$ be continuous, and let $f: [a, b] \to [a, b]$ be defined by $f(x) = x$. Show that if f Riemann integrable, then g is Riemann integrable.

Exercise Notes: For Exercise 1, use Definition 6.1.3 and Theorem 2.3.13. For Exercise 4, consider the function $f: [0, 1] \to \mathbb{R}$ in Exercise 7 on page 161. For Exercise 5, apply Theorem 6.2.14. For Exercise 7, see Exercise 3 on page 161. For Exercise 8, first note that $(yf + g)^2 \geq 0$ for any real number y. Thus, one can apply Theorem 6.2.7. Expand the integral $\int_c^d (yf + g)^2$. By algebra, if $ay^2 + by + c \geq 0$ for all y, then the equation $ay^2 + by + c = 0$ can have at most one real root. So, by the quadratic formula, we must have $b^2 - 4ac \leq 0$. For Exercise 9, expand the integral on the left side of $\leq$ and then use Exercise 8. Also note that $a + 2\sqrt{a}\sqrt{b} + b = (\sqrt{a} + \sqrt{b})^2$ when $a, b \geq 0$. For Exercise 11, use Theorem 6.2.14.

6.3 FAMILIES OF INTEGRABLE FUNCTIONS

It is useful to identify certain functional properties which, if satisfied, will ensure that a function is integrable. In this section, we show that continuous functions, monotone functions, and functions of bounded variation are Riemann integrable. We also derive a mean value theorem for integrals.

6.3.1 Continuous Functions

Theorem 6.2.14 indirectly implies that a continuous function defined on a closed interval is Riemann integrable (see Exercise 11 on page 169). Nevertheless, we will present a direct proof of this important result. In our proof, we will apply Theorem 6.1.18.

Theorem 6.3.1. If $f \colon [a, b] \to \mathbb{R}$ is continuous, then f is integrable on $[a, b]$.

Proof. Let $f \colon [a, b] \to \mathbb{R}$ be continuous. By Theorem 4.4.1 f is bounded. Let $\varepsilon > 0$. As f is uniformly continuous on $[a, b]$ by Theorem 4.5.5, there is a $\delta > 0$ such that

$$\text{for all } x, x' \in [a, b], \text{ if } |x - x'| < \delta, \text{ then } |f(x) - f(x')| < \frac{\varepsilon}{b - a}. \tag{6.34}$$

Let $P = \{x_0, x_1, x_2, \ldots, x_n\}$ be a partition of $[a, b]$ such that (▲) $\Delta x_i = (x_i - x_{i-1}) < \delta$ for each $i = 1, 2, \ldots, n$. Now, for each such i, let $M_i = M_i(f)$ and $m_i = m_i(f)$ be as in Definition 6.1.3. By Theorem 4.4.3, f attains its maximum and minimum values on each subinterval $[x_{i-1}, x_i]$. Thus, for each $i = 1, 2, \ldots, n$, let $t_i, s_i \in [x_{i-1}, x_i]$ be such that (▶) $M_i = f(t_i)$ and $m_i = f(s_i)$. Clearly, $|t_i - s_i| \leq (x_i - x_{i-1}) = \Delta x_i < \delta$ by (▲). Hence, by (6.34), we have (▼) $f(t_i) - f(s_i) < \frac{\varepsilon}{b-a}$ for all $i = 1, 2, \ldots, n$. Thus,

$$
\begin{aligned}
U(f, P) - L(f, P) &= \sum_{i=1}^n (M_i - m_i)\Delta x_i && \text{by Lemma 6.1.7} \\
&= \sum_{i=1}^n (f(t_i) - f(s_i))\Delta x_i && \text{by (▶)} \\
&< \sum_{i=1}^n \frac{\varepsilon}{b - a}\Delta x_i && \text{by (▼)} \\
&= \frac{\varepsilon}{b - a} \sum_{i=1}^n \Delta x_i && \text{by algebra} \\
&= \frac{\varepsilon}{b - a}(b - a) = \varepsilon && \text{by Lemma 6.1.8.}
\end{aligned}
$$

Theorem 6.1.18 now implies that f is Riemann integrable on $[a, b]$. □

Corollary 6.3.2. Let $k \in \mathbb{R}$ and $f \colon [a, b] \to \mathbb{R}$ be the constant function defined by $f(x) = k$ for all x in $[a, b]$. Then f is Riemann integrable and $\int_a^b f = k(b - a)$.

Proof. Since f is continuous, it is integrable. Because $k \le f(x) \le k$ for all x in $[a, b]$, Lemma 6.1.15 implies $k(b - a) \le \int_a^b f \le k(b - a)$ and therefore, $\int_a^b f = k(b - a)$. $\square$

We can now show that the integral of a positive continuous function f over $[a, b]$ is equal to the area of a rectangle with base $[a, b]$ and height $f(c)$ for some $c \in [a, b]$.

Theorem 6.3.3 (Mean Value Theorem for Integrals). If $f \colon [a, b] \to \mathbb{R}$ is continuous, then there is a point c in $[a, b]$ such that $\int_a^b f = f(c)(b - a)$.

Proof. Let $f \colon [a, b] \to \mathbb{R}$ be continuous. If f is a constant function, the result follows from Corollary 6.3.2. Assume that f is not constant. By Theorem 4.4.3, f attains a maximum value M and a minimum value m on $[a, b]$. Thus, $f(i) = m$ and $f(j) = M$ for some $i, j \in [a, b]$. As $m \le f(x) \le M$ for all x in $[a, b]$, Lemma 6.1.15 asserts that $m(b - a) \le \int_a^b f \le M(b - a)$. Hence, $m \le \frac{\int_a^b f}{(b-a)} \le M$. Theorem 4.4.6 implies that $f(c) = \frac{\int_a^b f}{(b-a)}$ for some c in $[a, b]$. So $\int_a^b f = f(c)(b - a)$. $\square$

We end this section with four applications of Theorem 6.3.3. The first three of these applications show how a strict inequality that holds for a function, affects the integral of the function.

Corollary 6.3.4. If $f \colon [a, b] \to \mathbb{R}$ is continuous and $f(x) > 0$ for all x in $[a, b]$, then $\int_a^b f > 0$.

Proof. Let $f \colon [a, b] \to \mathbb{R}$ be continuous such that $f(x) > 0$ for all x in $[a, b]$. By Theorem 6.3.3, $\int_a^b f = f(c)(b - a)$ for some c in $[a, b]$. Since $a < b$ and $f(c) > 0$, we conclude that $\int_a^b f > 0$. $\square$

Corollary 6.3.5. Let $f \colon [a, b] \to \mathbb{R}$ be continuous and $f(x) \ge 0$ for all x in $[a, b]$. If $f(x_0) > 0$ for some $x_0 \in [a, b]$, then $\int_a^b f > 0$.

Proof. Let $f \colon [a, b] \to \mathbb{R}$ be continuous, $f(x) \ge 0$ for all x in $[a, b]$, and $f(x_0) > 0$ for some $x_0 \in [a, b]$. Lemma 4.1.14 implies that there are real numbers c, d such that $a \le c < x_0 < d \le b$ and $f(x) > 0$ for all x in $[c, d]$. Corollary 6.2.10 asserts that f is integrable on $[c, d]$. Thus, $\int_c^d f > 0$ by Corollary 6.3.4. Corollary 6.2.13 implies that $\int_a^b f = \int_a^c f + \int_c^d f + \int_d^b f$. From Theorem 6.2.7 we infer that $\int_a^b f > 0$. $\square$

Corollaries 6.3.5 and 6.2.6 imply the following result.

Corollary 6.3.6. Let $f \colon [a, b] \to \mathbb{R}$ and $g \colon [a, b] \to \mathbb{R}$ be continuous such that $f(x) \le g(x)$ for all x in $[a, b]$. If $f(x_0) < g(x_0)$ for some $x_0 \in [a, b]$, then $\int_a^b f < \int_a^b g$.

Corollary 6.3.7. Let $f \colon [a, b] \to \mathbb{R}$ be continuous. If $\int_a^b f = 0$ and $f(x) \ge 0$ for each x in $[a, b]$, then $f(x) = 0$ for all x in $[a, b]$.

Proof. Let $f \colon [a, b] \to \mathbb{R}$ be continuous, $\int_a^b f = 0$, and $f(x) \ge 0$ for all x in $[a, b]$. Suppose, for a contradiction, that $f(x_0) > 0$ for some $x_0 \in [a, b]$. Corollary 6.3.5 implies that $\int_a^b f > 0$ which contradicts the fact that $\int_a^b f = 0$. $\square$

6.3.2 Monotone Functions

We now know that every continuous function, defined on a closed interval $[a, b]$, is Riemann integrable. The set of all monotone functions on an interval $[a, b]$ is another large collection of Riemann integrable functions. Moreover, there are many monotone functions that are not continuous. Recall that a monotone function is one that is either decreasing or increasing.

Theorem 6.3.8. If $f: [a, b] \to \mathbb{R}$ is monotone, then f is Riemann integrable on $[a, b]$.

Proof. Let $f: [a, b] \to \mathbb{R}$ be increasing. Thus, $f(a) \le f(x) \le f(b)$ for all $x \in [a, b]$. So f is bounded. Let $\varepsilon > 0$. Let $k > 0$ be such that (▲) $k(f(b) - f(a)) < \varepsilon$ and $P = \{x_0, x_1, x_2, \ldots, x_n\}$ be a partition of $[a, b]$ such that (►) $\Delta x_i = (x_i - x_{i-1}) \le k$ for each $i = 1, 2, \ldots, n$. Since f is increasing, we have that (▼) $m_i = f(x_{i-1})$ and $M_i = f(x_i)$ for all $i = 1, 2, \ldots, n$. Thus,

$$
\begin{aligned}
U(f, P) - L(f, P) &= \sum_{i=1}^{n}(M_i - m_i)\Delta x_i && \text{by Lemma 6.1.7} \\
&\le \sum_{i=1}^{n}(M_i - m_i)k && \text{by (►)} \\
&= \sum_{i=1}^{n}(f(x_i) - f(x_{i-1}))k && \text{by (▼)} \\
&= k\sum_{i=1}^{n}(f(x_i) - f(x_{i-1})) && \text{by algebra} \\
&= k(f(b) - f(a)) && \text{by Exercise 9 on page 27} \\
&< \varepsilon && \text{by (▲).}
\end{aligned}
$$

Thus, Theorem 6.1.18 implies that f is Riemann integrable. If f is decreasing, then $-f$ is increasing and hence, $-f$ is integrable by the above argument. Corollary 6.2.2 now implies that f is Riemann integrable. $\qquad\square$

6.3.3 Functions of Bounded Variation

We will now investigate the concept of a function having bounded variation. Such functions play a important role in the theory of integration and are closely related to monotone functions. The concept of a partition of an interval (see Definition 6.1.1) is needed in order to define a function of bounded variation. Our main goal in this section is to show that a function of bounded variation is Riemann integrable.

Definition 6.3.9. Let $f: [a, b] \to \mathbb{R}$ and let $[c, d]$ be a closed subinterval of $[a, b]$. Let P be a partition of $[c, d]$. The P-**variation** of f over $[c, d]$, denoted by $V(f, P, [c, d])$, is defined to be the real number

$$
V(f, P, [c, d]) = \sum_{1}^{n}|f(x_i) - f(x_{i-1})|
$$

where $P = \{x_i : 0 \le i \le n\}$.

One can increase a given variation of a function by using a refinement.

Lemma 6.3.10. Let $f: [a, b] \to \mathbb{R}$ and $[c, d]$ be a closed subinterval of $[a, b]$. Let P be a partition of $[c, d]$. If P^* is a refinement of P, then $V(f, P, [c, d]) \leq V(f, P^*, [c, d])$.

Proof. Let $f: [a, b] \to \mathbb{R}$, $[c, d]$, and P be as stated. Let $P = \{x_0, x_1, x_2, \ldots, x_n\}$ and P^* be a refinement of P. So, the P-variation of f over $[c, d]$ is

$$V(f, P, [c, d]) = \sum_{1}^{n} |f(x_i) - f(x_{i-1})|. \tag{6.35}$$

We will prove that $V(f, P, [c, d]) \leq V(f, P^*, [c, d])$. First suppose that P^* contains just one more point than P. So $P^* = P \cup \{x^*\}$ and $x_{k-1} < x^* < x_k$ for some natural number k where $1 \leq k \leq n$. In equation (6.2), let us "isolate" the value $|f(x_k) - f(x_{k-1})|$ in the sum (6.35) for $V(f, P, [c, d])$ by rewriting this sum as

$$V(f, P, [c, d]) = \left(\sum_{i \neq k}^{n} |f(x_i) - f(x_{i-1})| \right) + |f(x_k) - f(x_{k-1})|$$

$$= \left(\sum_{i \neq k}^{n} |f(x_i) - f(x_{i-1})| \right) + |f(x_k) - f(x^*) + f(x^*) - f(x_{k-1})|$$

$$\leq \left(\sum_{i \neq k}^{n} |f(x_i) - f(x_{i-1})| \right) + |f(x_k) - f(x^*)| + |f(x^*) - f(x_{k-1})|$$

$$= V(f, P^*, [c, d]).$$

Therefore, $V(f, P, [c, d]) \leq V(f, P^*, [c, d])$ when P^* contains just one more point than P. If a partition P^* contains $j > 1$ more points than P, then repeat the previous reasoning j times to arrive at the conclusion that $V(f, P, [c, d]) \leq V(f, P^*, [c, d])$. $\square$

Definition 6.3.11. Let $f: [a, b] \to \mathbb{R}$ and let $[c, d]$ be a closed subinterval of $[a, b]$. The function f is of **bounded variation** on $[c, d]$ if there is an $M \in \mathbb{R}$ such that $V(f, P, [c, d]) \leq M$ for all partitions P of $[c, d]$.

If a function is of bounded variation on an interval $[a, b]$, then it will be of bounded variation on any closed subinterval of $[a, b]$.

Lemma 6.3.12. Let $f: [a, b] \to \mathbb{R}$. Suppose that f is of bounded variation on $[a, b]$. Let $[c, d]$ be a subinterval of $[a, b]$. Then f is of bounded variation on $[c, d]$.

Proof. Assume that $f: [a, b] \to \mathbb{R}$ is of bounded variation on $[a, b]$. Let $M \in \mathbb{R}$ be such that $V(f, Q, [c, d]) \leq M$ for all partitions Q of $[a, b]$. Let $[c, d]$ be a subinterval of $[a, b]$ and let P be a partition of $[c, d]$. Hence, $Q = P \cup \{a, b\}$ is a partition of $[a, b]$. It easily follows that $V(f, P, [c, d]) \leq V(f, Q, [a, b]) \leq M$. Therefore, f is of bounded variation on $[c, d]$. $\square$

Definition 6.3.13. Let $f: [a, b] \to \mathbb{R}$ be of bounded variation on $[c, d]$, a subinterval of $[a, b]$. The **variation** of f over $[c, d]$, denoted by $V(f, [c, d])$, is defined by

$$V(f, [c, d]) = \sup\{V(f, P, [c, d]) : P \text{ is a partition of } [c, d]\}.$$

The variation of a function measures how much its graph "moves up and down." In other words, if a particle travels along the entire graph of the function, then the variation identifies the total *vertical* distance traveled by the particle.

Let $f\colon [a,b] \to \mathbb{R}$ be of bounded variation on $[a,b]$. Since $P = \{a,b\}$ is a partition of $[a,b]$ and $V(f,P,[a,b]) = |f(b) - f(a)|$, from Definition 6.3.13 we see that

$$|f(b) - f(a)| \le V(f,[a,b]). \tag{6.36}$$

Lemma 6.3.14. If $f\colon [a,b] \to \mathbb{R}$ is monotone, then f is of bounded variation on $[a,b]$ and $V(f,[a,b]) = |f(b) - f(a)|$.

Proof. Let $f\colon [a,b] \to \mathbb{R}$ be monotone. In particular, suppose that f is increasing. Let $P = \{x_0, x_1, x_2, \ldots, x_n\}$ be a partition of $[a,b]$. Then

$$
\begin{aligned}
V(f,P,[a,b]) &= \sum_{i=1}^{n} |f(x_i) - f(x_{i-1})| \\
&= \sum_{i=1}^{n} (f(x_i) - f(x_{i-1})) \quad \text{as } f(x_i) - f(x_{i-1}) \ge 0 \text{ for each } i \\
&= f(b) - f(a) \qquad\qquad \text{by Exercise 9 on page 27} \\
&= |f(b) - f(a)| \qquad\quad\ \text{as } f(b) - f(a) \ge 0.
\end{aligned}
$$

Thus, $V(f,P,[a,b]) = |f(b) - f(a)|$ for every partition P of $[a,b]$. We conclude that $V(f,[a,b]) = |f(b) - f(a)|$. A similar argument applies if f is decreasing. □

We will eventually show that a function of bounded variation is Riemann integral. The next result is a first step in this direction.

Lemma 6.3.15. If $f\colon [a,b] \to \mathbb{R}$ is of bounded variation on $[a,b]$, then f is bounded.

Proof. Assume that $f\colon [a,b] \to \mathbb{R}$ is of bounded variation on $[a,b]$. Thus, there is an $M \in \mathbb{R}$ such that such that $V(f,P,[a,b]) \le M$ for all partitions P of $[a,b]$. Let $x \in [a,b]$. Clearly, $P = \{a,x,b\}$ is a partition of $[a,b]$. Thus,

$$V(f,P,[a,b]) = |f(x) - f(a)| + |f(b) - f(x)| \le M.$$

Thus, $|f(x)| - |f(a)| \le |f(x) - f(a)| \le M$. Hence, $|f(x)| \le M + |f(a)|$ for all $x \in [a,b]$. Therefore, f is bounded. □

We now identify three algebraic closure properties which hold for functions of bounded variation.

Theorem 6.3.16. Let $f\colon [a,b] \to \mathbb{R}$ and $g\colon [a,b] \to \mathbb{R}$ be of bounded variation on $[a,b]$. Then

(1) kf is of bounded variation on $[a,b]$, for any $k \in \mathbb{R}$.
(2) $f + g$ is of bounded variation on $[a,b]$.
(3) fg is of bounded variation on $[a,b]$.

Let $c \in (a, b)$. Lemma 6.3.12 shows that if f is of bounded variation on $[a, b]$, then f is also of bounded variation on the subintervals of $[a, c]$ and $[c, b]$. The converse holds as well. Moreover, there is a simple equation that relates the total variation of f on these three intervals.

Theorem 6.3.17. Let $f \colon [a, b] \to \mathbb{R}$ and $c \in (a, b)$. If f is of bounded variation on $[a, c]$ and on $[c, b]$, then f is of bounded variation on $[a, b]$ and

$$V(f, [a, b]) = V(f, [a, c]) + V(f, [c, b]).$$

Proof. Let $f \colon [a, b] \to \mathbb{R}$ and $c \in (a, b)$. Assume that f is of bounded variation on $[a, c]$ and on $[c, b]$. Let P be a partition of $[a, b]$. Let $P^* = P \cup \{c\}$. So P^* is a refinement of P. Now let $Q = \{x \in P^* : x \leq c\}$ and $S = \{x \in P^* : x \geq c\}$. Clearly, $P^* = Q \cup S$, Q is a partition of $[a, c]$, and S is a partition of $[c, b]$. Therefore,

$$
\begin{aligned}
V(f, P, [a, b]) &\leq V(f, P^*, [a, b]) && \text{by Lemma 6.3.10} \\
&= V(f, Q, [a, c]) + V(f, S, [c, b]) && \text{by definition of } Q \text{ and } S \\
&\leq V(f, [a, c]) + V(f, [c, b]) && \text{by Definition 6.3.13.}
\end{aligned}
$$

So $V(f, [a, c]) + V(f, [c, b])$ is an upper bound for the set

$$\{V(f, P, [a, b]) : P \text{ is a partition of } [a, b]\}.$$

Thus, f is of bounded variation on $[a, b]$ and $V(f, [a, b]) \leq V(f, [a, c]) + V(f, [c, b])$.
Suppose, for a contradiction, that $V(f, [a, b]) < V(f, [a, c]) + V(f, [c, b])$. Then, as

$$V(f, [a, b]) - V(f, [c, b]) < V(f, [a, c]),$$

there is a partition Q' of $[a, c]$ such that

$$V(f, [a, b]) - V(f, [c, b]) < V(f, Q', [a, c]).$$

Since $V(f, [a, b]) - V(f, Q', [a, c]) < V(f, [c, b])$, there is a partition S' of $[c, b]$ such that

$$V(f, [a, b]) - V(f, Q', [a, c]) < V(f, S', [c, b]).$$

Hence,

$$V(f, [a, b]) < V(f, Q', [a, c]) + V(f, S', [c, b]) = V(f, Q' \cup S', [a, b]),$$

which contradicts Definition 6.3.13 because $Q' \cup S'$ is a partition of $[a, b]$. Therefore, we must have that $V(f, [a, b]) = V(f, [a, c]) + V(f, [c, b])$. $\qquad \square$

Corollary 6.3.18. If f is of bounded variation on $[a, b]$ and $a \leq c \leq d \leq b$, then $V(f, [a, c]) \leq V(f, [a, d])$.

Proof. Theorem 6.3.17, applied to the interval $[a, d]$, implies that

$$V(f, [a, c]) + V(f, [c, d]) = V(f, [a, d]).$$

Since $V(f, [c, d]) \geq 0$, we see that $V(f, [a, c]) \leq V(f, [a, d])$. $\qquad \square$

Corollary 6.3.19. If f is of bounded variation on $[a,b]$ and $a \le c \le d \le b$, then $V(f,[a,c]) - f(c) \le V(f,[a,d]) - f(d)$.

Proof. Theorem 6.3.17, applied to the interval $[a,d]$, implies that

$$V(f,[c,d]) = V(f,[a,d]) - V(f,[a,c]). \tag{6.37}$$

By (6.36) on page 174, we have

$$f(d) - f(c) \le |f(d) - f(c)| \le V(f,[c,d]). \tag{6.38}$$

Thus, (6.37) and (6.38) imply that $V(f,[a,c]) - f(c) \le V(f,[a,d]) - f(d)$. $\qquad\square$

Definition 6.3.20. Let f be of bounded variation on the interval $[a,b]$. Define the functions $V: [a,b] \to \mathbb{R}$ and $W: [a,b] \to \mathbb{R}$ by

- $V(x) = V(f,[a,x])$ and
- $W(x) = V(f,[a,x]) - f(x)$

for all $x \in [a,b]$.

Corollaries 6.3.18 and 6.3.19 imply that the above functions V and W are increasing and, since $f = V - W$, we have the following two theorems.

Theorem 6.3.21. If f is of bounded variation on $[a,b]$, then f is the difference of two increasing functions on $[a,b]$.

Theorem 6.3.22. If f is of bounded variation on $[a,b]$, then f is integrable on $[a,b]$.

Proof. This follows from Theorem 6.3.21, Theorem 6.3.8, and Corollary 6.2.6. $\qquad\square$

Exercises 6.3

1. Let f and g be continuous on $[a,b]$ where $\int_a^b f = \int_a^b g$. Prove that $f(c) = g(c)$ for some $c \in [a,b]$.

2. Let f and g be integrable on $[a,b]$. Suppose that $g(x) \ge 0$ for all $x \in [a,b]$. Prove that if $\int_a^b g = 0$, then $\int_a^b fg = 0$.

3. Let f be continuous on $[a,b]$ and let g be Riemann integrable on $[a,b]$. Suppose that $g(x) \ge 0$ for all $x \in [a,b]$ and $\int_a^b g > 0$. Prove that $\int_a^b fg = f(c) \int_a^b g$ for some $c \in [a,b]$.

4. Complete the proof of Theorem 6.3.8, by proving that the theorem holds when $f: [a,b] \to \mathbb{R}$ is decreasing.

5. Prove Corollary 6.3.6.

6. Suppose that $f: [a,b] \to \mathbb{R}$ and $g: [c,d] \to [a,b]$ are both monotone. Prove that $(f \circ g): [c,d] \to \mathbb{R}$ is Riemann integrable.

7. Let f be a function of bounded variation on $[a,b]$. Prove that $|f|$ is of bounded variation on $[a,b]$.

8. Prove part (1) of Theorem 6.3.16.

9. Prove part (2) of Theorem 6.3.16.

10. Prove part (3) of Theorem 6.3.16.

11. Let $f\colon [a,b] \to \mathbb{R}$ be of bounded variation on $[a,b]$, and let $V\colon [a,b] \to \mathbb{R}$ be as defined in Definition 6.3.20.

    (a) Prove the V is of bounded variation on $[a,b]$.
    (b) Let $c \in [a,b]$. Prove that if V is continuous at c, then f is continuous at c.
    (c) Prove that if f is continuous at $c \in [a,b]$, then V is continuous at c.
    (d) Suppose that f is continuous on $[a,b]$. Prove that f is the difference of two increasing continuous functions on $[a,b]$.

12. Let $f\colon [a,b] \to \mathbb{R}$ be a function of bounded variation on $[a,b]$. Prove that f has one-sided limits at each point $c \in [a,b]$

Exercise Notes: For Exercise 1, use the function $h = f - g$ and Theorem 6.3.3. For Exercise 2, use Corollary 6.2.8 and the fact that f is bounded. For Exercise 3, read and modify the proof of Theorem 6.3.3, using Corollary 6.2.8. For Exercise 10, see the note for Exercise 15 on page 109. For Exercise 11(c), let $c \in (a,b)$. Since V is increasing, we see that $\lim_{x \to c^-} V(x) \le V(c) \le \lim_{x \to c^+} V(x)$ (see Exercise 14 on page 118). Show that these inequalities are not strict. For Exercise 12, see Exercises 12 and 13 on page 118.

6.4 THE FUNDAMENTAL THEOREM OF CALCULUS

The Fundamental Theorem of Calculus, while taught early in elementary calculus courses, is actually a very deep result that identifies a key relationship between two central operations: differentiation and integration. Actually, there are two parts to this important theorem. The first part allows one to evaluate the definite integral of a function by using its antiderivative. The second part shows that continuous functions have antiderivatives.

Definition 6.4.1. Let $f\colon I \to \mathbb{R}$ where I is an interval. A function $F\colon I \to \mathbb{R}$ is an **antiderivative** of f if $F'(x) = f(x)$ for all $x \in I$.

Corollary 5.2.5 implies that any two antiderivatives of f differ by a constant on I. Moreover, if $f\colon I \to \mathbb{R}$ is differentiable on I, then f is an antiderivative of f'.

6.4.1 Evaluating Riemann Integrals

The definition of the Riemann integral of a function g is rather complicated. Moreover, in order to evaluate this integral one needs to evaluate the obscure value $U(g)$. Suppose that g has an antiderivative over a closed interval. Our first fundamental theorem shows that one can evaluate the Riemann integral of this function simply by evaluating its antiderivative at the endpoints of the interval and then perform a subtraction. It is remarkable that Riemann integration, a complex operation, can thus be reduced to antidifferentiation.

Theorem 6.4.2 (Fundamental Theorem of Calculus I). Let $f\colon [a,b] \to \mathbb{R}$ be differentiable on $[a,b]$. If f' is Riemann integrable on $[a,b]$, then $\int_a^b f' = f(b) - f(a)$.

Proof. Let $f\colon [a,b] \to \mathbb{R}$ be differentiable on $[a,b]$. By Corollary 5.1.3, f is continuous on $[a,b]$. Assume that f' is Riemann integrable on $[a,b]$. Thus, $\int_a^b f' = L(f') = U(f')$. We will show that $L(f') = U(f') = f(b) - f(a)$. Let $P = \{x_0, x_1, x_2, \ldots, x_n\}$ be any partition of $[a,b]$. Since f is continuous on each subinterval $[x_{i-1}, x_i]$, Theorem 5.2.3 implies that there exists a $t_i \in (x_{i-1}, x_i)$ such that $f'(t_i) = \frac{f(x_i) - f(x_{i-1})}{x_i - x_{i-1}}$; that is,

$$f(x_i) - f(x_{i-1}) = f'(t_i)(x_i - x_{i-1}).$$

By Exercise 9 on page 27, we see that $\sum_{i=1}^n (f(x_i) - f(x_{i-1})) = f(b) - f(a)$. Thus,

$$f(b) - f(a) = \sum_{i=1}^n (f(x_i) - f(x_{i-1})) = \sum_{i=1}^n f'(t_i)(x_i - x_{i-1}) = \sum_{i=1}^n f'(t_i)\Delta x_i. \quad (6.39)$$

Note that $m_i(f') \le f'(t_i) \le M_i(f')$ for each $i = 1, \ldots, n$. Hence,

$$L(f', P) = \sum_{i=1}^n m_i(f')\Delta x_i \le \sum_{i=1}^n f'(t_i)\Delta x_i \le \sum_{i=1}^n M_i(f')\Delta x_i = U(f', P). \quad (6.40)$$

Equation (6.39) yields $f(b) - f(a) = \sum_{i=1}^n f'(t_i)\Delta x_i$. So from (6.40) we infer that

$$L(f', P) \le f(b) - f(a) \le U(f', P). \quad (6.41)$$

Since P was arbitrary, we conclude that (6.41) holds for all partitions P. Thus,

$$L(f') \le f(b) - f(a) \le U(f'),$$

by Definition 6.1.12. Since f' is integrable, $L(f') = U(f')$ by Definition 6.1.13. Hence, $L(f') = f(b) - f(a) = U(f')$. Therefore, $\int_a^b f' = f(b) - f(a)$. □

6.4.2 Continuous Functions have Antiderivatives

If f is Riemann integrable on the closed interval $[a,b]$, then Theorem 6.2.9 implies that f is Riemann integrable on every subinterval of $[a,b]$. So, in particular, f is Riemann integrable on the subinterval $[a,x]$ for each $x \in (a,b]$. Thus, we can define a function $F\colon [a,b] \to \mathbb{R}$ by $F(x) = \int_a^x f$, where $F(a) = 0$ by Definition 6.2.12. We now show that the function F is uniformly continuous.

Theorem 6.4.3. Let $f\colon [a,b] \to \mathbb{R}$ be Riemann integrable. Define the function $F\colon [a,b] \to \mathbb{R}$ by $F(x) = \int_a^x f$. Then F is uniformly continuous on $[a,b]$.

Proof. Let $f\colon [a,b] \to \mathbb{R}$ be integrable. Let $\varepsilon > 0$. Since f is integrable, f is bounded (Definition 6.1.13). So let $B > 0$ be such that (▲) $|f(x)| \le B$ for all $x \in [a,b]$. Let $\delta = \frac{\varepsilon}{B}$. Let $c, x \in [a,b]$. Assume that $|x - c| < \delta$. Thus,

$$
\begin{aligned}
|F(x) - F(c)| &= \left| \int_a^x f - \int_a^c f \right| && \text{by definition of } F \\
&= \left| \left(\int_a^c f + \int_c^x f \right) - \int_a^c f \right| && \text{by Corollary 6.2.13} \\
&= \left| \int_c^x f \right| && \text{by algebra} \\
&\le B\,|x - c| && \text{by (▲) and Lemma 6.1.16} \\
&< B\delta = \varepsilon && \text{as } |x - c| < \delta \text{ and } \delta = \frac{\varepsilon}{B}.
\end{aligned}
$$

Therefore, $|F(x) - F(c)| < \varepsilon$ and F is uniformly continuous on $[a,b]$. □

Let f be continuous on $[a,b]$. Our second fundamental theorem implies that the function F defined in Lemma 6.4.3 is an antiderivative of f (see Theorem 6.4.6).

Theorem 6.4.4 (Fundamental Theorem of Calculus II). Suppose that $f\colon [a,b] \to \mathbb{R}$ is Riemann integrable. Define the function $F\colon [a,b] \to \mathbb{R}$ by $F(x) = \int_a^x f$. If f is continuous at $c \in [a,b]$, then F is differentiable at c and $F'(c) = f(c)$.

Proof. Let $f\colon [a,b] \to \mathbb{R}$ be integrable, and continuous at $c \in [a,b]$. We prove that

$$ \lim_{x \to c} \frac{F(x) - F(c)}{x - c} = f(c). $$

To do this, let $\varepsilon > 0$. Since f is continuous at c, there is a $\delta > 0$ such that

$$ |f(t) - f(c)| < \frac{\varepsilon}{2} \text{ when } |t - c| < \delta \text{ and } t \in [a,b]. \tag{6.42} $$

Let $x \in [a,b]$ and assume that $0 < |x - c| < \delta$. Thus,

$$
\begin{aligned}
\left| \frac{F(x) - F(c)}{x - c} - f(c) \right| &= \left| \frac{1}{(x-c)} \left(\int_a^x f - \int_a^c f \right) - f(c) \right| && \text{by definition of } F \\
&= \left| \frac{1}{(x-c)} \left(\int_a^x f + \int_c^a f \right) - f(c) \right| && \text{by Def. 6.2.12} \\
&= \left| \frac{1}{(x-c)} \left(\int_c^x f \right) - f(c) \right| && \text{by Cor. 6.2.13} \\
&= \left| \frac{1}{(x-c)} \int_c^x f(t)\,dt - f(c) \right| && \text{by Def. 6.1.13} \\
&= \left| \frac{1}{(x-c)} \int_c^x f(t)\,dt - \frac{1}{(x-c)} \int_c^x f(c)\,dt \right| && \text{by Cor. 6.3.2} \\
&= \frac{1}{|x-c|} \left| \int_c^x f(t)\,dt - \int_c^x f(c)\,dt \right| && \text{by algebra} \\
&= \frac{1}{|x-c|} \left| \int_c^x (f(t) - f(c))\,dt \right| && \text{by Thm. 6.2.5.}
\end{aligned}
$$

Therefore,

$$\left| \frac{F(x) - F(c)}{x - c} - f(c) \right| = \frac{1}{|x - c|} \left| \int_c^x (f(t) - f(c)) \, dt \right| \leq \frac{1}{|x - c|} \cdot \frac{\varepsilon}{2} \cdot |x - c| < \varepsilon$$

by (6.42) and Lemma 6.1.16. Thus, by Definition 5.1.1, $F'(c) = f(c)$. □

6.4.3 Techniques of Antidifferentiation

Let $f \colon I \to \mathbb{R}$ where I is an interval. Recall that a function $F \colon I \to \mathbb{R}$ is said to be an antiderivative of f if $F'(x) = f(x)$ for all $x \in I$. Before we continue, we will slightly extend the notion of being Riemann integrable.

Definition 6.4.5. Let $f \colon I \to \mathbb{R}$ where I is an interval. We say that f is **integrable over** I if f is Riemann integrable on $[a, b]$ for every $[a, b] \subseteq I$.

If $f \colon I \to \mathbb{R}$ is integrable over the interval I, then for each $a \in I$ we can define the function $F \colon I \to \mathbb{R}$ by $F(x) = \int_a^x f$. Our next theorem shows that if $f \colon I \to \mathbb{R}$ is continuous, then F is an antiderivative of f.

Theorem 6.4.6. Let $f \colon I \to \mathbb{R}$ be continuous on the interval I. Let $F \colon I \to \mathbb{R}$ be defined by $F(x) = \int_a^x f$, where $a \in I$. Then F is differentiable on I and $F'(x) = f(x)$ for all $x \in I$.

Proof. Let $f \colon I \to \mathbb{R}$ be continuous on I. By Theorem 6.3.1, f is integrable over I. Let $a \in I$ and define the function $F \colon I \to \mathbb{R}$ by $F(x) = \int_a^x f$. Let $x \in I$ and $b \in I$ be such that $b \neq a$ and x is between a and b. If $a \leq x \leq b$, then Theorem 6.4.4 implies that $F'(x) = f(x)$. If $b \leq x \leq a$, then Corollary 6.2.13 implies that $\int_a^b f = \int_a^x f + \int_x^b f$. Thus, $\int_a^b f = F(x) + \int_x^b f$ and so, $F(x) = \int_a^b f - \int_x^b f$. Hence, $F'(x) = -(\int_x^b f)' = f(x)$ by Exercise 1. □

So, by Theorem 6.4.6, if $f \colon I \to \mathbb{R}$ is continuous, then $\int_a^x f$ is an antiderivative of f whenever $a \in I$. Evaluating the antiderivative $\int_a^x f$ is often considerably more difficult than evaluating the derivative of f. Methods for evaluating antiderivatives, called *techniques of integration*, are covered in a calculus course. In this section we will establish three such techniques of integration.

Now suppose that $f \colon I \to \mathbb{R}$ is integrable over I but f is not continuous on I. Is the function $\int_a^x f$ an antiderivative of f? The following result shows that if f does have an antiderivative, then $\int_a^x f$ is also an antiderivative of f. On the other hand, there are integrable functions that do not have an antiderivative (see Exercise 4).

Lemma 6.4.7. Let $f \colon I \to \mathbb{R}$ be integrable over the interval I. Let $a \in I$. If G is an antiderivative of f, then for all $x \in I$

$$\int_a^x f = G(x) - G(a),$$

and $\int_a^x f$ is an antiderivative of f.

Proof. Let $f: I \to \mathbb{R}$ be integrable over I and let G be an antiderivative of f. Let $a \in I$. By Theorem 6.4.2, $\int_a^x f = G(x) - G(a)$ for all x in I. As $(G(x) - G(a))' = f(x)$ for each $x \in I$, we conclude that $\int_a^x f$ is an antiderivative of f. □

Our first technique of integration is often used to transform the antiderivative of a product of functions into an antiderivative that is easier to evaluate.

Theorem 6.4.8 (Integration by Parts). Let f and g be differentiable functions on an interval I such that f' and g' are both integrable over I. Then fg' and $f'g$ are integrable over I and for any $a \in I$,

$$\int_a^x fg' = f(x)g(x) - f(a)g(a) - \int_a^x f'g, \text{ for all } x \in I. \tag{6.43}$$

Proof. Let f and g be differentiable on I such that f' and g' are integrable over I. Corollaries 5.1.3, 6.2.16, and Theorem 6.3.1 imply that fg' and $f'g$ are integrable over I. By Theorem 5.1.4(3) (the product rule), we conclude that $fg' = (fg)' - gf'$. Let $a \in I$ and $x \in I$. Definition 6.2.12, Corollary 6.2.6, and Theorem 6.4.2 imply that

$$\int_a^x fg' = \int_a^x (fg)' - \int_a^x f'g = f(x)g(x) - f(a)g(a) - \int_a^x f'g. \quad □$$

If one can evaluate the integral of $f'g$ on the right in (6.43), then one can evaluate the integral of fg' on the left. This technique is applied when the integral of $f'g$ is easier to evaluate than that of fg'.

We now present another useful method for finding antiderivatives. This technique involves reversing the chain rule.

Theorem 6.4.9 (Integration by Substitution). Let $g: I \to J$ be differentiable on I and let $f: J \to \mathbb{R}$ be continuous on J, where I and J are intervals. If g' is integrable over I, then $(f \circ g)g'$ is integrable over I and for any $a \in I$,

$$\int_a^x (f \circ g)g' = \int_{g(a)}^{g(x)} f, \text{ for all } x \in I. \tag{6.44}$$

In particular, $\int_{g(a)}^{g(x)} f$ is an antiderivative of $(f \circ g)g'$.

Proof. Let $g: I \to J$ be differentiable on I, $f: J \to \mathbb{R}$ be continuous on J, and g' be integrable over I, where I and J are intervals. By Theorem 6.2.14, Corollaries 6.2.16, 5.1.3, and Theorem 6.3.1, $(f \circ g)g'$ is integrable over I. Let $a \in I$. Define $G: I \to \mathbb{R}$ by $G(x) = \int_{g(a)}^{g(x)} f$. The function G is an antiderivative of $(f \circ g)g'$ by Exercise 6. Since $G(a) = 0$, Lemma 6.4.7 implies that $\int_a^x (f \circ g)g' = \int_{g(a)}^{g(x)} f$ for all x in I. □

If one can evaluate the integral of f on the right in (6.44), then one can evaluate an antiderivative of $(f \circ g)g'$ on the left. In other words, Theorem 6.4.9 allows one to transform certain antiderivatives, which are difficult to evaluate, into ones that are easier to evaluate.

You may recall from calculus that in order to evaluate antiderivatives of certain functions that involve the square root of a sum, or difference, of squares, one would use integration by trigonometric substitution, which is an application of a general technique of integration called *inverse substitution*. Integration by inverse substitution essentially inverts the antidifferentiation equation (6.44), in Theorem 6.4.9. This is reason why the method is called "inverse substitution." This method is tacitly applied in calculus and real analysis books, but it is rarely stated and almost never verified. Recall that I^* denotes the set of points in an interval I that are not endpoints of I.

Theorem 6.4.10 (Integration by Inverse Substitution). Suppose that $f\colon J \to \mathbb{R}$ is continuous, $g\colon I \to J$ is differentiable, $J = g[I]$, and $g'(u) \neq 0$ for all $u \in I^*$, where I and J are intervals. Then $g^{-1}\colon J \to I$ exists and is continuous. Moreover, if g' is Riemann integrable over I, then for any $a \in J$,

$$\int_a^x f = \int_{g^{-1}(a)}^{g^{-1}(x)} (f \circ g)g', \text{ for all } x \in J. \tag{6.45}$$

Hence, $\int_{g^{-1}(a)}^{g^{-1}(x)} (f \circ g)g'$ is an antiderivative of f.

Proof. Let $f\colon J \to \mathbb{R}$ be continuous, $g\colon I \to J$ be differentiable, $J = g[I]$, and $g'(u) \neq 0$ for all $u \in I^*$. Corollaries 5.1.3 and 5.2.21 imply that the inverse function $g^{-1}\colon J \to I$ exists and is continuous. Given that g' is Riemann integrable over I, Theorem 6.4.9 implies that $(f \circ g)g'$ is Riemann integrable over I and for any $a \in J$,

$$\int_{g^{-1}(a)}^{g^{-1}(x)} (f \circ g)g' = \int_{g(g^{-1}(a))}^{g(g^{-1}(x))} f = \int_a^x f, \text{ for all } x \in J.$$

Thus, (6.45) holds, and $\int_{g^{-1}(a)}^{g^{-1}(x)} (f \circ g)g$ is an antiderivative of f by Theorem 6.4.6. □

Theorem 6.4.10 is implicitly applied in calculus books to evaluate an antiderivative of a difficult function f. This is done by introducing an appropriate invertible function g such that one can evaluate an antiderivative of $(f \circ g)g'$. Then, using equation (6.45), one can evaluate an antiderivative of f. The next two examples illustrate this method.

Example 6.4.11. Let $f\colon [-2,2] \to \mathbb{R}$ be defined by $f(x) = \sqrt{4 - x^2}$. Evaluate $\int_{-2}^x f$ using Theorem 6.4.10.

Solution. Clearly, f is continuous. Consider the right triangle in Figure 6.4. Thus, $x = 2\sin(\theta)$. Let $g\colon [-\frac{\pi}{2}, \frac{\pi}{2}] \to [-2,2]$ be defined by $g(\theta) = 2\sin(\theta)$. The function

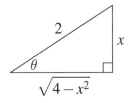

Figure 6.4: Trigonometric substitution triangle.

g is differentiable and has range $[-2, 2]$ which is also the domain of the function f. Moreover, $g'(\theta) \neq 0$ for all θ in $(-\frac{\pi}{2}, \frac{\pi}{2})$. (Observe that $g'(-\frac{\pi}{2}) = 0$ and $g'(\frac{\pi}{2}) = 0$.) Thus, $g^{-1} \colon [-2, 2] \to [-\frac{\pi}{2}, \frac{\pi}{2}]$ exists and is continuous. Now, using the substitution $x = g(\theta)$, where $\theta \in [-\frac{\pi}{2}, \frac{\pi}{2}]$, we obtain

$$(f \circ g)(\theta) = f(g(\theta)) = \sqrt{4 - 4\sin^2(\theta)} = 2\cos(\theta), \text{ and } g'(\theta) = 2\cos(\theta).$$

Thus, $f(g(\theta))g'(\theta) = 4\cos^2(\theta)$. By Theorem 6.4.10, we have that for all $x \in [-2, 2]$,

$$\int_{-2}^{x} f = \int_{g^{-1}(-2)}^{g^{-1}(x)} (f \circ g)g' = \int_{g^{-1}(-2)}^{g^{-1}(x)} 4\cos^2(\theta) \, d\theta. \tag{6.46}$$

As $4\cos^2(\theta)$ has antiderivative $2(\sin(\theta)\cos(\theta) + \theta)$ and $g^{-1}(-2) = -\frac{\pi}{2}$, we conclude from Theorem 6.4.2 that the equation

$$\int_{-2}^{x} f = \int_{-\frac{\pi}{2}}^{g^{-1}(x)} 4\cos^2(\theta) \, d\theta = 2(\sin(g^{-1}(x))\cos(g^{-1}(x)) + g^{-1}(x)) - \pi \tag{6.47}$$

holds for all $x \in [-2, 2]$. Since $g(\theta) = 2\sin(\theta)$, by solving the equation $x = 2\sin(\theta)$ for θ, we can obtain a formula for $g^{-1}(x)$. Then one can replace $g^{-1}(x)$ in (6.47) with this formula and obtain a explicit solution; however, there is another equivalent method. Since $x = g(\theta)$ implies that $g^{-1}(x) = \theta$, we can use Figure 6.4 to infer that

$$\sin(g^{-1}(x)) = \frac{x}{2}, \quad \cos(g^{-1}(x)) = \frac{\sqrt{4 - x^2}}{2}, \text{ and } g^{-1}(x) = \sin^{-1}\left(\frac{x}{2}\right). \tag{6.48}$$

Therefore, using (6.47) and (6.48), we obtain

$$\int_{-2}^{x} f = 2\left(\frac{x}{2}\frac{\sqrt{4 - x^2}}{2} + \sin^{-1}\left(\frac{x}{2}\right)\right) - \pi, \text{ for all } x \in [-2, 2].$$

Example 6.4.12. Let $f \colon (-\frac{\pi}{2}, \frac{\pi}{2}) \to \mathbb{R}$ be defined by $f(x) = \frac{\sin(x)+1}{(1-\sin(x))\cos(x)}$. Evaluate $\int_0^x f$ using Theorem 6.4.10.

Solution. The function f is continuous. Now let $g \colon (-1, 1) \to (-\frac{\pi}{2}, \frac{\pi}{2})$ be defined by $g(t) = \sin^{-1}(t)$, the inverse sine function. The function g is differentiable, has range $(-\frac{\pi}{2}, \frac{\pi}{2})$, and $g'(t) \neq 0$ for all t in $(-1, 1)$. Thus, $g^{-1} \colon (-\frac{\pi}{2}, \frac{\pi}{2}) \to (-1, 1)$ exists and is continuous. From calculus, we have the identities

$$\cos(\sin^{-1}(t)) = \sqrt{1 - t^2}$$

$$\frac{d}{dx}[\sin^{-1}(t)] = \frac{1}{\sqrt{1 - t^2}}.$$

Using the substitution $x = g(t)$ and the above two identities, we obtain

$$f(g(t)) = \frac{t + 1}{(1 - t)\sqrt{1 - t^2}} \text{ and } g'(t) = \frac{1}{\sqrt{1 - t^2}}.$$

Hence, $f(g(t))g'(t) = \frac{1}{(1-t)^2}$ which has antiderivative $\frac{1}{1-t}$. Since $g^{-1}(x) = \sin(x)$ and $g^{-1}(0) = 0$, Theorem 6.4.10 and Theorem 6.4.2 imply that

$$\int_0^x f = \int_0^{\sin(x)} \frac{1}{(1 - t)^2} \, dt = \frac{1}{1 - \sin(x)} - 1.$$

6.4.4 Improper Integrals

The definition of the Riemann integral is based on a bounded function that is defined on a closed and bounded interval. Under these conditions, if the integral exists, then the integral is said to be **proper**. There are cases, however, in which these conditions can be relaxed by relying on the limit concept.

Unbounded Intervals

In the definition of Riemann integral $\int_a^b f$, we assume that the domain of f has the form $[a, b]$. However, in many applications in physics, economics, and probability, one must integrate over unbounded intervals. We must therefore give meaning to operations like the following:

$$\int_c^\infty f \text{ and } \int_{-\infty}^c f. \tag{6.49}$$

Any integral having one of the forms in (6.49) is called an **improper integral**.

Definition 6.4.13. Let c be a real number. Given a function $f \colon [c, \infty) \to \mathbb{R}$ such that f Riemann integrable on $[c, \ell]$ for every $\ell > c$, we define

$$\int_c^\infty f = \lim_{\ell \to \infty} \int_c^\ell f. \tag{I}$$

For any function $g \colon (-\infty, c] \to \mathbb{R}$ such that g Riemann integrable on $[\ell, c]$ for every $\ell < c$, we define

$$\int_{-\infty}^c g = \lim_{\ell \to -\infty} \int_\ell^c g. \tag{II}$$

In (I) and (II), if the limit on the right exists and is finite, then the corresponding improper integral **converges**. Otherwise, the improper integral **diverges**.

Example 6.4.14. Let $f \colon [0, \infty) \to \mathbb{R}$ be defined by $f(x) = e^{-x}$. Since $F(x) = -e^{-x}$ is an antiderivative of f, we have

$$\int_0^\infty f = \lim_{\ell \to \infty} \int_0^\ell f = \lim_{\ell \to \infty} (F(\ell) - F(0)) = 0 - (-1) = 1.$$

Thus, the improper integral $\int_0^\infty f$ converges.

Unbounded Integrands

In the definition of Riemann Integral $\int_a^b f$, we assumed that the **integrand** f was defined for all x in the interval $[a, b]$. However, in some cases of interest, we may have that $f(a)$ or $f(b)$ are undefined. In either of these cases, we also call $\int_a^b f$ an **improper integral**. We can give meaning to such integrals.

Definition 6.4.15. Given a function f and an finite interval $[a, b]$, if $f(b)$ is undefined and f Riemann integrable on $[a, \ell]$ whenever $a < \ell < b$, we define

$$\int_a^b f = \lim_{\ell \to b^-} \int_a^\ell f. \tag{i}$$

If $f(a)$ is undefined and f Riemann integrable on $[\ell, b]$ whenever $a < \ell < b$, we define

$$\int_a^b f = \lim_{\ell \to a^+} \int_\ell^b f. \qquad (ii)$$

In (i) and (ii), if the limit on the right exists and is finite, then corresponding improper integral **converges**. Otherwise, the improper integral is said to **diverge**.

Example 6.4.16. Let $f: [0, 1) \to \mathbb{R}$ be defined by $f(x) = \frac{1}{1-x}$. Clearly, $f(1)$ is undefined. Since $F(x) = -\ln(1 - x)$ is an antiderivative of f, we have

$$\int_0^1 f = \lim_{\ell \to 1^-} \int_0^\ell f = \lim_{\ell \to 1^-} (F(\ell) - F(0)) = \infty.$$

Thus, the improper integral $\int_0^1 f$ diverges.

Exercises 6.4 ───

*1. Let f be continuous on $[a, b]$. Let $G: [a, b] \to \mathbb{R}$ be defined by $G(x) = \int_x^b f$. Prove that $G'(x) = -f(x)$.

2. Let $g: [0, 1] \to \mathbb{R}$ be defined by

$$g(x) = \begin{cases} 1, & \text{if } x < 1; \\ 0, & \text{if } x = 1. \end{cases}$$

   The function g is Riemann integrable (see Example 6.1.20). Show that $\int_0^c g = c$ for all $c \in [0, 1]$.

3. Let f be continuous on $[0, 1]$. Suppose that $\int_0^x f = x$ for all $x \in [0, 1]$. Prove that $f(x) = 1$ for all $x \in [0, 1]$. Using Exercise 2, show that this conclusion may not hold if f is not continuous.

*4. The function $g: [0, 1] \to \mathbb{R}$ in Exercise 2 is Riemann integrable. Show that g does not have an antiderivative. (See Theorem 5.2.17.)

5. Let $f: [a, b] \to \mathbb{R}$ be continuous and let $g: [a, b] \to \mathbb{R}$ be differentiable on $[a, b]$. Define $H: [a, b] \to \mathbb{R}$ by $H(x) = \int_a^x g(x)f(t)\, dt$. Find $H'(x)$.

*6. Let f be continuous on an interval J. Suppose that $g: I \to J$ is differentiable on the interval I. Let $a \in J$. Define the function $G: I \to \mathbb{R}$ be

$$G(x) = \int_a^{g(x)} f.$$

   Prove that G is differentiable and that $G'(x) = f(g(x))g'(x)$ for all $x \in I$.

7. Let f be continuous on an interval J and $h: I \to J$ be differentiable on the interval I. Let $b \in J$. Define the function $k: I \to \mathbb{R}$ by

$$k(x) = \int_{h(x)}^b f.$$

   Prove that k is differentiable and that $k'(x) = -f(h(x))h'(x)$ for all $x \in I$.

8. Let f be continuous on an interval J. Let $g\colon I \to J$ and $h\colon I \to J$ be differentiable on the interval I. Define $k\colon I \to \mathbb{R}$ by

$$k(x) = \int_{g(x)}^{h(x)} f.$$

Prove that k' exists and that $k'(x) = f(g(x))g'(x) - f(h(x))h'(x)$ for each $x \in I$.

9. Let f be continuous on $[a, b]$ such that $f(x) > 0$ for all $x \in [a, b]$. Define $F\colon [a, b] \to \mathbb{R}$ by $F(x) = \int_a^x f$. Prove that F is strictly increasing on $[a, b]$.

10. Let $f\colon [0, \infty) \to \mathbb{R}$ be continuous with $f(x) > 0$ for all $x \in [0, \infty)$. Suppose that $(f(x))^2 = 2 \int_0^x f$ for all $x \in [0, \infty)$. Show that $f(x) = x$ for all $x \in [0, \infty)$.

11. Let $f\colon [a, b] \to \mathbb{R}$ be continuous such that $\int_a^x f = \int_x^b f$ for all x in $[a, b]$. Show that $f(x) = 0$ for all x in $[a, b]$.

12. Use Theorem 6.4.8 to evaluate $\int_\pi^x fg'$ when $f(x)g'(x) = x\sin(x)$.

13. Use Theorem 6.4.9 to evaluate $\int_1^x (f \circ g)g'$ when $f(x) = \frac{1}{1+x^2}$ and $g(x) = \sqrt{x}$.

14. Use Theorem 6.4.10 to evaluate $\int_1^x f$ when $f(x) = \frac{1}{(1+x)2\sqrt{x}}$ and $g(x) = x^2$.

15. Let $f\colon [a, b] \to \mathbb{R}$ be Riemann integrable. Show that $\lim_{\ell \to b^-} \int_a^\ell f = \int_a^b f$.

16. Evaluate $\int_0^\infty xe^{-x^2}\, dx$.

17. Show that $\int_1^\infty \frac{1}{x^p}\, dx$ converges for $p > 1$ and diverges for $p \le 1$.

18. For what values of $p < 0$ does $\int_0^1 x^p\, dx$ converge?

19. Assume that $\int_{-\infty}^c f$ and $\int_c^\infty f$ converge for any real number c. Show that the sum $\int_{-\infty}^c f + \int_c^\infty f$ is independent of the choice of c.

20. Evaluate $\int_0^{16} \frac{1}{\sqrt[4]{x}}\, dx$ and $\int_0^2 \frac{1}{\sqrt{2-x}}\, dx$.

21. Define $f\colon (0, \infty) \to \mathbb{R}$ by $f(x) = \frac{1}{x}$ and $L\colon (0, \infty) \to \mathbb{R}$ by $L(x) - \int_1^x f$. Let $a \in \mathbb{R}^+$ and $b \in \mathbb{R}^+$. Validate, in order, the following properties of L.

   (a) L is continuous on $(0, \infty)$.
   (b) L is strictly increasing on $(0, \infty)$. Thus, $L(2) > L(1) = 0$.
   (c) L is differentiable on $(0, \infty)$.
   (d) $(L(ax))' = \frac{1}{x}$ for all $x \in (0, \infty)$. Thus, $L(ax) = L(x) + C$ for a constant C.
   (e) $L(ab) = L(a) + L(b)$.
   (f) $L(a^n) = nL(a)$, for all integers $n \ge 0$.
   (g) $(L(\frac{a}{x}))' = -\frac{1}{x}$ for all $x \in (0, \infty)$. Thus, $L(\frac{a}{x}) = -L(x) + C$ for a constant C.
   (h) $L(\frac{a}{b}) = L(a) - L(b)$.
   (i) $L(a^n) = nL(a)$, for all integers $n < 0$.
   (j) Let $\gamma \in \mathbb{R}$. Let $m, n \in \mathbb{Z}$ be such that $m < \frac{\gamma}{L(2)} < n$. So $L(2^m) < \gamma < L(2^n)$.
   (k) $\mathrm{ran}(L) = \mathbb{R}$, and there is a unique real number $e > 0$ such that $L(e) = 1$.

22. Let $L\colon (0,\infty) \to \mathbb{R}$ be as defined in Exercise 21. As L is strictly increasing, the inverse function $L^{-1}\colon \mathbb{R} \to (0,\infty)$ exists. Let $E(x) = L^{-1}(x)$ for all $x \in \mathbb{R}$. Thus, $E(x) = y$ if and only if $L(y) = x$ for all real numbers x and y. Let $c \in \mathbb{R}$ and $d \in \mathbb{R}$. Validate the following:

   (a) E is differentiable and $E'(x) = E(x)$, for all $x \in \mathbb{R}$.
   (b) E is strictly increasing on $\mathbb{R}$.
   (c) $E(c)E(d) = E(c+d)$.
   (d) $\frac{E(c)}{E(d)} = E(c-d)$.
   (e) $E(nc) = (E(c))^n$, for all integers n.

23. Let $L\colon (0,\infty) \to \mathbb{R}$ and e be as defined in Exercise 21.

   (a) Show that $\lim\limits_{x\to 0} \frac{L(1+x)}{x} = 1$.   (c) Show that $\lim\limits_{n\to\infty} (1 + \frac{1}{n})^n = e$.

   (b) Show that $\lim\limits_{n\to\infty} L((1 + \frac{1}{n})^n) = 1$.

Exercise Notes: For Exercise 1, Corollary 6.2.13 implies that $\int_a^b f = \int_a^x f + \int_x^b f$. For Exercise 2, one can use Theorem 6.4.3 to show that $\int_0^1 g = 1$. For Exercise 5, see Theorem 6.2.5(a). For Exercise 6, use the chain rule and Theorem 6.4.6. For Exercise 7, use Definition 6.2.12 and Exercise 6. For Exercise 8, use Corollary 6.2.13, Exercise 6, and Exercise 7. For Exercise 9, apply Theorem 5.2.8. Exercises 13 and 14 offer two methods for finding an antiderivative of $\frac{1}{(1+x)2\sqrt{x}}$. Exercise 21 presents a formal definition of the *natural logarithm function* $L(x)$, which is often denoted by $\ln(x)$. Exercise 22 presents a formal definition of the *natural exponential function* $E(x)$, which is usually denoted by e^x. For Exercise 23(b), use part (a) and Theorem 4.3.4. For Exercise 23(c), see Theorems 3.4.8 and 4.2.1.

CHAPTER 7

Infinite Series

Around 1665, Newton discovered that any binomial of the form $(1+x)^r$, where $r \notin \mathbb{N}$ and $|x| < 1$, can be expressed as an infinite series. In 1673, Leibniz discovered that π can be written as an infinite series. For Newton and Leibniz, infinite series were of fundamental importance in their development of calculus; however, during these times, questions of rigor and convergence were of secondary importance.

One typically first studies infinite series, in a standard calculus course, without going deeply into the theory of infinite series. In this chapter, using the results in Chapter 3 on the convergence of sequences, we will prove the important theorems that concern an infinite series of real numbers. In Section 7.1, we introduce the concept of a convergent infinite series, and discuss a variety of different kinds of infinite series. We then examine in Section 7.2 a range of tests for convergence. Finally, in Section 7.3 we investigate the effect of regrouping and rearranging the terms of an infinite series.

7.1   CONVERGENCE AND DIVERGENCE

As we have seen, the completeness axiom and the limit operation are critical components in real analysis. In particular, these concepts are used to develop the derivative and the Riemann integral. The limit concept was also used in our study of sequences, and it will again be used in our examination of infinite series.

Definition 7.1.1 (Infinite Series). Given a sequence $\langle a_n \rangle$ of real numbers, the nth **partial sum** s_n is defined by

$$s_n = \sum_{k=1}^{n} a_k = a_1 + a_2 + a_3 + \cdots + a_n.$$

We refer to the sequence $\langle s_n \rangle$ of partial sums as the **infinite series** $\sum_{k=1}^{\infty} a_k$. The real numbers $a_1, a_2, a_3, \ldots$ are called the **terms** of the infinite series.

The notion of convergence or divergence of an infinite series is defined in terms of the convergence or divergence of the sequence of its partial sums.

Definition 7.1.2. An infinite series $\sum_{k=1}^{n} a_k$ **converges** if the sequence of partial sums

$\langle s_n \rangle$ converges to a real number L, and we write $\sum_{k=1}^{\infty} a_k = \lim_{n \to \infty} s_n = \lim_{n \to \infty} \sum_{k=1}^{n} a_k = L$.

We refer to L as the **sum** of the series $\sum_{k=1}^{\infty} a_k$. If the sequence $\langle s_n \rangle$ does not converge, then the series $\sum_{k=1}^{\infty} a_k$ is said to **diverge**.

It will often be convenient to write $\sum_{k=1}^{\infty} a_k = a_1 + a_2 + a_3 + \cdots$. However, the notation $\sum_{k=1}^{\infty} a_k$ does not denote an "infinite sum." The notation $\sum_{k=1}^{\infty} a_k$ represents the sequence $\langle s_n \rangle$ of partial sums and it also represents the limit of the sequence $\langle s_n \rangle$ of these partial sums, if the limit exists. The context in which the notation $\sum_{k=1}^{\infty} a_k$ is being used should make its meaning clear.

Example. The nth term of the sequence of partial sums $\langle s_n \rangle$ of the series $\sum_{k=1}^{\infty} \frac{1}{2^k}$ is

$$s_n = \sum_{k=1}^{n} \frac{1}{2^k} = \sum_{k=0}^{n} \frac{1}{2^{k+1}} = \frac{1}{2} \sum_{k=0}^{n-1} \frac{1}{2^k} = 1 - \frac{1}{2^n},$$

by the Shift Rule 1.4.8 and Theorem 1.4.7. Since $\lim_{n \to \infty} s_n = 1$, we see that $\sum_{k=1}^{\infty} \frac{1}{2^k} = 1$.

Example. The sequence of partial sums $\langle s_n \rangle$ of the series $\sum_{k=1}^{\infty} (-1)^{k+1}$ is given by

$$s_1 = 1 = 1$$
$$s_2 = 1 - 1 = 0$$
$$s_3 = 1 - 1 + 1 = 1$$
$$s_4 = 1 - 1 + 1 - 1 = 0$$
$$\vdots$$

Thus, we see that the sequence $\langle s_n \rangle$ does not converge. Thus, the series diverges.

As illustrated in the above two examples, the convergence of an infinite series depends on the convergence of the sequence of its partial sums. In many cases, we will not be able to evaluate the limit of a convergent infinite series; but, we will be interested in the question: *Does the series converge or diverge?* It is this question that makes the theory of infinite series most interesting.

Theorem 7.1.3. Let $\sum_{k=1}^{\infty} a_k$ be a series such that $a_k \geq 0$ for all k. The series $\sum_{k=1}^{\infty} a_k$ converges if and only if the sequence $\langle s_n \rangle$ of partial sums is bounded above.

Proof. Let $\sum_{k=1}^{\infty} a_k$ be a series with nonnegative terms. For each $n \in \mathbb{N}$, let s_n be the nth partial sum of this series. It follows that $\langle s_n \rangle$ is an increasing sequence.

($\Rightarrow$). Assume that $\sum\limits_{k=1}^{\infty} a_k$ converges and let $\sum\limits_{k=1}^{\infty} a_k = L$ for some $L \in \mathbb{R}$. Since $\langle s_n \rangle$ is increasing, Lemma 3.4.2 implies that $s_n \leq L$ for all n. So $\langle s_n \rangle$ is bounded above.

($\Leftarrow$). Suppose that $\langle s_n \rangle$ is bounded above. As $\langle s_n \rangle$ is increasing, Theorem 3.4.6 implies that $\langle s_n \rangle$ converges, that is, $\sum\limits_{k=1}^{\infty} a_k$ converges. □

Remark 7.1.4. It is possible for an infinite series to start at 0 or a natural number $m > 1$. Any such series can be "reindexed" to an equal series with a new starting integer n. Consider the series $\sum\limits_{k=m}^{\infty} a_k$ with starting value m. To rewrite this series as one with the new starting value n, first compute

$$s = \text{old start} - \text{new start} = m - n.$$

Then $\sum\limits_{k=m}^{\infty} a_k = \sum\limits_{k=n}^{\infty} a_{k+s}$. For example, $\sum\limits_{k=3}^{\infty} \frac{1}{k^5} = \sum\limits_{k=1}^{\infty} \frac{1}{(k+2)^5}$ and $\sum\limits_{k=1}^{\infty} b_k = \sum\limits_{k=0}^{\infty} b_{k+1}$.

Definition 7.1.5. A series $\sum\limits_{k=1}^{\infty} b_k$ is a **telescoping series**, if there is a sequence $\langle a_k \rangle$ such that one of the following holds:

(i) Each term b_k can be written as $b_k = a_k - a_{k+1}$.

(ii) Each term b_k can be written as $b_k = a_{k+1} - a_k$.

Example. Since $\frac{1}{k^2+k} = \frac{1}{k} - \frac{1}{k+1}$, the series $\sum\limits_{k=1}^{\infty} \frac{1}{k^2+k}$ is a telescoping series.

The partial sums of a telescoping series can be easily simplified by cancellation and thereby, there is a simple way to determine it's sum, if it converges.

Theorem 7.1.6 (Telescoping Series). Suppose that $\sum\limits_{k=1}^{\infty} b_k$ is a telescoping series.

1. If $b_k = a_k - a_{k+1}$ for all $k \in \mathbb{N}$ and $\lim\limits_{n \to \infty} a_{n+1}$ exists, then $\sum\limits_{k=1}^{\infty} b_k = a_1 - \left(\lim\limits_{n \to \infty} a_{n+1} \right)$.

2. If $b_k = a_{k+1} - a_k$ for all $k \in \mathbb{N}$ and $\lim\limits_{n \to \infty} a_{n+1}$ exists, then $\sum\limits_{k=1}^{\infty} b_k = \left(\lim\limits_{n \to \infty} a_{n+1} \right) - a_1$.

Proof. We prove item 1. The proof of 2 is similar. Suppose that each term b_k can be written as $b_k = a_k - a_{k+1}$ and $\lim\limits_{n \to \infty} a_{n+1}$ exists. The sequence of partial sums $\langle s_n \rangle$ is given by (see Exercise 9 on page 27)

$$s_n = \sum_{k=1}^{n} b_k = \sum_{k=1}^{n} (a_1 - a_{k+1}) = a_1 - a_{n+1}.$$

Thus, $\sum\limits_{k=1}^{\infty} b_k = \lim\limits_{n \to \infty} s_n = a_1 - \left(\lim\limits_{n \to \infty} a_{n+1} \right)$. □

Example. The series $\sum\limits_{k=1}^{\infty} \frac{1}{k^2+k}$ is a telescoping series as $\frac{1}{k^2+k} = \frac{1}{k} - \frac{1}{(k+1)} = a_k - a_{k+1}$.

Since $a_1 = 1$ and $\lim\limits_{n \to \infty} a_{n+1} = \lim\limits_{n \to \infty} \frac{1}{(n+1)} = 0$, $\sum\limits_{k=1}^{\infty} \frac{1}{k^2+k} = 1$ by Theorem 7.1.6(1).

Let $\sum_{k=1}^{\infty} b_k$ be a telescoping series as in the statement of Theorem 7.1.6. If $\lim_{n\to\infty} a_{n+1}$ **does not** exist, then $\sum_{k=1}^{\infty} b_k$ diverges.

Since an infinite series converges if and only if the sequence of its partial sums converges, Definition 7.1.2, Theorem 3.2.2, and Exercise 12 on page 67 easily imply the following theorem.

Theorem 7.1.7 (Linearity). If $\sum_{k=1}^{\infty} a_k$ and $\sum_{k=1}^{\infty} b_k$ converge and c is a constant, then

(1) $\sum_{k=1}^{\infty} (a_k + b_k) = \sum_{k=1}^{\infty} a_k + \sum_{k=1}^{\infty} b_k$;

(2) $\sum_{k=1}^{\infty} ca_k = c \sum_{k=1}^{\infty} a_k$.

As alluded to earlier, the theory of infinite series is mainly concerned with the convergence or divergence of a series. We now show that a series cannot converge if its terms do not approach 0.

Theorem 7.1.8. If $\sum_{k=1}^{\infty} a_k$ converges, then $\lim_{k\to\infty} a_k = 0$.

Proof. Let s_n be the nth partial sum of the convergent series $\sum_{k=1}^{\infty} a_k$. Let $\lim_{n\to\infty} s_n = L$. Note that $a_n = s_n - s_{n-1}$ for each $n \geq 2$. Thus,

$$\lim_{n\to\infty} a_n = \lim_{n\to\infty} s_n - \lim_{n\to\infty} s_{n-1} = L - L = 0. \qquad \square$$

The contrapositive of Theorem 7.1.8 is identified in our next theorem

Theorem 7.1.9 (Divergence Test). If $\langle a_k \rangle$ does not converge to 0, then the series $\sum_{k=1}^{\infty} a_k$ diverges.

For example, consider the series $\sum_{k=1}^{\infty} \frac{k}{k+1}$. As $\lim_{k\to\infty} \frac{k}{k+1} = 1$, Theorem 7.1.9 implies that $\sum_{k=1}^{\infty} \frac{k}{k+1}$ diverges. The converse of Theorem 7.1.8 is false; that is, the condition $\lim_{k\to\infty} a_k = 0$ does not imply that $\sum_{k=1}^{\infty} a_k$ converges. This is verified by out next result.

Theorem 7.1.10 (Harmonic Series). The *harmonic series* $\sum_{k=1}^{\infty} \frac{1}{k}$ diverges.

Proof. Consider the nth partial sum:

$$s_n = 1 + \frac{1}{2} + \frac{1}{3} + \frac{1}{4} + \frac{1}{5} + \cdots + \frac{1}{n}.$$

Imagine n to be very large and write

$$
\begin{aligned}
s_n &= 1 + \frac{1}{2} + \frac{1}{3} + \frac{1}{4} + \frac{1}{5} + \cdots + \frac{1}{n} \\
&= 1 + \frac{1}{2} + \left(\frac{1}{3} + \frac{1}{4}\right) + \left(\frac{1}{5} + \frac{1}{6} + \frac{1}{7} + \frac{1}{8}\right) + \left(\frac{1}{9} + \cdots + \frac{1}{16}\right) + \cdots + \frac{1}{n} \\
&> 1 + \frac{1}{2} + \left(\frac{1}{4} + \frac{1}{4}\right) + \left(\frac{1}{8} + \frac{1}{8} + \frac{1}{8} + \frac{1}{8}\right) + \left(\frac{1}{16} + \cdots + \frac{1}{16}\right) + \cdots + \frac{1}{n} \\
&= 1 + \frac{1}{2} + \frac{1}{2} + \frac{1}{2} + \frac{1}{2} + \cdots + \frac{1}{n}.
\end{aligned}
$$

By taking n sufficiently large we can introduce as many $\frac{1}{2}$'s in the above sum as we wish. In fact, one can prove by induction that $s_{2^n} \geq 1 + n \cdot \frac{1}{2}$ for all $n \in \mathbb{N}$. Thus, the sequence of partial sums $\langle s_n \rangle$ is not bounded above. Hence, Theorem 7.1.3 implies that the harmonic series diverges. □

A **geometric series** is any series that can be put into the form

$$
\sum_{k=0}^{\infty} ar^k = a + ar + ar^2 + \cdots + ar^k + \cdots ,
$$

where $a \neq 0$ and r are constants. Geometric series have applications in many areas outside of mathematics; for example, physics, engineering, biology, economics, and computer science.

Theorem 7.1.11 (Geometric Series). Let a be nonzero. The geometric series $\sum_{k=0}^{\infty} ar^k$ converges if $|r| < 1$ and diverges if $|r| \geq 1$. Moreover, if $|r| < 1$, then $\sum_{k=0}^{\infty} ar^k = \frac{a}{1-r}$.

Proof. Let $a \neq 0$. If $|r| \geq 1$, then the sequence $\langle ar^k \rangle$ does not converge to 0. Thus, by Theorem 7.1.9, the series $\sum_{k=0}^{\infty} ar^k$ diverges. Suppose that $|r| < 1$. For each $n \geq 0$, the nth partial sum is (▲) $s_n = \sum_{k=0}^{n} ar^k = a\frac{r^{n+1}-1}{r-1}$, by Theorem 1.4.7. Since $|r| < 1$, $\lim_{n \to \infty} r^{n+1} = 0$ by Corollary 3.1.12. Thus, $\lim_{n \to \infty} s_n = \frac{a}{1-r}$ by (▲). So $\sum_{k=0}^{\infty} ar^k = \frac{a}{1-r}$. □

Any series of the form $\sum_{k=m}^{\infty} ar^k$ where $m > 0$ is also a geometric series. However, the formula in Theorem 7.1.11 for the sum of a geometric series is valid only when the index starts at 0. Nevertheless, using Remark 7.1.4, one can rewrite the series as one with a starting value 0. For example,

$$
\sum_{k=3}^{\infty} 5\left(\frac{1}{3}\right)^k = \sum_{k=0}^{\infty} 5\left(\frac{1}{3}\right)^{k+3} = \sum_{k=0}^{\infty} \frac{5}{27}\left(\frac{1}{3}\right)^k = \frac{\frac{5}{27}}{1 - \frac{1}{3}} = \frac{5}{18}.
$$

Recall that a sequence converges if and only if it is a Cauchy sequence. This leads to our next theorem. Let $\sum\limits_{k=1}^{\infty} a_k$ be an infinite series and, for each $n \in \mathbb{N}$, let $s_n = \sum\limits_{k=1}^{n} a_k$ be the nth partial sum. Whenever $n > m \geq 1$, we have that

$$s_n - s_m = a_{m+1} + \cdots + a_n = \sum_{k=m+1}^{n} a_k.$$

Thus, Theorem 3.6.5 implies our next result.

Theorem 7.1.12 (Cauchy Series Criterion). An infinite series $\sum\limits_{n=1}^{\infty} a_n$ converges if and only if for each $\varepsilon > 0$, there is an N such that for all $n > m > N$,

$$\left| \sum_{k=m+1}^{n} a_k \right| = |a_{m+1} + \cdots + a_n| < \varepsilon.$$

Given a series $\sum\limits_{k=1}^{\infty} a_k = a_1 + a_2 + a_3 + \cdots$ and an integer $j \geq 0$, consider the series $\sum\limits_{k=1}^{\infty} a_{j+k} = a_{j+1} + a_{j+2} + a_{j+3} + \cdots$ which is obtained by eliminating the first j terms of the series $\sum\limits_{k=1}^{\infty} a_k$. The series $\sum\limits_{k=1}^{\infty} a_{j+k}$ is called a **tail** of the series $\sum\limits_{k=1}^{\infty} a_k$. Moreover, for $n \geq 1$, the nth partial sum t_n of this tail series is

$$t_n = \sum_{k=1}^{n} a_{j+k} = a_{j+1} + a_{j+2} + a_{j+3} + \cdots + a_{j+n}.$$

Example. For the series $\sum\limits_{k=1}^{\infty} \frac{1}{k^2} = 1 + \frac{1}{4} + \frac{1}{9} + \frac{1}{16} + \frac{1}{25} + \cdots$, let $j = 3$. Then the tail series is

$$\sum_{k=1}^{\infty} \frac{1}{(k+3)^2} = \frac{1}{16} + \frac{1}{25} + \frac{1}{36} + \frac{1}{49} + \frac{1}{64} + \cdots$$

with the third partial sum $t_3 = \frac{1}{16} + \frac{1}{25} + \frac{1}{36}$.

Theorem 7.1.13 (Tail Convergence Test). Consider the series $\sum\limits_{k=1}^{\infty} a_k$ and let $j \geq 1$ be an integer. Then $\sum\limits_{k=1}^{\infty} a_k$ converges if and only if $\sum\limits_{k=1}^{\infty} a_{j+k}$ converges.

Proof. Let $j \geq 1$ and let $s_n = \sum\limits_{k=1}^{n} a_k$ and $t_n = \sum\limits_{k=1}^{n} a_{j+k}$, for each $n \in \mathbb{N}$. Thus, for all $n \in \mathbb{N}$, we have the identity (▲) $s_{j+n} = \left(\sum\limits_{k=1}^{j} a_k \right) + t_n$. Therefore,

$$
\begin{array}{lll}
\sum\limits_{k=1}^{\infty} a_k \text{ converges} & \text{iff } \langle s_n \rangle \text{ converges} & \text{by Definition 7.1.1} \\
& \text{iff } \langle s_{j+n} \rangle \text{ converges} & \text{by Corollary 3.3.11} \\
& \text{iff } \langle t_n \rangle \text{ converges} & \text{by (▲)} \\
& \text{iff } \sum\limits_{k=1}^{\infty} a_{j+k} \text{ converges} & \text{by Definition 7.1.1.} \qquad \square
\end{array}
$$

Exercises 7.1

1. Prove Theorem 7.1.7.

2. Prove that if $\sum_{k=1}^{\infty}(a_k + b_k)$ and $\sum_{k=1}^{\infty} b_k$ converge, then $\sum_{k=1}^{\infty} a_k$ converges.

3. Let $\sum_{k=1}^{\infty} a_k$ be a convergent series. For each $k \in \mathbb{N}$, let

$$\rho_k = \begin{cases} a_k, & \text{if } a_k > 0; \\ 0, & \text{if } a_k \leq 0. \end{cases} \qquad \eta_k = \begin{cases} a_k, & \text{if } a_k < 0; \\ 0, & \text{if } a_k \geq 0. \end{cases}$$

   (a) Show that $a_k = \rho_k + \eta_k$ for all $k \in \mathbb{N}$.

   (b) Prove that $\sum_{k=1}^{\infty} \rho_k$ converges if and only if $\sum_{k=1}^{\infty} \eta_k$ converges.

4. Prove Theorem 7.1.12.

5. Show that $\sum_{n=1}^{\infty} \frac{1}{\sqrt{n+1}+\sqrt{n}}$ diverges.

6. Show that $\sum_{k=0}^{\infty} \frac{1}{7}\left(\frac{3}{2}\right)^k$ diverges.

7. Show that $\sum_{k=0}^{\infty} 3\left(-\frac{1}{5}\right)^k$ converges and find its sum.

8. Given that
$$3 + \sqrt{3} + 1 + \frac{1}{\sqrt{3}} + \frac{1}{3} + \cdots$$
   is a geometric series, find its sum. [Hint: $a = 3$ and $ar = \sqrt{3}$.]

9. Show that $\sum_{k=1}^{\infty} \frac{1}{k(k+2)}$ converges and evaluate its sum.

10. Let $\langle x_n \rangle$ be a sequence of nonnegative real numbers, and let $y_n = x_n - x_{n+1}$ for each $n \in \mathbb{N}$.

   (a) Prove that $\sum_{n=1}^{\infty} y_n$ converges if and only if the sequence $\langle x_n \rangle$ converges.

   (b) If $\sum_{n=1}^{\infty} y_n$ converges, then what is the sum?

11. Let $a \neq 0$ and $|r| < 1$. Find the sum of $\sum_{k=0}^{\infty} ar^{2k}$.

12. Let $a \neq 0$ and $|r| < 1$. Find the sum of $\sum_{k=0}^{\infty} ar^{2k+1}$.

13. Suppose that $\sum_{n=1}^{\infty} |a_n|$ converges and that $\langle b_n \rangle$ is a bounded sequence. Using Theorem 7.1.12, prove that $\sum_{n=1}^{\infty} a_n b_n$ converges.

7.2 CONVERGENCE TESTS

Given any infinite series $\sum_{k=1}^{\infty} a_k$, there are two significant questions:

1. Does $\sum_{k=1}^{\infty} a_k$ converge?

2. If $\sum_{k=1}^{\infty} a_k$ does converge, can we evaluate its sum?

In Section 7.1, we presented methods for evaluating the sum of convergent geometric series and telescoping series. In general, however, it is difficult to determine the exact sum of a convergent series. Thus, in this section, our focus will be on establishing several tests that may be used to determine whether or not a series converges.

7.2.1 Comparison Tests

If you know that a given series converges (diverges) and you want to know if another series converges (diverges), then you may be able to use a *comparison test* to resolve your inquiry. Our first comparison test follows.

Theorem 7.2.1 (Direct Comparison Test). Let $\sum_{k=1}^{\infty} a_k$ and $\sum_{k=1}^{\infty} b_k$ be series such that $0 \le a_k \le b_k$ for all $k > N$, for some integer $N \ge 0$.

(1) If $\sum_{k=1}^{\infty} b_k$ converges, then $\sum_{k=1}^{\infty} a_k$ converges.

(2) If $\sum_{k=1}^{\infty} a_k$ diverges, then $\sum_{k=1}^{\infty} b_k$ diverges. (This is the contrapositive of (1).)

Proof. Let $\sum_{k=1}^{\infty} a_k$ and $\sum_{k=1}^{\infty} b_k$ be such that (▲) $0 \le a_k \le b_k$ for all $k > N \ge 0$. For each $n \in \mathbb{N}$, let $s_n = \sum_{k=1}^{n} a_{N+k}$ and $t_n = \sum_{k=1}^{n} b_{N+k}$. Assume that $\sum_{k=1}^{\infty} b_k$ converges. Theorem 7.1.13 implies that the tail series $\sum_{k=1}^{n} b_{N+k}$ converges. So, by Theorem 7.1.3, there is a real number B such that $t_n \le B$ for all $n \in \mathbb{N}$. From (▲), it follows that $s_n \le t_n \le B$, for all $n \in \mathbb{N}$. So $\langle s_n \rangle$ is bounded above. Theorem 7.1.3 implies that $\sum_{k=1}^{\infty} a_{N+k}$ converges. Hence, $\sum_{k=1}^{\infty} a_k$ converges by Theorem 7.1.13. □

Given a series with nonnegative terms, to apply Theorem 7.2.1, one should find a convergent (divergent) series whose terms are nonnegative and are eventually larger (smaller) than the corresponding terms of the given series. To find this relevant series, it often requires using properties of inequality. For example, consider the series $\sum_{k=1}^{\infty} \frac{3k+2}{k^2-k+5}$. On can verify that $\frac{3k+2}{k^2-k+5} > 0$ for all $k \ge 1$. Observe that for $k \ge 5$, we have that $\frac{3k+2}{k^2-k+5} \ge \frac{k}{k^2-k+5} \ge \frac{k}{k^2-k+k} = \frac{k}{k^2} = \frac{1}{k}$. By Theorems 7.1.10 and 7.2.1, we conclude that $\sum_{k=1}^{\infty} \frac{3k+2}{k^2-k+5}$ diverges.

Our next comparison test is often easier to apply than the direct comparison test, because it removes the need to verify that the required inequality holds between the terms of the two sequences. In our next theorem, we write $x < \infty$ to mean the x is a real number.

Theorem 7.2.2 (Limit Comparison Test). Let $\sum\limits_{k=1}^{\infty} a_k$ and $\sum\limits_{k=1}^{\infty} b_k$ be such that $a_k > 0$ and $b_k > 0$ for all $k \in \mathbb{N}$.

(1) If $0 < \lim\limits_{k\to\infty} \frac{a_k}{b_k} < \infty$, then $\sum\limits_{k=1}^{\infty} a_k$ converges if and only if $\sum\limits_{k=1}^{\infty} b_k$ converges.

(2) If $\lim\limits_{k\to\infty} \frac{a_k}{b_k} = 0$ and $\sum\limits_{k=1}^{\infty} b_k$ converges, then $\sum\limits_{k=1}^{\infty} a_k$ converges.

(3) If $\lim\limits_{k\to\infty} \frac{a_k}{b_k} = \infty$ and $\sum\limits_{k=1}^{\infty} a_k$ converges, then $\sum\limits_{k=1}^{\infty} b_k$ converges.

Proof. Let $\sum\limits_{k=1}^{\infty} a_k$ and $\sum\limits_{k=1}^{\infty} b_k$ be series where $a_k > 0$ and $b_k > 0$ for all $k \in \mathbb{N}$. To prove item (1), let $L = \lim\limits_{k\to\infty} \frac{a_k}{b_k}$ be a positive real number. Since $0 < L$ and $\frac{a_k}{b_k} > 0$ for all $k \in \mathbb{N}$, Theorem 3.1.24 and Lemma 3.2.5 imply that there are positive real numbers M and N such that $M < \frac{a_k}{b_k} < N$ for all $k \in \mathbb{N}$. Since $b_k > 0$ for each $k \in \mathbb{N}$, it follows that

$$M b_k < a_k < N b_k \text{ for all } k. \tag{7.1}$$

If $\sum\limits_{k=1}^{\infty} b_k$ converges, then $\sum\limits_{k=1}^{\infty} N b_k$ converges by Theorem 7.1.7. Thus, (7.1) implies that $\sum\limits_{k=1}^{\infty} a_k$ converges by Theorem 7.2.1, the Direct Comparison Test.

If $\sum\limits_{k=1}^{\infty} a_k$ converges, then (7.1) implies that $\sum\limits_{k=1}^{\infty} M b_k$ converges by Theorem 7.2.1. Since $M \neq 0$, Theorem 7.1.7(2) implies that $\sum\limits_{k=1}^{\infty} b_k$ converges.

The proofs of items (2) and (3) are left to Exercises 1 and 2, respectively. □

Given a series, to apply the Limit Comparison Test, one needs to find a relevant *companion* series to compare with the given series. If a sum or difference of powers of k appear in the numerator or denominator of the given series, delete all but the highest power of k in the numerator and in denominator. For example, consider once again the series $\sum\limits_{k=1}^{\infty} \frac{3k+2}{k^2-k+5}$. As instructed, we obtain the series $\sum\limits_{k=1}^{\infty} \frac{k}{k^2} = \sum\limits_{k=1}^{\infty} \frac{1}{k}$. One can verify that $\lim\limits_{k\to\infty} \frac{\frac{3k+2}{k^2-k+5}}{\frac{1}{k}} = 3$. Thus, by Theorems 7.1.10 and 7.2.2(1), the series $\sum\limits_{k=1}^{\infty} \frac{3k+2}{k^2-k+5}$ diverges.

Geometric series and p-series (see Theorem 7.2.5) offer a collection of companion series that are frequently used in applications of Theorem 7.2.2.

7.2.2 The Integral Test

Suppose that a nonnegative decreasing continuous function can be used to determine the terms of a series. Then there is a convergence test for the series that requires one to evaluate an improper integral of the function.

Theorem 7.2.3 (Integral Test). Let $f\colon [1,\infty) \to \mathbb{R}$ be a decreasing and continuous function such that $f(x) \geq 0$, for all x in $[1,\infty)$. Let $a_k = f(k)$ for all $k \in \mathbb{N}$. Then

$$\sum_{k=1}^{\infty} a_k \text{ converges if and only if } \int_1^{\infty} f \text{ converges.}$$

Proof. Suppose that $a_k = f(k)$ for all k, and $f\colon [1,\infty) \to \mathbb{R}$ is a decreasing continuous function such that $f(x) \geq 0$, for all x in $[1,\infty)$. Note that each $a_k \geq 0$. Since f is decreasing, we have that

$$a_{k+1} = f(k+1) \leq f(x) \leq f(k) = a_k, \text{ for all } k \in \mathbb{N} \text{ and all } x \in [k, k+1]. \quad (7.2)$$

Since $k + 1 - k = 1$, (7.2) and Lemma 6.1.15 imply that

$$a_{k+1} \leq \int_k^{k+1} f \leq a_k, \text{ for all } k \in \mathbb{N}. \quad (7.3)$$

Consider the sequences $\langle a_{k+1} \rangle$, $\left\langle \int_k^{k+1} f \right\rangle$, and $\langle a_k \rangle$. Theorem 7.2.1 and (7.3) imply that $\sum_{k=1}^{\infty} a_k$ converges if and only if $\sum_{k=1}^{\infty} \left(\int_k^{k+1} f\right)$ converges. By Theorem 6.2.11 and Definition 6.4.13, we also have that

$$\sum_{k=1}^{\infty} \int_k^{k+1} f = \lim_{n \to \infty} \sum_{k=1}^{n} \int_k^{k+1} f = \lim_{n \to \infty} \int_1^{n+1} f = \int_1^{\infty} f,$$

where the last equality holds because $f(x) \geq 0$, for all x in $[1,\infty)$. Therefore, $\sum_{k=1}^{\infty} a_k$ converges if and only if $\int_1^{\infty} f$ converges. □

The following definition identifies a generalization of the harmonic series.

Definition 7.2.4 (p-series). A **p-series** is any series of the form $\sum_{k=1}^{\infty} \frac{1}{k^p}$, where $p > 0$.

In the definition of a p-series, the value p is a fixed positive real number. The integral test is used to establish the following p-series test.

Theorem 7.2.5 (p-Series Test). The p-series $\sum_{k=1}^{\infty} \frac{1}{k^p}$ converges if $p > 1$ and diverges if $0 < p \leq 1$.

Proof. If $p = 1$, then $\sum_{k=1}^{\infty} \frac{1}{k^p}$ is the harmonic series, which diverges by Theorem 7.1.10. So assume that $p \neq 1$ and $p > 0$. Define $f\colon [1,\infty) \to \mathbb{R}$ by $f(x) = x^{-p}$. Then

$\frac{1}{k^p} = f(k)$ for all $k \in \mathbb{N}$, and f is positive, continuous, and decreasing on $[1, \infty)$. We now evaluate the improper integral

$$\int_1^{\infty} x^{-p}\, dx = \lim_{\ell \to \infty} \left[\frac{x^{1-p}}{1-p} \right]_1^{\ell} = \lim_{\ell \to \infty} \frac{\ell^{1-p} - 1}{1-p} = \begin{cases} \frac{1}{p-1}, & \text{if } p > 1; \\ \infty, & \text{if } p < 1. \end{cases}$$

Thus, by Theorem 7.2.3, $\sum\limits_{k=1}^{\infty} \frac{1}{k^p}$ converges if $p > 1$ and diverges if $0 < p < 1$. $\quad\square$

7.2.3 Alternating Series

The tests in the above two sections involve series with nonnegaitve terms. We will now discuss series whose terms alternate between positive and negative values.

Definition 7.2.6 (Alternating Series). An **alternating series** is a series having one of the forms $\sum\limits_{k=1}^{\infty} (-1)^{k+1} a_k$ or $\sum\limits_{k=1}^{\infty} (-1)^k a_k$ where $a_k > 0$ for all k.

For example, $\sum\limits_{k=1}^{\infty} (-1)^{k+1} \frac{1}{k} = 1 - \frac{1}{2} + \frac{1}{3} - \cdots + (-1)^{k+1} \frac{1}{k} + \cdots$ is alternating series.

Theorem 7.2.7 (Alternating Series Test). Let $\sum\limits_{k=1}^{\infty} (-1)^{k+1} a_k$ be a series such that $a_k \geq a_{k+1} > 0$ for all k, and $\lim\limits_{k \to \infty} a_k = 0$. Then $\sum\limits_{k=1}^{\infty} (-1)^{k+1} a_k$ converges to a sum S such that $|S - s_n| \leq a_{n+1}$ for all n, where s_n is the nth partial sum of the series.

Proof. Suppose that the series $\sum\limits_{k=1}^{\infty} (-1)^{k+1} a_k$ is such that (▲) $a_k \geq a_{k+1} > 0$ for all k, and (▼) $\lim\limits_{k \to \infty} a_k = 0$. Consider the "even" partial sums $s_2, s_4, s_6, \ldots, s_{2n}, \ldots$ where, for each $n \in \mathbb{N}$,

$$s_{2n} = a_1 - a_2 + a_3 - a_4 + \cdots + a_{2n-1} - a_{2n};$$
$$s_{2(n+1)} = a_1 - a_2 + a_3 - a_4 + \cdots + a_{2n-1} - a_{2n} + a_{2n+1} - a_{2(n+1)}$$
$$= s_{2n} + a_{2n+1} - a_{2(n+1)}.$$

So, for any $n \in \mathbb{N}$, $s_{2(n+1)} - s_{2n} = a_{2n+1} - a_{2(n+1)} \geq 0$, by (▲). Thus, $\langle s_{2n} \rangle$ is an increasing sequence. Note that, for every $n \in \mathbb{N}$,

$$s_{2n} = a_1 - (a_2 - a_3) - (a_4 - a_5) - \cdots - (a_{2n-2} - a_{2n-1}) - a_{2n} \leq a_1 \qquad (7.4)$$

as $a_{2n} > 0$ and $a_k - a_{k+1} \geq 0$ for all $k \in \mathbb{N}$. Hence, the sequence $\langle s_{2n} \rangle$ is bounded above by a_1 and thus, Theorem 3.4.6 implies that $\langle s_{2n} \rangle$ converges. Let $\lim\limits_{n \to \infty} s_{2n} = S$. Now we consider the "odd" partial sums $s_1, s_3, s_5, \ldots, s_{2n-1}, \ldots$. Observe that for each natural number n,

$$s_{2n-1} = a_1 - a_2 + a_3 - a_4 + \cdots + a_{2n-1};$$
$$s_{2n} = a_1 - a_2 + a_3 - a_4 + \cdots + a_{2n-1} - a_{2n}$$
$$= s_{2n-1} - a_{2n}.$$

Thus, $s_{2n-1} = s_{2n} + a_{2n}$, for all $n \in \mathbb{N}$. By ($\blacktriangledown$), we have that $\lim\limits_{n\to\infty} a_{2n} = 0$. Since $\lim\limits_{n\to\infty} s_{2n} = S$, Theorem 3.2.2 implies that

$$\lim_{n\to\infty} s_{2n-1} = \lim_{n\to\infty} s_{2n} + \lim_{n\to\infty} a_{2n} = S + 0 = S.$$

So, by Exercise 14 on page 79, $\lim\limits_{n\to\infty} s_n = S$. Therefore, $\sum\limits_{k=1}^{\infty} (-1)^{k+1} a_k = S$.

We will now prove that the *error estimate* $|S - s_n| \leq a_{n+1}$ holds for all n. By Exercise 3, $\langle s_{2n-1} \rangle$ is a decreasing sequence. Since $\langle s_{2n} \rangle$ is an increasing sequence, $\langle s_{2n-1} \rangle$ is a decreasing sequence, and $\lim\limits_{n\to\infty} s_{2n} = S = \lim\limits_{n\to\infty} s_{2n-1}$, it follows (see Corollaries 3.4.3 and 3.4.5) that

$$s_{2n} \leq S \leq s_{2n+1} \leq s_{2n-1},$$

for each $n \in \mathbb{N}$. Thus, for all natural numbers n, we have that

$$|S - s_{2n}| \leq |s_{2n+1} - s_{2n}| \text{ and } |S - s_{2n-1}| \leq |s_{2n} - s_{2n-1}|.$$

Therefore, $|S - s_n| \leq |s_{n+1} - s_n|$ for all $n \in \mathbb{N}$. Since $s_{n+1} - s_n = a_{n+1} > 0$, we conclude that $|S - s_n| \leq a_{n+1}$ for every natural number n. □

Since the alternating series $\sum\limits_{k=1}^{\infty} (-1)^{k+1} \frac{1}{k}$ satisfies the conditions of Theorem 7.2.7, it converges. We note that since $\sum\limits_{k=1}^{\infty} (-1)^{k+1} a_k = (-1) \sum\limits_{k=1}^{\infty} (-1)^k a_k$, Theorem 7.2.7 also applies to alternating series of the form $\sum\limits_{k=1}^{\infty} (-1)^k a_k$.

7.2.4 Absolute Convergence

The partial sums s_n of a series with nonnegative terms forms an increasing sequence $\langle s_n \rangle$. For this reason, such a series is typically easier to investigate than a series with a varied arrangement of positive and negative terms. Given a series $\sum\limits_{k=1}^{\infty} a_k$ with positive and negative terms, consider the series $\sum\limits_{k=1}^{\infty} |a_k|$ with nonnegative terms. Can we use the series $\sum\limits_{k=1}^{\infty} |a_k|$ to determine whether or not $\sum\limits_{k=1}^{\infty} a_k$ converges? Our next theorem, addresses this question.

Theorem 7.2.8. Let $\sum\limits_{k=1}^{\infty} a_k$ be a series of real numbers. If $\sum\limits_{k=1}^{\infty} |a_k|$ converges, then the series $\sum\limits_{k=1}^{\infty} a_k$ converges.

Proof. Let $n > m$ be natural numbers. By the triangle inequality, we have that

$$\left| \sum_{k=m+1}^{n} a_k \right| \leq \sum_{k=m+1}^{n} |a_k|.$$

Thus, the result follows from Theorem 7.1.12. □

Definition 7.2.9. Let $\sum\limits_{k=1}^{\infty} a_k$ be a series of real numbers.

1. If $\sum\limits_{k=1}^{\infty} |a_k|$ converges, then the series $\sum\limits_{k=1}^{\infty} a_k$ is said to be **absolutely convergent**.

2. If $\sum\limits_{k=1}^{\infty} a_k$ converges and $\sum\limits_{k=1}^{\infty} |a_k|$ diverges, then $\sum\limits_{k=1}^{\infty} a_k$ is said to be **conditionally convergent**.

Moreover, if a series is absolutely convergent, then we will say that it *converges absolutely*. Similarly, if a series is conditionally convergent, then we will say that it *converges conditionally*.

Clearly, by Theorem 7.2.8, if a series is absolutely convergent, then the series itself converges. So, suppose that $\sum\limits_{k=1}^{\infty} a_k$ is a series with positive and negative terms. Since the tests in Sections 7.2.1 and 7.2.2 all concern series with nonnegative terms, some of these tests may determine that $\sum\limits_{k=1}^{\infty} |a_k|$ converges, and if it does, then Theorem 7.2.8 would imply that $\sum\limits_{k=1}^{\infty} a_k$ converges. On the other hand, the converse of Theorem 7.2.8 does not hold, that is, if a series $\sum\limits_{k=1}^{\infty} a_k$ converges, we cannot conclude that $\sum\limits_{k=1}^{\infty} |a_k|$ converges. For example, $\sum\limits_{k=1}^{\infty} (-1)^{k+1} \frac{1}{k}$ converges, but $\sum\limits_{k=1}^{\infty} \left|(-1)^{k+1} \frac{1}{k}\right| = \sum\limits_{k=1}^{\infty} \frac{1}{k}$ does not converge. Thus, $\sum\limits_{k=1}^{\infty} (-1)^{k+1} \frac{1}{k}$ is conditionally convergent.

7.2.5 The Ratio and Root Tests

As mentioned above, one may be able to use some of the tests already introduced to determine if a series converges absolutely. But these tests do not work for all series. In this section we shall derive two new tests which often prove to be useful in showing that a series is absolutely convergent. These tests are called the Ratio Test and the Root Test. These tests deal with the terms of a given series by measuring the rate at which the terms converge to 0. Of course, we know that if the terms of a series converge to 0, this does not imply that the series converges. However, the proofs of these new tests show that if this rate of decrease is comparable to that of a convergent geometric series, then this will ensure convergence.

The Ratio Test

The Ratio Test measures the rate of decline of the terms of a series $a_1, a_2, \ldots, a_k, \ldots$ by investigating the long-term behavior of the ratio $\left|\frac{a_{k+1}}{a_k}\right|$. If this rate is less than 1, then the series will converge. The proof of the test relies on the direct comparison test and a geometric series.

Theorem 7.2.10 (Ratio Test I). Let $\sum\limits_{k=1}^{\infty} a_k$ be a series of nonzero real numbers.

(1) If there is real number $\rho < 1$ and a natural number N such that $\left|\frac{a_{k+1}}{a_k}\right| \leq \rho$ for all $k \geq N$, then $\sum\limits_{k=1}^{\infty} a_k$ converges absolutely.

(2) If there is an $N \in \mathbb{N}$ such that $\left|\frac{a_{k+1}}{a_k}\right| \geq 1$ for all $k \geq N$, then $\sum\limits_{k=1}^{\infty} a_k$ diverges.

Proof. Let $\sum\limits_{k=1}^{\infty} a_k$ be series of nonzero real numbers. To prove (1), let $\rho < 1$ and let $N \in \mathbb{N}$ be such that $\left|\frac{a_{k+1}}{a_k}\right| \leq \rho$ for all $k \geq N$. It follows that

$$|a_{k+1}| \leq \rho |a_k| \text{ for all } k \geq N. \tag{7.5}$$

Using (7.5), one can prove by induction on k that

$$|a_{N+k}| \leq \rho^k |a_N| \text{ for all } k \in \mathbb{N}. \tag{7.6}$$

Since $0 < \rho < 1$, the geometric series $\sum\limits_{k=1}^{\infty} \rho^k |a_N|$ clearly converges by Theorem 7.1.11. Thus, (7.6) and Theorem 7.2.1 imply that $\sum\limits_{k=1}^{\infty} |a_{N+k}|$ converges. Since $\sum\limits_{k=1}^{\infty} |a_{N+k}|$ is a tail of the series $\sum\limits_{k=1}^{\infty} |a_k|$, we conclude from Theorem 7.1.13 that the series $\sum\limits_{k=1}^{\infty} |a_k|$ converges. Therefore, $\sum\limits_{k=1}^{\infty} a_k$ converges absolutely.

To prove (2), let N be a natural number such that $\left|\frac{a_{k+1}}{a_k}\right| \geq 1$ for all $n \geq N$. Thus,

$$|a_{k+1}| \geq |a_k| \text{ for all } k \geq N.$$

Since $|a_N| > 0$, the sequence $\langle a_k \rangle$ does not converge to 0. Hence, $\sum\limits_{k=1}^{\infty} a_k$ diverges by Theorem 7.1.9. □

Theorem 3.8.16 and Theorem 7.2.10 imply the following more familiar appearing version of the ratio test (also see Exercise 12.)

Corollary 7.2.11 (Ratio Test II). Let $\sum\limits_{k=1}^{\infty} a_k$ be a series of nonzero real numbers.

(1) If $\limsup\limits_{k \to \infty} \left|\frac{a_{k+1}}{a_k}\right| < 1$, then $\sum\limits_{k=1}^{\infty} a_k$ converges absolutely.

(2) If $\liminf\limits_{k \to \infty} \left|\frac{a_{k+1}}{a_k}\right| > 1$, then $\sum\limits_{k=1}^{\infty} a_k$ diverges.

The Root Test

The Root Test is similar to the Ratio Test, except that it measures the rate of decrease of the terms of a series $a_1, a_2, \ldots, a_k, \ldots$ by examining the long-term behavior of the value $\sqrt[k]{|a_k|}$. Again, if this rate is less than 1, then the series will converge. The proof of this test also depends on the direct comparison test and a geometric series.

Theorem 7.2.12 (Root Test I). Let $\sum_{k=1}^{\infty} a_k$ be an infinite series.

(1) If there is a positive real number $\rho < 1$ and a natural number N such that $\sqrt[k]{|a_k|} \leq \rho$ for all $k \geq N$, then $\sum_{k=1}^{\infty} a_k$ converges absolutely.

(2) If $\sqrt[k]{|a_k|} \geq 1$ for infinitely many natural numbers k, then $\sum_{k=1}^{\infty} a_k$ diverges.

Proof. To prove (1), let $\rho \in \mathbb{R}^+$ and $N \in \mathbb{N}$ be such that $\sqrt[k]{|a_k|} \leq \rho < 1$ for all $k \geq N$. It follows, by induction, that

$$|a_{N+k}| \leq \rho^N \rho^k \text{ for all } k \in \mathbb{N}. \qquad (7.7)$$

Since $0 < \rho < 1$, the series $\sum_{k=1}^{\infty} \rho^N \rho^k$ converges, by Theorem 7.1.11. Thus, (7.7) and Theorem 7.2.1 imply that $\sum_{k=1}^{\infty} |a_{N+k}|$ converges. Since $\sum_{k=1}^{\infty} |a_{N+k}|$ is a tail of the series $\sum_{k=1}^{\infty} |a_k|$, Theorem 7.1.13 and Theorem 7.2.8 imply that $\sum_{k=1}^{\infty} a_k$ converges absolutely.

To prove (2), assume that $\sqrt[k]{|a_k|} \geq 1$ for infinitely many natural numbers k. Thus, $|a_k| \geq 1$ for infinitely many k. So the sequence $\langle a_k \rangle$ does not converge to 0. Hence, the series $\sum_{k=1}^{\infty} a_k$ diverges by Theorem 7.1.9. □

Theorem 3.8.16 and Theorem 7.2.12 imply the following modified version of the above root test (also see Exercise 13.)

Corollary 7.2.13 (Root Test II). Let $\sum_{k=1}^{\infty} a_k$ be an infinite series.

(1) If $\limsup_{k \to \infty} \sqrt[k]{|a_k|} < 1$, then $\sum_{k=1}^{\infty} a_k$ converges absolutely.

(2) If $\limsup_{k \to \infty} \sqrt[k]{|a_k|} > 1$ or $\left\langle \sqrt[k]{|a_k|} \right\rangle$ is not bounded above, then $\sum_{k=1}^{\infty} a_k$ diverges.

Theorem 3.8.17 implies that if ratio test II decides either the convergence or the divergence of a series, then root test II will arrive at the same conclusion. So root test II is considered to be a stronger test. However, ratio test II is often easier to apply.

Recall that if a sequence $\langle s_n \rangle$ of partial sums diverges, then the corresponding series $\sum_{k=1}^{\infty} a_k$ is said to diverge. Moreover, if $\langle s_n \rangle$ diverges to ∞ (see Definition 3.7.1), then we will write $\sum_{k=1}^{\infty} a_k = \infty$. We also write $\sum_{k=1}^{\infty} a_k = -\infty$, if $\langle s_n \rangle$ diverges to $-\infty$.

Exercises 7.2

*1. Prove Theorem 7.2.2(2). (See Theorem 3.1.24.)

*2. Prove Theorem 7.2.2(3).

*3. Under the hypothesis of Theorem 7.2.7, show that the "odd" sequence $\langle s_{2n-1}\rangle$ is decreasing and bounded below.

4. Let $\sum_{k=1}^{\infty}(-1)^k a_k$ be a series such that $a_k \geq a_{k+1} > 0$ for all k, and $\lim_{k\to\infty} a_k = 0$. Show that Theorem 7.2.7 implies that $\sum_{k=1}^{\infty}(-1)^k a_k$ converges to a sum T such that $|T - s_n| \leq a_{n+1}$ for all n, where s_n is the nth partial sum of the series.

*5. Let $\sum_{k=1}^{\infty} a_k$ be a series. For each $k \in \mathbb{N}$, let

$$\rho_k = \begin{cases} a_k, & \text{if } a_k > 0; \\ 0, & \text{if } a_k \leq 0. \end{cases} \qquad \eta_k = \begin{cases} a_k, & \text{if } a_k < 0; \\ 0, & \text{if } a_k \geq 0. \end{cases}$$

(a) Show that $|a_k| = \rho_k - \eta_k$ and $a_k = \rho_k + \eta_k$, for all $k \in \mathbb{N}$.

(b) Prove: $\sum_{k=1}^{\infty} a_k$ converges absolutely if and only if $\sum_{k=1}^{\infty} \rho_k$ and $\sum_{k=1}^{\infty} \eta_k$ converge.

(c) Suppose that $\sum_{k=1}^{\infty} a_k$ converges conditionally. Prove that $\sum_{k=1}^{\infty} \rho_k = \infty$ and $\sum_{k=1}^{\infty} \eta_k = -\infty$ (see Exercise 3 on page 195). Then prove that $\sum_{k=m}^{\infty} \rho_k = \infty$ and $\sum_{k=m}^{\infty} \eta_k = -\infty$, for each $m \in \mathbb{N}$.

6. Prove that if $\sum_{k=1}^{\infty} a_k^2$ and $\sum_{k=1}^{\infty} b_k^2$ converge, then $\sum_{k=1}^{\infty} a_k b_k$ converges absolutely. [Hint: $(a-b)^2 \geq 0$.]

7. Suppose that $\sum_{k=1}^{\infty} a_k$ is absolutely convergent and $\langle b_k\rangle$ is a convergent sequence. Prove that $\sum_{k=1}^{\infty} a_k b_k$ converges absolutely.

8. Using the integral test, show that $\sum_{k=1}^{\infty} \frac{k}{e^{k^2}}$ converges.

9. Test each series for convergence: (a) $\sum_{k=1}^{\infty} \frac{1}{\sqrt{k^3}}$, (b) $\sum_{k=1}^{\infty} \frac{1}{\sqrt[3]{k^2}}$.

10. Using a comparison test, test each series for convergence:
(a) $\sum_{k=1}^{\infty} \frac{1}{3^k+1}$, (b) $\sum_{k=1}^{\infty} \frac{1}{\sqrt{k}-.5}$, (c) $\sum_{k=1}^{\infty} \frac{1}{2^k+5}$, (d) $\sum_{k=1}^{\infty} \frac{3k+2}{\sqrt{k}(3k+5)}$, (e) $\sum_{k=1}^{\infty} \frac{1}{\ln(k)}$.

11. Determine whether or not the following alternating series converge.
(a) $\sum_{k=1}^{\infty}(-1)^{k+1}\frac{1}{k!}$, (b) $\sum_{k=1}^{\infty}(-1)^{k+1}\frac{1}{k}$, (c) $\sum_{k=1}^{\infty}(-1)^{k+1}\frac{k+3}{k^2+k}$.

*12. Let $\sum_{k=1}^{\infty} b_k$ be an series of nonzero terms. Show that Theorem 7.2.10 implies:

(1) If $\lim_{n\to\infty}\left|\frac{b_{n+1}}{b_n}\right| < 1$, then $\sum_{k=1}^{\infty} b_k$ converges absolutely.

(2) If $\lim_{n\to\infty}\left|\frac{b_{n+1}}{b_n}\right| > 1$ or $\lim_{n\to\infty}\left|\frac{b_{n+1}}{b_n}\right| = \infty$, then $\sum_{k=1}^{\infty} b_k$ diverges.

*13. Let $\sum\limits_{k=1}^{\infty} a_k$ be an infinite series. Show that Theorem 7.2.12 implies the following:

    (1) If $\lim\limits_{n\to\infty} \sqrt[k]{|a_k|} < 1$, then $\sum\limits_{k=1}^{\infty} a_k$ converges absolutely.

    (2) If $\lim\limits_{n\to\infty} \sqrt[k]{|a_k|} > 1$ or $\lim\limits_{n\to\infty} \sqrt[k]{|a_k|} = \infty$, then $\sum\limits_{k=1}^{\infty} a_k$ diverges.

14. Use an absolute convergence test to show that the following series converge.

    (a) $\sum\limits_{k=1}^{\infty} (-1)^k \frac{1}{k^2}$, (b) $\sum\limits_{k=1}^{\infty} \frac{\sin(k)}{2^k}$.

15. Test each of these series for convergence:

    (a) $\sum\limits_{k=1}^{\infty} \frac{2^k}{k!}$, (b) $\sum\limits_{k=1}^{\infty} \frac{k^k}{k!}$, (c) $\sum\limits_{k=1}^{\infty} \frac{1}{2k-3}$.

16. Show that the series

    (a) $\sum\limits_{k=1}^{\infty} (-1)^k \frac{1}{k!}$ is absolutely convergent

    (b) $\sum\limits_{k=1}^{\infty} (-1)^{k+1} \frac{1}{k}$ conditionally convergent.

17. Test each of these series for convergence:

    (a) $\sum\limits_{k=2}^{\infty} \frac{1}{(\ln k)^k}$, (b) $\sum\limits_{k=1}^{\infty} \frac{1}{k^2}$.

18. Suppose that the series $\sum\limits_{k=1}^{\infty} a_k$ is conditionally convergent where $a_k \neq 0$ for all $k \in \mathbb{N}$. Show that $\sum\limits_{k=1}^{\infty} \eta_k$ diverges, where for each $k \in \mathbb{N}$, η_k is defined by

$$\eta_k = \begin{cases} a_k, & \text{if } a_k < 0; \\ 0, & \text{if } a_k > 0. \end{cases}$$

7.3 REGROUPING AND REARRANGING TERMS OF A SERIES

Given a finite sum of real numbers, say $a_1 + a_2 + a_3 + a_4 + a_5$, by the associative law we see that

$$a_1 + a_2 + a_3 + a_4 + a_5 = a_1 + (a_2 + a_3) + (a_4 + a_5).$$

So we can regroup the terms of the finite sum and get the same answer. Clearly, any regrouping of the terms in a finite sum will yield the same sum. Moreover, by the commutative law, we have that

$$a_1 + a_2 + a_3 + a_4 + a_5 = a_4 + a_2 + a_5 + a_3 + a_1.$$

Thus, by associativity and commutativity, we can rearrange the terms of a finite sum get the same result; that is, the order in which one performs the addition does not affect the sum. In the same way that one can change the terms of a finite sum, one can alter the terms of an infinite series. Can we reach the same conclusions for an infinite series that we did for a finite sum? In this section, we address the following question: What happens to the convergence or divergence of an infinite series if the terms of the series are regrouped or rearranged?

7.3.1 Regrouping

A regrouping of an infinite series is obtained simply by "inserting non-overlapping parentheses" in which each term is enclosed within one pair of a parentheses. For example, consider the infinite series

$$\sum_{k=1}^{\infty} a_k = a_1 + a_2 + a_3 + a_4 + a_5 + a_6 + \cdots .$$

We can construct a new series by regrouping the terms of the series as follows

$$\sum_{k=1}^{\infty} (a_{2k-1} + a_{2k}) = (a_1 + a_2) + (a_3 + a_4) + (a_5 + a_6) + \cdots . \tag{7.8}$$

By using different regroupings, one can construct many other new series as well. Suppose that a given series converges. Does a regrouping of the series also converge? The answer is yes; but, before proving this fact, we must first present a precise definition of a regrouping of a given series.

Definition 7.3.1. Let $\sum_{k=1}^{\infty} a_k$ be a given series. Let $\psi \colon \mathbb{N} \to \mathbb{N}$ be strictly increasing; that is, for all m and n in $\mathbb{N}$, if $m < n$, then $\psi(m) < \psi(n)$. Define a new series $\sum_{k=1}^{\infty} b_k$ as follows:

$$b_1 = \Big(a_1 + a_2 + \cdots + a_{\psi(1)} \Big)$$

$$b_2 = \Big(a_{\psi(1)+1} + a_{\psi(1)+2} + \cdots + a_{\psi(2)} \Big)$$

$$\vdots \tag{7.9}$$

$$b_k = \Big(a_{\psi(k-1)+1} + a_{\psi(k-1)+2} + \cdots + a_{\psi(k)} \Big)$$

$$\vdots$$

The series $\sum_{k=1}^{\infty} b_k$ is said to be a **regrouping** of the series $\sum_{k=1}^{\infty} a_k$.

The regrouping in (7.8) can be formally defined by using the function $\psi \colon \mathbb{N} \to \mathbb{N}$ defined by $\psi(k) = 2k$. The next theorem shows that if a series converges, then any regrouping of the series will converge to the same value. So a convergent infinite series satisfies a generalized associative law.

Theorem 7.3.2. Let $\sum_{k=1}^{\infty} a_k$ be a series that converges to the real number α, and let $\sum_{k=1}^{\infty} b_k$ be a regrouping of $\sum_{k=1}^{\infty} a_k$ using the strictly increasing function $\psi \colon \mathbb{N} \to \mathbb{N}$. Then $\sum_{k=1}^{\infty} b_k$ converges to α.

Proof. For each $n \in \mathbb{N}$, let $s_n = \sum_{k=1}^{n} a_k$ and $t_n = \sum_{k=1}^{n} b_k$ be the respective nth partial sums. Assume that $\langle s_n \rangle$ converges to α. Observe that $t_n = s_{\psi(n)}$ for each $n \in \mathbb{N}$ (see (7.9)). Thus, $\langle t_n \rangle = \langle s_{\psi(n)} \rangle$ and so, $\langle t_n \rangle$ is a subsequence of $\langle s_n \rangle$. Theorem 3.3.4 implies that $\langle t_n \rangle$ converges to α. $\square$

In a manner similar to Definition 7.3.1, one can define the regrouping of any series having the form $\sum_{k=0}^{\infty} a_k$ (see Exercise 1). Theorem 7.3.2 also holds for such a regrouping.

7.3.2 Rearrangements

Suppose that $A = a_1 + a_2 + a_3 + \cdots + a_n$, where $a_1, a_2, a_3, \ldots, a_n$ are real numbers and n is a natural number. Thus, A is a finite sum of real numbers. Since addition satisfies the commutative law, if one evaluates a rearrangement of the terms in this sum, say, $a_4 + a_1 + a_n + \cdots + a_7$, then one will obtain, again, the original value A. Now, suppose that the infinite series $\sum_{k=1}^{\infty} a_n$ converges to A. What happens if we rearrange the terms in this infinite series? Will the rearranged series converge to A, as well? We will now confront this question. First, we must precisely define what it means to rearrange the terms of an infinite series.

If $\sigma \colon \mathbb{N} \to \mathbb{N}$ be one-to-one and onto $\mathbb{N}$, then σ is a **permutation** of $\mathbb{N}$. If σ is such a permutation, then

$$\sigma(1), \sigma(2), \sigma(3), \sigma(4), \sigma(5), \ldots$$

produces a reordering of all the natural numbers, when σ is not the identity function. For example, let $\sigma \colon \mathbb{N} \to \mathbb{N}$ be defined by

$$\sigma(n) = \begin{cases} n+1, & \text{if } n \text{ is odd;} \\ n-1, & \text{if } n \text{ is even.} \end{cases} \tag{7.10}$$

The function σ is one-to-one and onto $\mathbb{N}$. By evaluating $\sigma(1), \sigma(2), \sigma(3), \sigma(4), \sigma(5), \ldots$, we obtain the following simple reordering of the natural numbers:

$$2, 1, 4, 3, 6, 5, 8, 7, \ldots.$$

Definition 7.3.3. Let $\sum_{k=1}^{\infty} a_k$ be an infinite series and let $\sigma \colon \mathbb{N} \to \mathbb{N}$ be a permutation of $\mathbb{N}$. The series $\sum_{k=1}^{\infty} a_{\sigma(k)}$ is called a **rearrangement** of the series $\sum_{k=1}^{\infty} a_k$.

To illustrate this notion, let $\sigma \colon \mathbb{N} \to \mathbb{N}$ be the permutation of $\mathbb{N}$ given by (7.10). Applying σ to the indices of $\sum_{k=1}^{\infty} a_k$, we obtain the rearrangement

$$\sum_{k=1}^{\infty} a_{\sigma(k)} = a_2 + a_1 + a_4 + a_3 + a_6 + a_5 + \cdots.$$

Now, suppose that a series $\sum_{k=1}^{\infty} a_k$ converges to A and $\sum_{k=1}^{\infty} a_{\sigma(k)}$ is a rearrangement of $\sum\limits_{k=1}^{\infty} a_k$. Will the rearrangement converge? If so, will it converge to A? Lemma 7.3.4 and Corollary 7.3.5, below, provide a partial answer.

The proof of Lemma 7.3.4 takes advantage of the following observation: Whenever σ is a permutation of $\mathbb{N}$, then Theorem 1.3.9 implies that σ^{-1} is also permutation of $\mathbb{N}$ such that $\sigma^{-1}(\sigma(k)) = k$ for all $k \in \mathbb{N}$. Thus, $\sum\limits_{k=1}^{\infty} a_k$ is a rearrangement of $\sum\limits_{k=1}^{\infty} a_{\sigma(k)}$.

Lemma 7.3.4. If the series $\sum\limits_{k=1}^{\infty} p_k$ converges to A where $p_k \geq 0$ for all $k \in \mathbb{N}$, then any rearrangement of this series converges to A.

Proof. Suppose that $\sum\limits_{k=1}^{\infty} p_k = A$ and that (▲) $p_k \geq 0$ for all $k \in \mathbb{N}$. Let σ be a permutation of $\mathbb{N}$. So, $\sum\limits_{k=1}^{\infty} p_{\sigma(k)}$ is a rearrangement of $\sum\limits_{k=1}^{\infty} p_k$. For each $n \in \mathbb{N}$, let

$$s_n = \sum_{k=1}^{n} p_k \text{ and } t_n = \sum_{k=1}^{n} p_{\sigma(k)}.$$

Since $p_k \geq 0$ for all $k \in \mathbb{N}$, the sequences $\langle s_n \rangle$ and $\langle t_n \rangle$ are increasing. As $\sum\limits_{k=1}^{\infty} p_k = A$, Corollary 3.4.3 implies that $s_n \leq A$ for all $n \in \mathbb{N}$. We will now show that $t_n \leq A$ for all $n \in \mathbb{N}$. Let $n \in \mathbb{N}$. Since $\{\sigma(1), \sigma(2), \ldots, \sigma(n)\}$ is a finite set of natural numbers, there is a natural number N such that $\sigma(k) \leq N$ for all $k \leq n$. So for each $k \leq n$, there is an $i \leq N$ such that $p_{\sigma(k)} = p_i$. Hence, by (▲),

$$t_n = \sum_{k=1}^{n} p_{\sigma(k)} \leq \sum_{k=1}^{N} p_k = s_N \leq A.$$

Thus, $t_n \leq A$ for all $n \in \mathbb{N}$. Lemma 3.4.2 implies that there is a $B \in \mathbb{R}$ such that $\lim\limits_{n \to \infty} t_n = B$ and $B \leq A$, that is, $\sum\limits_{k=1}^{\infty} p_{\sigma(k)} = B \leq A$.

As $\sum\limits_{k=1}^{\infty} p_{\sigma(k)} = B$ and $\sum\limits_{k=1}^{\infty} p_k$ is a rearrangement of $\sum\limits_{k=1}^{\infty} p_{\sigma(k)}$, the above argument implies that $\sum\limits_{k=1}^{\infty} p_k = A \leq B$. Hence, $A = B$. Therefore, $\sum\limits_{k=1}^{\infty} p_{\sigma(k)} = A$. $\square$

Corollary 7.3.5. If $\sum\limits_{k=1}^{\infty} q_k$ converges to B where $q_k \leq 0$ for all $k \in \mathbb{N}$, then any rearrangement of this series converges to B.

Proof. Let $\sum\limits_{k=1}^{\infty} q_k = B$ and $q_k \leq 0$ for all $k \in \mathbb{N}$, and let σ be a permutation of $\mathbb{N}$. Theorem 7.1.7 implies that $\sum\limits_{k=1}^{\infty} (-q_k) = -B$ where $-q_k \geq 0$ for all $k \in \mathbb{N}$. Thus, $\sum\limits_{k=1}^{\infty} -q_{\sigma(k)} = -B$ by Lemma 7.3.4. Hence, again by Theorem 7.1.7, $\sum\limits_{k=1}^{\infty} q_{\sigma(k)} = B$. $\square$

So, if a series has only positive terms, or only negative terms, and it converges to a value, then any rearrangement of this series will converge to this same value. Suppose that a series has an "infinite" mix of positive and negative terms. Will the same hold for such a series? Yes, if the series is absolutely convergent.

Theorem 7.3.6. If $\sum\limits_{k=1}^{\infty} a_k$ is an absolutely convergent series that converges to A, then any rearrangement of this series converges to A.

Proof. Assume that $\sum\limits_{k=1}^{\infty} a_k$ is an absolutely convergent and converges to A, that is,

(▲) $\sum\limits_{k=1}^{\infty} a_k = A$. For each $k \in \mathbb{N}$, let

$$\rho_k = \begin{cases} a_k, & \text{if } a_k > 0; \\ 0, & \text{if } a_k \leq 0. \end{cases} \qquad \eta_k = \begin{cases} a_k, & \text{if } a_k < 0; \\ 0, & \text{if } a_k \geq 0. \end{cases}$$

Clearly, $\rho_k \geq 0$ and $\eta_k \leq 0$, for all $k \in \mathbb{N}$. Note that (▼) $a_k = \rho_k + \eta_k$ for all $k \in \mathbb{N}$. Since $\sum\limits_{k=1}^{\infty} a_k$ is absolutely convergent, Exercise 5 on page 204 implies that (►) both $\sum\limits_{k=1}^{\infty} \rho_k$ and $\sum\limits_{k=1}^{\infty} \eta_k$ converge. Let σ be a permutation of $\mathbb{N}$. We now have the following:

$$\begin{aligned}
A = \sum_{k=1}^{\infty} a_k = \sum_{k=1}^{\infty} (\rho_k + \eta_k) & \qquad \text{by (▲) and (▼)} \\
= \sum_{k=1}^{\infty} \rho_k + \sum_{k=1}^{\infty} \eta_k & \qquad \text{by (►) and Theorem 7.1.7} \\
= \sum_{k=1}^{\infty} \rho_{\sigma(k)} + \sum_{k=1}^{\infty} \eta_{\sigma(k)} & \qquad \text{by Lemma 7.3.4 \& Corollary 7.3.5} \\
= \sum_{k=1}^{\infty} (\rho_{\sigma(k)} + \eta_{\sigma(k)}) & \qquad \text{by Theorem 7.1.7} \\
= \sum_{k=1}^{\infty} a_{\sigma(k)} & \qquad \text{by (▼).} \qquad \square
\end{aligned}$$

We next prove a theorem which shows that the conclusion of Theorem 7.3.6 fails for any conditionally convergent series, a surprising result. The theorem states that one can rearrange a conditionally convergent series to one that converges to any real number of one's choosing. The key idea behind the proof of this theorem is not complicated; however, the formal proof is a bit technical, in part, because one has to inductively define three sequences before one can identify the relevant rearrangement of the given conditionally convergent series.

Before presenting the proof, we offer the following summary of the proof: Since the series is conditionally convergent, it must have an infinite number of positive terms and an infinite number of negative terms (see Exercise 5(c) on page 204). Let $L \geq 0$. First we choose successive positive terms from the conditionally convergent series until their sum is greater than L. Then we choose successive negative terms, from the series, until the sum of *all* the chosen positive and negative terms is less than L. We then inductively repeat this process by adding, in the same manner, more positive and negative terms to the preceding sum. During this process, we also record how this rearrangement is being constructed. In the end, we will have a rearrangement of the series that converges to L.

Theorem 7.3.7. If $\sum_{k=1}^{\infty} a_k$ is a conditionally convergent series, then for any real number L, there is a rearrangement of this series that converges to L.

Proof. Let $\sum_{k=1}^{\infty} a_k$ be a conditionally convergent series. Without loss of generality, we will assume that $a_k \neq 0$ for all $k \in \mathbb{N}$. Let $\rho_1, \rho_2, \rho_3, \ldots$ be all the positive terms in the series $\sum_{k=1}^{\infty} a_k$, ordered as they occur in the series; and let $\eta_1, \eta_2, \eta_3, \ldots$ be all the negative terms in $\sum_{k=1}^{\infty} a_k$, also ordered as in the series. Since $\sum_{k=1}^{\infty} a_k$ converges, Theorem 7.1.8 implies that the sequences $\langle \rho_k \rangle$ and $\langle \eta_k \rangle$ converge to 0. Moreover, Exercise 5(c) on page on 204 implies that $\sum_{k=m}^{\infty} \rho_k = \infty$ and $\sum_{k=m}^{\infty} \eta_k = -\infty$, for each $m \in \mathbb{N}$.

Let $L \geq 0$. (The proof for $L < 0$ is Exercise 10.) We inductively define sequences $\langle i_k \rangle$, $\langle n_k \rangle$, and $\langle s_{n_k} \rangle$. These sequences will allow us to identify a rearrangement of $\sum_{k=1}^{\infty} a_k$ and a sequence of partial sums of this rearrangement. At the odd stages of the induction we put some of the positive terms of $\sum_{k=1}^{\infty} a_k$ into the rearrangement, and at the even stages we place a some of the negative terms, of the original series, into the rearrangement. Let $\rho_0 = 0$ and $\eta_0 = 0$. Since the partial sums of $\sum_{k=1}^{\infty} \rho_k$ are not bounded above and $L \geq 0$, there is an $i_1 \in \mathbb{N}$ such that

$$\rho_1 + \rho_2 + \rho_3 + \cdots + \rho_{i_1-1} \leq L < \rho_1 + \rho_2 + \rho_3 + \cdots + \rho_{i_1-1} + \rho_{i_1},$$

and let $n_1 = i_1$ and $s_{n_1} = \rho_1 + \rho_2 + \cdots + \rho_{i_1}$. Thus, $L < s_{n_1}$ and $|s_{n_1} - L| \leq |\rho_{i_1}|$ (Corollary 2.2.10(1)). Since the partial sums of $\sum_{k=1}^{\infty} \eta_k$ are not bounded below and $L < s_{n_1}$, there is an $i_2 \in \mathbb{N}$ such that

$$s_{n_1} + \eta_1 + \eta_2 + \eta_3 + \cdots + \eta_{i_2-1} + \eta_{i_2} < L \leq s_{n_1} + \eta_1 + \eta_2 + \eta_3 + \cdots + \eta_{i_2-1}.$$

Let $n_2 = n_1 + i_2$ and $s_{n_2} = s_{n_1} + \eta_1 + \eta_2 + \cdots + \eta_{i_2}$. So $s_{n_2} < L$ and $|s_{n_2} - L| \leq |\eta_{i_2}|$ (Corollary 2.2.10(2)).

Let $k \geq 2$. Assume that i_{k-1}, n_k, s_{n_k} are defined and that $s_{n_k} < L$ if k is even, and $s_{n_k} > L$ if k is odd. We shall now define $i_{k+1}, n_{k+1}, s_{n_{k+1}}$. There are two cases to consider.

CASE 1: $k + 1$ is odd. Since the partial sums of $\sum_{k=(i_{k-1}+1)}^{\infty} \rho_k$ are not bounded above and $s_{n_k} < L$, there is a natural number $i_{k+1} > i_{k-1}$ such that

$$s_{n_k} + \rho_{(i_{k-1}+1)} + \cdots + \rho_{(i_{k+1}-1)} \leq L < s_{n_k} + \rho_{(i_{k-1}+1)} + \cdots + \rho_{(i_{k+1}-1)} + \rho_{i_{k+1}},$$

and let $n_{k+1} = n_k + i_{k+1}$ and

$$s_{n_{k+1}} = s_{n_k} + \rho_{(i_{k-1}+1)} + \cdots + \rho_{(i_{k+1}-1)} + \rho_{i_{k+1}}.$$

So $L < s_{n_{k+1}}$ and $|s_{n_{k+1}} - L| \leq |\rho_{i_{k+1}}|$.

CASE 2: $k + 1$ is even. Since the partial sums of $\sum_{k=(i_{k-1}+1)}^{\infty} \eta_k$ are not bounded below and $L < s_{n_k}$, there is a natural number $i_{k+1} > i_{k-1}$ such that

$$s_{n_k} + \eta_{(i_{k-1}+1)} + \cdots + \eta_{(i_{k+1}-1)} + \eta_{i_{k+1}} < L \leq s_{n_k} + \eta_{(i_{k-1}+1)} + \cdots + \eta_{(i_{k+1}-1)},$$

and let $n_{k+1} = n_k + i_{k+1}$ and

$$s_{n_{k+1}} = s_{n_k} + \eta_{i_{k-1}+1} + \cdots + \eta_{i_{k+1}-1} + \eta_{i_{k+1}}.$$

Hence, $s_{n_{k+1}} < L$ and $|s_{n_{k+1}} - L| \leq |\eta_{i_{k+1}}|$.

The inductive definition of the sequences $\langle i_k \rangle$, $\langle n_k \rangle$, and $\langle s_{n_k} \rangle$ is now complete. As a result, for all $k \in \mathbb{N}$, we have that

$$\begin{aligned} |s_{n_k} - L| &\leq |\rho_{i_k}| \text{ if } k \text{ is odd;} \\ |s_{n_k} - L| &\leq |\eta_{i_k}| \text{ if } k \text{ is even.} \end{aligned} \tag{7.11}$$

The constructed series is

$$\rho_1 + \cdots + \rho_{i_1} + \eta_1 + \cdots + \eta_{i_2} + \rho_{i_1+1} + \cdots + \rho_{i_3} + \eta_{i_2+1} + \cdots + \eta_{i_4} + \cdots, \tag{7.12}$$

which is a rearrangement of $\sum_{k=1}^{\infty} a_k$, denoted by $\sum_{k=1}^{\infty} a_{\sigma(k)}$. Let $\langle s_n \rangle$ be the sequence of partial sums of the series (7.12), and let $\langle s_{n_k} \rangle$ be the subsequence of $\langle s_n \rangle$ that was defined in the above inductive definition. For example, $s_{n_1} = \rho_1 + \cdots + \rho_{i_1}$ and $s_{n_2} = \rho_1 + \cdots + \rho_{i_1} + \eta_1 + \cdots + \eta_{i_2}$.

Since the sequences $\langle \rho_k \rangle$ and $\langle \eta_k \rangle$ converge to 0, the subsequence $\langle s_{n_k} \rangle$ converges to L by (7.11). From the definition of $\langle s_{n_k} \rangle$ and the terms in (7.12), it follows that for all $k \in \mathbb{N}$ and $n \in \mathbb{N}$, if $n_k \leq n \leq n_{k+1}$, then $s_{n_k} \leq s_n \leq s_{n_{k+1}}$ or $s_{n_{k+1}} \leq s_n \leq s_{n_k}$. Lemma 3.3.7 implies that $\langle s_n \rangle$ converges to L, that is, $\sum_{k=1}^{\infty} a_{\sigma(k)} = L$. □

Thus, given any specific real number, the terms of a conditionally convergent series can be rearranged to obtain a series that converges to this given number. Hence, the sum of a conditionally convergent series depends on the arrangement of its terms. Whereas, a rearrangement of the terms in a absolutely convergent series, or in a finite sum, will have no effect on the resulting sum. Thus, the properties of addition that hold for finite sums do not always hold for infinite series.

Exercises 7.3

*1. Let $\mathbb{N}^* = \{0, 1, 2, 3, 4, \dots\}$. Define a strictly increasing function $\psi: \mathbb{N}^* \to \mathbb{N}^*$ that formally verifies that $\sum_{k=0}^{\infty} (a_{2k} + a_{2k+1})$ is a regrouping of $\sum_{k=0}^{\infty} a_k$.

2. Show that $\sum\limits_{k=0}^{\infty} (4^{2k}r^{2k} + 2^{2k+1}r^{2k+1})$ is a regrouping of $\sum\limits_{k=0}^{\infty} (3 + (-1)^k)^k r^k$, using Exercise 1. Let $|r| < \frac{1}{4}$. Given that $\sum\limits_{k=0}^{\infty} (3 + (-1)^k)^k r^k$ converges, find its sum.

3. Find a regrouping of $\sum\limits_{k=1}^{\infty} (-1)^{k+1}$ that converges to 0. Now find another regrouping of this series that converges to 1.

4. Let $\sum\limits_{k=1}^{\infty} a_k$ be a series of positive terms. Let $\psi \colon \mathbb{N} \to \mathbb{N}$ be strictly increasing and let $\sum\limits_{k=1}^{\infty} b_k$ be the resulting regrouping of $\sum\limits_{k=1}^{\infty} a_k$, as in Definition 7.3.1. Suppose that $\sum\limits_{k=1}^{\infty} b_k$ converges to α. Using Lemma 3.3.7, show that $\sum\limits_{k=1}^{\infty} a_k$ converges to α.

5. Suppose that $\sum\limits_{k=1}^{\infty} a_k$ is absolutely convergent, and let $\sum\limits_{k=1}^{\infty} a_{\sigma(k)}$ be a rearrangement of $\sum\limits_{k=1}^{\infty} a_k$. Prove that $\sum\limits_{k=1}^{\infty} |a_k| = \sum\limits_{k=1}^{\infty} |a_{\sigma(k)}|$.

6. Let $\sum\limits_{k=1}^{\infty} a_k$ be a series and let σ be a permutation of $\mathbb{N}$. Prove that $\sum\limits_{k=1}^{\infty} a_k$ is absolutely convergent if and only if $\sum\limits_{k=1}^{\infty} a_{\sigma(k)}$ is absolutely convergent.

7. Let σ be a permutation of $\mathbb{N}$. Evaluate the sum of $\sum\limits_{k=1}^{\infty} \frac{1}{2^{\sigma(k)}}$.

8. Let $\sum\limits_{k=1}^{\infty} a_k$ be divergent with nonnegative terms. Show that $\sum\limits_{k=1}^{\infty} a_{\sigma(k)}$ also diverges for any rearrangement.

9. Let $\sum\limits_{k=1}^{\infty} a_k$ be a conditionally convergent series.

   (a) Show that there is a rearrangement of $\sum\limits_{k=1}^{\infty} a_k$ such that the sequence of partial sums of the rearranged series is not bounded above.

   (b) Show that there is a rearrangement of $\sum\limits_{k=1}^{\infty} a_k$ such that the sequence of partial sums of the rearranged series is bounded but does not converge.

*10. Let $\sum\limits_{k=1}^{\infty} b_k$ be a conditionally convergent series.

   (a) Show that $\sum\limits_{k=1}^{\infty} (-b_k)$ is conditionally convergent.

   (b) Assume that Theorem 7.3.7 holds whenever $L \geq 0$. Show that this implies that for any $J < 0$, there is a permutation σ of $\mathbb{N}$ such that $\sum\limits_{k=1}^{\infty} b_{\sigma(k)} = J$.

CHAPTER 8

Sequences and Series of Functions

In this chapter, we will investigate sequences and series of functions. We have already investigated sequences and series of real numbers. Each real number number that appears in a sequence is viewed as a distinct object. A function f is also a distinct mathematical object. Recall that $f(x)$ denotes only the value of f at x. So $f(x)$ is a real number, whereas the object f is not a real number; it is an operation that produces a real number value $f(x)$ for each x in the domain of f.

8.1 POINTWISE AND UNIFORM CONVERGENCE

8.1.1 Sequences of Functions

In Definition 3.0.1, we defined a sequence to be a function $s\colon \mathbb{N} \to \mathbb{R}$. We will now extend the notion of a sequence to be any function whose domain is the set of natural numbers. So to obtain a sequence of functions, we need to associate a function f_n to each natural number n. We will then let $\langle f_n \rangle$ denote this **sequence of functions**. For example, for each $n \in \mathbb{N}$, let $f_n\colon \mathbb{R} \to \mathbb{R}$ be defined by $f_n(x) = nx^n$. Then $\langle f_n \rangle$ is a sequence of functions. In most cases, the functions in a sequence will have the same domain.

Let $\langle f_n \rangle$ be a sequence of functions each of which has domain D. For each $x \in D$, we have that $\langle f_n(x) \rangle$ is a sequence of real numbers, and it is not not a sequence of functions. Moreover, the sequence $\langle f_n(x) \rangle$ of real numbers may or may not converge. However, when $\langle f_n(x) \rangle$ converges for all $x \in D$, we can define a new function.

Definition 8.1.1. Let $\langle f_n \rangle$ be a sequence of functions that are defined on $D \subseteq \mathbb{R}$. Then $\langle f_n \rangle$ **converges pointwise** on D if for each $x \in D$, the sequence $\langle f_n(x) \rangle$ converges. Whenever $\langle f_n \rangle$ converges pointwise on D, we can define the **limit function** $f\colon D \to \mathbb{R}$ by

$$f(x) = \lim_{n \to \infty} f_n(x)$$

for each $x \in D$, and we say that $\langle f_n \rangle$ converges pointwise to f on D.

Example 8.1.2. Let $f_n\colon [0,1] \to \mathbb{R}$ be defined by $f_n(x) = x^n$, for each $n \in \mathbb{N}$. Then for each $x \in [0,1]$, recalling Corollary 3.1.12, we have that $\lim_{n \to \infty} f_n(x) = f(x)$ where

$$f(x) = \begin{cases} 0, & \text{if } 0 \le x < 1; \\ 1, & \text{if } x = 1. \end{cases}$$

Observe that the limit function f is not continuous while all of the functions in the sequence $\langle f_n \rangle$ are continuous. In addition, each f_n is differentiable on $[0,1]$, but f is not differentiable at 1.

As Example 8.1.2 demonstrates, there are sequences of continuous functions whose pointwise limit is not continuous. There are also sequences of bounded functions whose pointwise limit is not bounded. We now define a stronger notion of convergence which does not have these particular shortcomings.

Definition 8.1.3. Let $\langle f_n \rangle$ be a sequence of functions that are defined on $D \subseteq \mathbb{R}$, and let $f\colon D \to \mathbb{R}$. Then $\langle f_n \rangle$ **converges uniformly** to f on D if for every $\varepsilon > 0$, there is an $N \in \mathbb{N}$ such that for all $n \in \mathbb{N}$, if $n > N$, then $|f_n(x) - f(x)| < \varepsilon$ for all $x \in D$. A sequence $\langle f_n \rangle$ is said to *converge uniformly* on D if there is a function f to which $\langle f_n \rangle$ converges uniformly on D.

There is a key difference between pointwise convergence and uniform convergence. For uniform convergence, given any $\varepsilon > 0$, there is an $N \in \mathbb{N}$ so that whenever $n > N$, we have that $|f_n(x) - f(x)| < \varepsilon$, not just for a specific point in D, but *for all $x \in D$*; that is, the entire graph of f_n must lie within the graphs of $f - \varepsilon$ and $f + \varepsilon$ (see Figure 8.1).

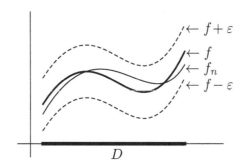

Figure 8.1: If $n > N$, then $|f_n(x) - f(x)| < \varepsilon$ for all $x \in D$.

Remark. If $\langle f_n \rangle$ converges uniformly to f on D, then

1. $\langle f_n \rangle$ converges pointwise to f on D, and
2. $\langle f_n \rangle$ converges uniformly to f on any subset of D.

In logical terms, $\langle f_n \rangle$ converges uniformly to f on D means that

$$(\forall \varepsilon > 0)(\exists N \in \mathbb{N})(\forall n \in \mathbb{N})(\forall x \in D)(n > N \to |f_n(x) - f(x)| < \varepsilon),$$

and it is this logical form that motivates the following proof strategy.

Proof Strategy 8.1.4. To prove that $\langle f_n \rangle$ converges uniformly to f on D, use the proof diagram:

Let $\varepsilon > 0$ be arbitrary.
Let $N = $ (the value in $\mathbb{N}$ you found).
Let $n > N$ be a natural number.
Let $x \in D$ be arbitrary.
Prove $|f_n(x) - f(x)| < \varepsilon$.

In a proof we may be assuming that a given sequence of functions converges uniformly. Our next strategy will then be useful.

Assumption Strategy 8.1.5. When assuming that $\langle f_n \rangle$ converges uniformly to f on D, then for any $\varepsilon > 0$, there is an $N \in \mathbb{N}$ such that $|f_n(x) - f(x)| < \varepsilon$, for all $n > N$ and all $x \in D$.

In the proof of our next theorem, we shall apply both of the above strategies.

Theorem 8.1.6. Suppose that $\langle f_n \rangle$ converges uniformly to f on D. Let $c \in \mathbb{R}$ be nonzero. Then $\langle cf_n \rangle$ converges uniformly to cf on the set D.

Proof. Assume that $\langle f_n \rangle$ converges uniformly to f on D and $c \neq 0$. Let $\varepsilon > 0$. Thus, there is an $N \in \mathbb{N}$ such that for all $n > N$ and all $x \in D$,

$$|f_n(x) - f(x)| < \frac{\varepsilon}{|c|}. \tag{8.1}$$

We will now prove that $|cf_n(x) - cf(x)| < \varepsilon$ for all $n > N$ and all $x \in D$. To do this, let $n > N$ and let $x \in D$. Hence,

$$|cf_n(x) - cf(x)| = |c|\,|f_n(x) - f(x)| \quad \text{by property of abs. value}$$
$$< |c|\frac{\varepsilon}{|c|} = \varepsilon \qquad \text{by (8.1)}.$$

Therefore, the sequence $\langle cf_n \rangle$ converges uniformly to cf. $\qquad \square$

The above proof is fairly straightforward. However, to prove that a sequence of functions converges uniformly to a specific function using Definition 8.1.3 is typically difficult. Fortunately, our next theorem offers an alternative method.

Theorem 8.1.7. Let $\langle f_n \rangle$ be a sequence of functions that converges pointwise to f on D. For each $n \in \mathbb{N}$, let $\beta_n = \sup\{|f_n(x) - f(x)| : x \in D\}$. Then $\langle f_n \rangle$ converges uniformly to f on D if and only if $\lim_{n \to \infty} \beta_n = 0$.

Proof. See Exercise 3. $\qquad \square$

Theorem 8.1.7 can also be used to show that a sequence of functions does not converge uniformly to a given function. This is illustrated in our next example.

Example 8.1.8. For each $n \in \mathbb{N}$ define $f_n \colon (0,1) \to \mathbb{R}$ by $f_n(x) = x^n$. The sequence of functions $\langle f_n \rangle$ converges pointwise to the zero function on $(0,1)$. Observe that

$$\beta_n = \sup\{|x^n - 0| : x \in (0,1)\} = 1$$

for each $n \in \mathbb{N}$. Since $\lim_{n \to \infty} \beta_n = 1 \neq 0$, Theorem 8.1.7 implies that $\langle f_n \rangle$ does not converge uniformly to the zero function on $(0,1)$.

So by Example 8.1.8, pointwise convergence does not imply uniform convergence. We now extend the notion of a Cauchy sequence of real numbers to a sequence of functions.

Definition 8.1.9. Let $\langle f_n \rangle$ be a sequence of functions that are defined on $D \subseteq \mathbb{R}$. Then $\langle f_n \rangle$ is **uniformly Cauchy** on D if for every $\varepsilon > 0$, there is an $N \in \mathbb{N}$ such that for all $m, n \in \mathbb{N}$, if $n > N$ and $m > N$, then $|f_n(x) - f_m(x)| < \varepsilon$ for all $x \in D$.

One can use our next theorem to prove that a sequence of functions converges uniformly without knowing the purported limit function.

Theorem 8.1.10 (Cauchy Criterion). Let $\langle f_n \rangle$ be a sequence of functions that are defined on $D \subseteq \mathbb{R}$. Then $\langle f_n \rangle$ converges uniformly on D if and only if $\langle f_n \rangle$ is uniformly Cauchy on D.

Proof. See Exercises 11 and 12. □

8.1.2 Series of Functions

Recall that an infinite series of real numbers is viewed as a sequence of its partial sums, where each partial sum is a real number. Similarly, an infinite series of functions is viewed as a sequence of its partial sums, where each partial sum is a function. Let $\langle f_n \rangle$ be a sequence of functions. For each $n \in \mathbb{N}$, the nth **partial sum function** $s_n \colon D \to \mathbb{R}$ is defined by $s_n(x) = \sum_{k=1}^{n} f_k(x)$ for all $x \in D$. The infinite series of functions $\sum_{n=1}^{\infty} f_n$ denotes the sequence $\langle s_n \rangle$ of partial sum functions.

Definition 8.1.11. Let $\langle f_n \rangle$ be a sequence of functions defined on $D \subseteq \mathbb{R}$ and let $f \colon D \to \mathbb{R}$. The series $\sum_{n=1}^{\infty} f_n$ **converges pointwise** to f on D if the corresponding sequence of partial sum functions $\langle s_n \rangle$ converges pointwise to f on D. Moreover, $\sum_{n=1}^{\infty} f_n$ **converges uniformly** to f on D if $\langle s_n \rangle$ converges uniformly to f on D.

We now present an important condition which will ensure the uniform convergence of a series of functions. The proof of this result applies Theorem 8.1.10, the Cauchy Criterion.

Theorem 8.1.12 (Weierstrass M-test). Let $\langle f_n \rangle$ be a sequence of functions defined on $D \subseteq \mathbb{R}$. Let $\langle M_n \rangle$ be a sequence of real numbers such that $|f_n(x)| \leq M_n$ for all $x \in D$ and all $n \in \mathbb{N}$. If $\sum_{n=1}^{\infty} M_n$ converges, then $\sum_{n=1}^{\infty} f_n$ converges uniformly on D.

Proof. Let $\langle M_n \rangle$ be a sequence of real numbers such that $\sum_{n=1}^{\infty} M_n$ converges and

$$|f_n(x)| \leq M_n, \text{ for all } x \in D \text{ and all } n \in \mathbb{N}. \tag{8.2}$$

Let $\langle s_n \rangle$ be the sequence of partial sum functions for the series $\sum_{n=1}^{\infty} f_n$. We prove that $\sum_{n=1}^{\infty} f_n$ is uniformly Cauchy on D. Let $\varepsilon > 0$. Since $\sum_{n=1}^{\infty} M_n$ converges, Theorem 7.1.12 implies that there is a natural number N such that

$$\sum_{k=m+1}^{n} M_k < \varepsilon, \text{ for all } n > m > N. \tag{8.3}$$

Let $n > m > N$ and $x \in D$. We shall prove that $|s_n(x) - s_m(x)| < \varepsilon$ as follows:

$$|s_n(x) - s_m(x)| = \left| \sum_{k=m+1}^{n} f_k(x) \right| \quad \text{by algebra}$$

$$\leq \sum_{k=m+1}^{n} |f_k(x)| \quad \text{by the triangle inequality}$$

$$\leq \sum_{k=m+1}^{n} M_k < \varepsilon \quad \text{by (8.2) and (8.3)}.$$

Thus, by Theorem 8.1.10 and Definition 8.1.11, $\sum_{n=1}^{\infty} f_n$ converges uniformly on D. □

Example. For each $n \in \mathbb{N}$, let $f_n : \mathbb{R} \to \mathbb{R}$ be defined by $f_n(x) = \frac{1}{x^2 + n^3}$. Observe that $|f_n(x)| \leq \frac{1}{n^3}$ for all $x \in \mathbb{R}$ and all $n \in \mathbb{N}$. Since $\sum_{n=1}^{\infty} \frac{1}{n^3}$ converges by Theorem 7.2.5, the Weierstrass M-test implies that series $\sum_{n=1}^{\infty} f_n$ converges uniformly on $\mathbb{R}$.

Exercises 8.1

*1. For each $n \in \mathbb{N}$, let $f_n(x) = \frac{\sin(nx)}{\sqrt{n}}$ for all $x \in \mathbb{R}$.

   (a) Show that $\langle f_n \rangle$ converges uniformly to the zero function on $\mathbb{R}$.
   (b) Show that $\langle f_n' \rangle$ does not converge pointwise on $\mathbb{R}$ to the function f'.

2. For each $n \in \mathbb{N}$, let $f_n(x) = \frac{x}{x+n}$ for all $x \geq 0$. Let $f(x) = 0$ for all $x \geq 0$.

   (a) Show that $\langle f_n \rangle$ converges pointwise to f on $[0, \infty)$.
   (b) Let $b > 0$. Show that $\langle f_n \rangle$ converges uniformly to f on $[0, b]$.
   (c) Show that $\langle f_n \rangle$ does not converge uniformly to f on $[0, \infty)$.

*3. Prove Theorem 8.1.7

4. For each $n \in \mathbb{N}$, let $f_n : D \to \mathbb{R}$ be increasing on D. Show that if $\langle f_n \rangle$ converges pointwise to f, then f is increasing on D. [See Theorem 3.2.13.]

5. Suppose that $\langle f_n \rangle$ and $\langle g_n \rangle$ converge uniformly, respectively, to f and g on D. Prove that $\langle f_n + g_n \rangle$ converges uniformly to $f + g$ on D.

6. Suppose that $\langle f_n \rangle$ converge uniformly to f on D, and that $g \colon D \to \mathbb{R}$ is bounded. Prove that $\langle gf_n \rangle$ converges uniformly to gf on D.

7. Let $\langle f_n \rangle$ be a sequence of functions that converges uniformly to f on D. Suppose that there is an $m > 0$ such that each $|f(x)| \geq m$ and $|f_n(x)| \geq m$ for all $x \in D$ and all $n \in \mathbb{N}$. Prove that $\left\langle \frac{1}{f_n} \right\rangle$ converges uniformly to $\left\langle \frac{1}{f} \right\rangle$ on D.

8. Let $\langle f_n \rangle$ be a sequence of functions that converges uniformly to f on D. Suppose that f is bounded on D. Prove that there is an $N \in \mathbb{N}$ such that f_n is bounded on D, for each $n > N$.

*9. Let $\langle f_n \rangle$ be a sequence of functions that converges uniformly to f on D. Suppose that each f_n is bounded on D.

   (a) Prove that f is bounded on D.
   (b) Prove that $\langle f_n \rangle$ is uniformly bounded on D, that is, prove that there exists a real number M such that $|f_n(x)| \leq M$ for all $x \in D$ and all $n \in \mathbb{N}$.
   (c) Let M be as in (b). Show that $|f(x)| \leq M$ for all $x \in D$. [Apply Corollary 3.2.16.]

10. Suppose $\langle f_n \rangle$ and $\langle g_n \rangle$ are sequences of bounded functions with domain D that converge uniformly on D to f and g, respectively.

   (a) Show that $f_n(x)g_n(x) - f(x)g(x) = f_n(x)(g_n(x) - g(x)) + g(x)(f_n(x) - f(x))$.
   (b) Using Exercise 9, prove that $\langle f_n g_n \rangle$ converges uniformly to fg on D.

*11. Suppose that $\langle f_n \rangle$ converges uniformly to f on D. Prove that $\langle f_n \rangle$ is uniformly Cauchy. [See the proof of Lemma 3.6.3.]

*12. Let $\langle f_n \rangle$ be a sequence of functions on D that is uniformly Cauchy.

   (a) Prove that there is a function $f \colon D \to \mathbb{R}$ such that $\langle f_n \rangle$ converges pointwise to f. [Apply Theorem 3.6.5.]
   (b) Let $\varepsilon > 0$. Since $\langle f_n \rangle$ is uniformly Cauchy, there exists an $N \in \mathbb{N}$ such that for all $x \in D$, if $n > N$ and $m > N$, then $|f_n(x) - f_m(x)| < \frac{\varepsilon}{2}$. Let $x \in D$. Since $\lim\limits_{m \to \infty} f_m(x) = f(x)$, use Corollary 3.2.16 to show that $|f_n(x) - f(x)| \leq \frac{\varepsilon}{2} < \varepsilon$ whenever $n > N$.

13. For each $n \in \mathbb{N}$, let $f_n(x) = (\sin(x))^{2n}$ for all $x \in [0, \frac{\pi}{3}]$. Prove that the series $\sum\limits_{n=1}^{\infty} f_n$ converges uniformly on $[0, \frac{\pi}{3}]$.

8.2 PRESERVATION THEOREMS

As shown in the previous section, pointwise convergence does not ensure that the limit function will possess the same properties as the functions in a sequence. Suppose that a sequence of functions converges uniformly to f. What properties that are shared by all of the functions in the sequence will be preserved and thus hold for f? In this section, we will pursue this question.

We will first show that uniform convergence is sufficient to ensure that the limit function will be continuous if all the functions in the sequence are continuous. This is established, in the proof of our next theorem, by applying Strategy 4.1.2 on page 102.

Theorem 8.2.1. Let $\langle f_n \rangle$ be a sequence such that $f_n \colon D \to \mathbb{R}$ is continuous on D, for all $n \in \mathbb{N}$. If $\langle f_n \rangle$ converges uniformly to f on D, then f is continuous on D.

Proof. Let $\langle f_n \rangle$ be a sequence such that $f_n \colon D \to \mathbb{R}$ is continuous on D, for all $n \in \mathbb{N}$. Assume that $\langle f_n \rangle$ converges uniformly to f on D. To prove that f is continuous on D, let $c \in D$ be arbitrary and let $\varepsilon > 0$. Since $\langle f_n \rangle$ converges uniformly to f, there is an N such that

$$|f_n(x) - f(x)| < \frac{\varepsilon}{3} \text{ for } all \ x \in D \text{ and all } n > N. \qquad (8.4)$$

So, let $n > N$. Since f_n is continuous and $c \in D$, there is a $\delta > 0$ such that for all $x \in D$, we have

$$|f_n(x) - f_n(c)| < \frac{\varepsilon}{3} \text{ when } |x - c| < \delta. \qquad (8.5)$$

Now, let $x \in D$ be such that $|x - c| < \delta$. We prove that $|f(x) - f(c)| < \varepsilon$ as follows:

$$
\begin{aligned}
|f(x) - f(c)| &= |f(x) - f(c) + f_n(x) - f_n(x) + f_n(c) - f_n(c)| && \text{by algebra} \\
&= |(f(x) - f_n(x)) + (f_n(c) - f(c)) + (f_n(x) - f_n(c))| && \text{by algebra} \\
&\leq |f(x) - f_n(x)| + |f_n(c) - f(c)| + |f_n(x) - f_n(c)| && \text{by triangle ineq.} \\
&< \frac{\varepsilon}{3} + \frac{\varepsilon}{3} + |f_n(x) - f_n(c)| && \text{by (8.4)} \\
&< \frac{\varepsilon}{3} + \frac{\varepsilon}{3} + \frac{\varepsilon}{3} = \varepsilon && \text{by (8.5).}
\end{aligned}
$$

Thus, $|f(x) - f(c)| < \varepsilon$, if $|x - c| < \delta$. Therefore, f is continuous on D. □

Since a series of functions is a sequence of partial sum functions, Theorem 8.2.1 applies to a series of functions as well. Consequently, Theorems 4.1.6 and 8.2.1 imply the following corollary.

Corollary 8.2.2. Let $\langle f_n \rangle$ be such that $f_n \colon D \to \mathbb{R}$ is continuous on D, for all $n \in \mathbb{N}$. If $\sum_{n=1}^{\infty} f_n$ converges uniformly to f on D, then f is continuous on D.

Example 8.1.2 presents a pointwise convergent sequence of continuous functions that converges to a discontinuous function. Theorem 8.2.1 thus implies that this convergence is not uniform. Furthermore, Example 8.1.8 identifies a sequence of continuous functions that does converge pointwise to a continuous function and yet, this convergence is not uniform. So the converse of Theorem 8.2.1 does not hold. Under certain additional conditions, however, a converse does hold (see Theorem 8.2.3).

Let $\langle f_n \rangle$ be a sequence of functions defined on a closed interval $[a, b]$. Then $\langle f_n \rangle$ is said to be an **increasing sequence of functions** if $f_n(x) \leq f_{n+1}(x)$ for all $n \in \mathbb{N}$ and all x in $[a, b]$.

Theorem 8.2.3 (Dini's Theorem). Let $\langle f_n \rangle$ be an increasing sequence of continuous functions, where $f_n \colon [a, b] \to \mathbb{R}$ for all $n \in \mathbb{N}$. If $\langle f_n \rangle$ converges pointwise to f on $[a, b]$ and f is continuous, then $\langle f_n \rangle$ converges uniformly to f on $[a, b]$.

Proof. See Exercises 11 and 12. ◻

Suppose that each f_n is Riemann integrable on $[a, b]$, and $\langle f_n \rangle$ converges pointwise to f on $[a, b]$. In addition, suppose that f is Riemann integrable on $[a, b]$. Can one conclude that

$$\lim_{n \to \infty} \int_a^b f_n = \int_a^b f?$$

Exercise 10 shows that the answer, in general, is "no." On the other hand, if $\langle f_n \rangle$ converges uniformly to f on $[a, b]$, then the answer is "yes." Our next theorem shows that uniform convergence implies that the limit of Riemann integrable functions is Riemann integrable and, in addition, it implies that the "limit of the integrals is the integral of the limit."

Theorem 8.2.4. Let $\langle f_n \rangle$ be a sequence where $f_n \colon [a, b] \to \mathbb{R}$ and f_n is Riemann integrable on $[a, b]$, for all $n \in \mathbb{N}$. If $\langle f_n \rangle$ converges uniformly to f on $[a, b]$, then f is Riemann integrable on $[a, b]$ and

$$\lim_{n \to \infty} \int_a^b f_n = \int_a^b f.$$

Proof. Let $\langle f_n \rangle$ be a sequence of functions where $f_n \colon [a, b] \to \mathbb{R}$ and f_n is integrable for each $n \in \mathbb{N}$. Thus, each f_n is bounded. Suppose that $\langle f_n \rangle$ converges uniformly to f on $[a, b]$. Exercise 9(a) on page 218 implies that f is bounded. We apply Theorem 6.1.18 to prove that f is Riemann integrable. Let $\varepsilon > 0$. Now let $\varepsilon^* = \frac{\varepsilon}{3(b-a)}$. Since $\langle f_n \rangle$ converges uniformly to f on $[a, b]$, there is an $n \in \mathbb{N}$ such that

$$|f_n(x) - f(x)| < \varepsilon^* \text{ for all } x \in [a, b].$$

Thus,

$$f(x) - f_n(x) < \varepsilon^* \text{ and } f_n(x) - f(x) < \varepsilon^*, \text{ for all } x \in [a, b].$$

Therefore,

(▲) $f(x) < f_n(x) + \varepsilon^*$, and (▶) $f_n(x) < f(x) + \varepsilon^*$, for all $x \in [a, b]$.

Since f_n is Riemann integrable, there is a partition P of $[a, b]$ such that

$$U(f_n, P) - L(f_n, P) < \frac{\varepsilon}{3} \tag{8.6}$$

by Theorem 6.1.18. Inequality (▲) and Exercises 3(a), 4(a) on page 161 imply that

$$U(f, P) \le U(f_n, P) + \varepsilon^*(b - a). \tag{8.7}$$

In addition, inequality (▶) and Exercises 3(b), 4(b) on page 161 imply that

$$L(f_n, P) \le L(f, P) + \varepsilon^*(b - a) \text{ and thus,} \tag{8.8}$$

$$-L(f, P) \le -L(f_n, P) + \varepsilon^*(b - a). \tag{8.9}$$

Adding inequalities (8.7) and (8.9), we obtain from (8.6) the following result

$$U(f,P) - L(f,P) \le U(f_n,P) - L(f_n,P) + 2\varepsilon^*(b-a) < \frac{\varepsilon}{3} + 2\varepsilon^*(b-a) = \frac{\varepsilon}{3} + 2\frac{\varepsilon}{3} = \varepsilon.$$

So $U(f,P) - L(f,P) < \varepsilon$ and Theorem 6.1.18 implies that f is Riemann integrable. We now prove that $\lim_{n\to\infty} \int_a^b f_n = \int_a^b f$. Let $\varepsilon > 0$ be given. Let ε^* be such that (▼) $0 < \varepsilon^* < \frac{\varepsilon}{b-a}$. Since $\langle f_n \rangle$ converges uniformly to f on $[a,b]$, there is an $N \in \mathbb{N}$ such that for all $n > N$,

$$|f_n(x) - f(x)| < \varepsilon^* \text{ for all } x \in [a,b]. \tag{8.10}$$

We conclude that

$$\left| \int_a^b f_n - \int_a^b f \right| = \left| \int_a^b (f_n - f) \right| \qquad \text{by Corollary 6.2.6}$$

$$\le \int_a^b |f_n - f| \qquad \text{by Corollary 6.2.17}$$

$$\le \int_a^b \varepsilon^* \qquad \text{by (8.10) and Corollary 6.2.8}$$

$$= \varepsilon^* \cdot (b-a) < \varepsilon \qquad \text{by Exercise 2 on page 161, and (▼).}$$

Hence, $\left| \int_a^b f_n - \int_a^b f \right| < \varepsilon$ for all $n > N$. Therefore, $\lim_{n\to\infty} \int_a^b f_n = \int_a^b f$. □

Lemma 6.2.4 and Theorem 8.2.4 imply the following corollary.

Corollary 8.2.5. Let $\langle f_n \rangle$ be a sequence such that $f_n \colon [a,b] \to \mathbb{R}$ and f_n is Riemann integrable, for each $n \in \mathbb{N}$. If $\sum_{n=1}^{\infty} f_n$ converges uniformly to f on $[a,b]$, then f is Riemann integrable and

$$\int_a^b f = \sum_{n=1}^{\infty} \int_a^b f_n.$$

Exercise 1 on page 217 identifies a sequence $\langle f_n \rangle$ of differentiable functions that uniformly converges to a differentiable function f; however, as stated in the exercise, the "derivative sequence" $\langle f_n' \rangle$ does not converge pointwise to f'. One way to avoid such results, is to deal with derivative sequences that converge uniformly and consist of continuous functions.

Theorem 8.2.6. Let $\langle f_n \rangle$ be a sequence such that $f_n \colon [a,b] \to \mathbb{R}$ for all $n \in \mathbb{N}$. Suppose for each $n \in \mathbb{N}$ we have that

(1) f_n is differentiable on $[a,b]$, and
(2) f_n' is continuous on $[a,b]$.

If $\langle f_n \rangle$ converges pointwise to f on $[a,b]$ and $\langle f_n' \rangle$ converges uniformly on $[a,b]$, then f is differentiable on $[a,b]$ and $\langle f_n' \rangle$ converges uniformly to f'.

Proof. Let $\langle f_n \rangle$ be a sequence such that $f_n \colon [a, b] \to \mathbb{R}$ for each $n \in \mathbb{N}$. Suppose that the above (1) and (2) hold for every $n \in \mathbb{N}$. Assume that $\langle f_n \rangle$ converges pointwise to f on $[a, b]$ and that $\langle f_n' \rangle$ converges uniformly to $g \colon [a, b] \to \mathbb{R}$. As f_n' is continuous for all $n \in \mathbb{N}$, Theorem 6.3.1 implies that each f_n' is Riemann integrable on $[a, x]$ for each $x \in [a, b]$. Thus, Theorem 8.2.4 implies that g is Riemann integrable on $[a, x]$ for every $x \in [a, b]$. Moreover, Theorem 8.2.4 implies that

$$\int_a^x g = \lim_{n \to \infty} \int_a^x f_n' \qquad (8.11)$$

for all $x \in [a, b]$. Because $\langle f_n \rangle$ converges pointwise to f, equation (8.11) and Theorem 6.4.2 (Fundamental Theorem of Calculus I) imply that

$$\int_a^x g = \lim_{n \to \infty} (f_n(x) - f_n(a)) = f(x) - f(a)$$

for all $x \in [a, b]$. So,

$$f(x) = \left(\int_a^x g \right) + f(a), \text{ for each } x \in [a, b].$$

Theorem 8.2.1 implies that g is continuous on $[a, b]$. Theorem 6.4.4 (Fundamental Theorem of Calculus II) thus implies that f is differentiable and that $f'(x) = g(x)$ for all $x \in [a, b]$, as $f(a)$ is a constant. Since $\langle f_n' \rangle$ converges uniformly to g and $f' = g$, we conclude that $\langle f_n' \rangle$ converges uniformly to f', as required. □

Corollary 8.2.7. Let $\langle f_n \rangle$ be a sequence such that $f_n \colon [a, b] \to \mathbb{R}$ for all $n \in \mathbb{N}$. Suppose for each $n \in \mathbb{N}$ we have that

(1) f_n is differentiable on $[a, b]$, and

(2) f_n' is continuous on $[a, b]$.

If $\sum\limits_{n=1}^{\infty} f_n$ converges pointwise to f on $[a, b]$ and $\sum\limits_{n=1}^{\infty} f_n'$ converges uniformly to f' on $[a, b]$, then f is differentiable on $[a, b]$, and for all $x \in [a, b]$, $f'(x) = \sum\limits_{n=1}^{\infty} f_n'(x)$.

Exercises 8.2

1. For each $n \in \mathbb{N}$, define $f_n \colon [0, 1] \to \mathbb{R}$ by $f_n(x) = \frac{\sin(nx) + 1}{n^2}$.

   (a) Show that $\langle f_n \rangle$ converges pointwise to f, the zero function on $[0, 1]$.

   (b) Using Theorem 8.1.7, show that $\langle f_n \rangle$ converges uniformly to f on $[0, 1]$.

   (c) Does $\langle f_n' \rangle$ converges uniformly to f' on $[0, 1]$?

   (d) Using Theorem 8.1.12, show that $\sum\limits_{n=1}^{\infty} f_n$ converges uniformly on $[0, 1]$.

   (e) Is $\sum\limits_{n=1}^{\infty} f_n$ continuous on $[0, 1]$?

   (f) Does $\sum\limits_{n=1}^{\infty} f_n'$ converge uniformly on $[0, 1]$?

2. Prove Corollary 8.2.5.

3. Prove Corollary 8.2.2.

4. Let $f_n: D \to \mathbb{R}$ be uniformly continuous on D, for all $n \in \mathbb{N}$. Prove that if $\langle f_n \rangle$ converges uniformly to f on D, then f is uniformly continuous on D.

5. Let $\langle f_n \rangle$ be a sequence of functions that are continuous on $[a, b]$. For each $n \in \mathbb{N}$, define $F_n: [a, b] \to \mathbb{R}$ by $F_n(x) = \int_a^x f$. Suppose that $\langle F_n \rangle$ converges uniformly on $[a, b]$. Prove that $\langle f_n \rangle$ converges uniformly on $[a, b]$.

6. For each $n \in \mathbb{N}$, define $f_n: [0, 1] \to \mathbb{R}$ by $f_n(x) = \frac{x^n}{n} + 1$.

   (a) Show that $\langle f_n \rangle$ converges pointwise to f, the constant 1 function on $[0, 1]$.

   (b) Using Theorem 8.1.7, show that $\langle f_n \rangle$ converges uniformly to f on $[0, 1]$.

   (c) Does $\langle f_n' \rangle$ converges uniformly to f' on $[0, 1]$?

7. Let $\langle f_n \rangle$ be such that $f_n: D \to \mathbb{R}$ is continuous on D, for all $n \in \mathbb{N}$. Suppose that $\langle f_n \rangle$ converges uniformly to f on D. Let $x \in D$ and let $\langle x_n \rangle$ be a sequence of points in D. Prove that if $\lim_{n \to \infty} x_n = x$, then $\lim_{n \to \infty} f_n(x_n) = f(x)$. [Theorem 8.2.1 implies that f is continuous and thus, $\lim_{n \to \infty} f(x_n) = f(x)$.]

8. Prove Corollary 8.2.7.

9. Let $\langle f_n \rangle$ be a sequence of functions that are Riemann integrable on $[a, b]$. Suppose that $\langle f_n \rangle$ converges uniformly to f on $[a, b]$. For each $n \in \mathbb{N}$, define $F_n: [a, b] \to \mathbb{R}$ by $F_n(x) = \int_a^x f$. Prove that $\langle F_n \rangle$ converges uniformly on $[a, b]$.

*10. For each $n \in \mathbb{N}$ define $f_n: [0.1] \to \mathbb{R}$ by

$$
f_n(x) = \begin{cases} n(n+1), & \text{if } \frac{1}{n+1} \leq x \leq \frac{1}{n}; \\ 0, & \text{otherwise,} \end{cases}
$$

   for all $x \in [0, 1]$. Show that $\langle f_n \rangle$ converges pointwise on $[0, 1]$ to f, the zero function on $[0, 1]$. Each of the functions f_n is Riemann integrable on $[0, 1]$, and so is the function f. Show that $\lim_{n \to \infty} \int_0^1 f_n \neq \int_0^1 f$.

*11. Let $\langle f_n \rangle$ be an increasing sequence where $f_n: [a, b] \to \mathbb{R}$ for all $n \in \mathbb{N}$. Suppose that $\langle f_n \rangle$ converges pointwise to f on $[a, b]$.

   (a) Show that $f_n(x) \leq f(x)$ for all $n \in \mathbb{N}$ and all $x \in [a, b]$.

   (b) Let $n \in \mathbb{N}$. Show that $f(x) - f_n(x) \geq f(x) - f_k(x)$ for all $x \in [a, b]$ and all $k \geq n$.

   (c) Prove that $\langle f_n \rangle$ converges uniformly to f on $[a, b]$ if and only if for every $\varepsilon > 0$, there is an $n \in \mathbb{N}$ such that $f(x) - f_n(x) < \varepsilon$ for all $x \in [a, b]$.

*12. Let $\langle f_n \rangle$ be an increasing sequence where $f_n: [a, b] \to \mathbb{R}$ is continuous on $[a, b]$, for all $n \in \mathbb{N}$. Suppose that $\langle f_n \rangle$ converges pointwise to f, where f is continuous on $[a, b]$. Prove Theorem 8.2.3 as follows: Suppose, for a contradiction, that $\langle f_n \rangle$ does not converge uniformly to f on $[a, b]$. Now, justify the following steps:

(a) There is an $\varepsilon > 0$ so that for every $n \in \mathbb{N}$, there is an $x \in [a, b]$ such that $f(x) - f_n(x) \geq \varepsilon$ (see Exercise 11(c)).

(b) There is a sequence $\langle x_n \rangle$ such that $x_n \in [a, b]$ and $f(x_n) - f_n(x_n) \geq \varepsilon$, for every $n \in \mathbb{N}$.

(c) There exists a subsequence $\langle x_{n_k} \rangle$ of $\langle x_n \rangle$ that converges to a point $x^* \in [a, b]$.

(d) For each $n \in \mathbb{N}$, $f(x_{n_k}) - f_n(x_{n_k}) \geq f(x_{n_k}) - f_{n_k}(x_{n_k}) \geq \varepsilon$ for all $k \geq n$. (see Exercise 11(b)).

(e) For each $n \in \mathbb{N}$, $f(x^*) - f_n(x^*) \geq \varepsilon$ (see Exercise 9 on page 74). Therefore, $\langle f_n \rangle$ does not converge pointwise to f.

8.3 POWER SERIES

We will now continue our study of infinite series of functions; however, in this section we will examine the infinite series of power functions, that is, functions that have the form $a_n(x - c)^n$. Such a series is said to be a *power series*. A given power series may converge for some values of x and it may diverge for other values. In this section will present tools that will allow us to identify the particular values for which the power series converges. In the next section, we will address the question of representing a specific function as an infinite series of power functions.

Definition 8.3.1. Let $\langle a_n \rangle_{n=0}^{\infty}$ be a sequence of real numbers and let $c \in \mathbb{R}$. For each $n \in \mathbb{N}$, let $f_n \colon I \to \mathbb{R}$ be defined by $f_n(x) = a_n(x - c)^n$, where I is an interval. Then $\sum_{n=1}^{\infty} f_n$ is called a **power series** in $(x - c)$ and is also denoted by

$$\sum_{n=1}^{\infty} f_n(x) = \sum_{k=0}^{\infty} a_k(x - c)^k = a_0 + a_1(x - c) + a_2(x - c)^2 + a_3(x - c)^3 + \cdots . \quad (8.12)$$

The numbers $a_0, a_1, a_2, \ldots$ are called the **coefficients** of the power series, and the number c is referred to as the **center** of the power series.

A power series in $(x - c)$ can be viewed as an "infinite" polynomial in $(x - c)$. When $c = 0$, a power series takes on the form $\sum_{k=0}^{\infty} a_k x^k = a_0 + a_1 x + a_2 x^2 + a_3 x^3 + \cdots$.

Intervals of Convergence

Given a power series in $(x - c)$, for what values of x will the series converge? It seems likely that the answer to this question depends on the coefficients of the power series. Our next theorem will address this question. First, we need to slightly extend the definition of the limit superior of a (bounded) sequence given in Section 3.8.1. If a sequence $\langle x_k \rangle$ is not bounded above, then we shall write $\limsup_{n \to \infty} x_k = \infty$, in this specific case. In addition, we shall write $x < \infty$ to mean the x is a real number. The term $\sqrt[n]{x}$ is undefined when $x \geq 0$. So whenever a term of the form $\sqrt[k]{x}$ appears below, we will be implicitly assuming that $k \geq 1$.

Theorem 8.3.2. Let $\sum\limits_{k=0}^{\infty} a_k(x-c)^k$ be a power series, $\ell = \limsup\limits_{k\to\infty} \sqrt[k]{|a_k|}$, and let

$$\rho = \begin{cases} 0, & \text{if } \ell = \infty; \\ \frac{1}{\ell}, & \text{if } 0 < \ell < \infty; \\ \infty, & \text{if } \ell = 0. \end{cases} \tag{8.13}$$

(1) If $\rho = 0$, then the series converges only for $x = c$.

(2) If $0 < \rho < \infty$ and $|x - c| < \rho$, then $\sum\limits_{k=0}^{\infty} a_k(x-c)^k$ converges absolutely.

(3) If $0 < \rho < \infty$ and $|x - c| > \rho$, then $\sum\limits_{k=0}^{\infty} a_k(x-c)^k$ diverges.

(4) If $\rho = \infty$, then $\sum\limits_{k=0}^{\infty} a_k(x-c)^k$ converges absolutely for all real numbers x.

Moreover, if $0 < \zeta < \rho$, then $\sum\limits_{k=0}^{\infty} a_k(x-c)^k$ converges uniformly on $[c-\zeta, c+\zeta]$.

Proof. Let $\ell = \limsup\limits_{k\to\infty} \sqrt[k]{|a_k|}$. First we note that for any $x \in \mathbb{R}$ and $k \in \mathbb{N}$,

$$\sqrt[k]{|a_k(x-c)^k|} = |x - c| \sqrt[k]{|a_k|}. \tag{8.14}$$

If $\rho = 0$, then $\limsup\limits_{k\to\infty} \sqrt[k]{|a_k|} = \infty$, by (8.13). Thus, $\left\langle \sqrt[k]{|a_k|} \right\rangle$ is not bounded above. If $x \neq c$, then from (8.14) we see that $\left\langle \sqrt[k]{|a_k(x-c)^k|} \right\rangle$ is also not bounded above. By Theorem 7.2.12(2), $\sum\limits_{k=0}^{\infty} a_k(x-c)^k$ diverges. Thus, the series converges only for $x = c$.

Suppose that $0 < \rho < \infty$. So $\rho = \frac{1}{\ell}$. By Exercise 10 on page 98, we have

$$\limsup\limits_{k\to\infty} \sqrt[k]{|a_k(x-c)^k|} = |x - c| \limsup\limits_{k\to\infty} \sqrt[k]{|a_k|} = |x - c| \frac{1}{\rho} \tag{8.15}$$

for any $x \in \mathbb{R}$. By Corollary 7.2.13, $\sum\limits_{k=0}^{\infty} a_k(x-c)^k$ converges absolutely if $|x - c| < \rho$ and diverges if $|x - c| > \rho$.

Suppose that $\rho = \infty$. Thus, $\limsup\limits_{k\to\infty} \sqrt[k]{|a_k|} = 0$. So, $\limsup\limits_{k\to\infty} \sqrt[k]{|a_k(x-c)^k|} = 0$ for all $x \in \mathbb{R}$. Thus, the series converges absolutely for all $x \in \mathbb{R}$, by Corollary 7.2.13(1).

Let $0 < \zeta < \rho$. If $x \in [c-\zeta, c+\zeta]$, then $|x - c| \leq \zeta$ and $|a_k(x-c)^k| \leq |a_k|\zeta^k$, for all $k \in \mathbb{N}$. Moreover, as in (8.15), we have that $\limsup\limits_{k\to\infty} \sqrt[k]{|a_k|\zeta^k} = \frac{\zeta}{\rho} < 1$.[1] Hence, by Corollary 7.2.13(1), $\sum\limits_{k=0}^{\infty} |a_k|\zeta^k$ converges. Theorem 8.1.12 implies that the series $\sum\limits_{k=0}^{\infty} a_k(x-c)^k$ converges uniformly on the closed interval $[c-\zeta, c+\zeta]$. □

---

[1] Replace $\frac{\zeta}{\rho}$ with 0, if $\rho = \infty$.

The value ρ, defined in (8.13), is called the **radius of convergence**. If $\rho \neq 0$, then by Theorem 8.3.2 the set of points x such that $\sum_{k=0}^{\infty} a_k(x-c)^k$ converges is an interval. This interval, called the **interval of convergence**, is determined as follows:

1. If $0 < \rho < \infty$, then the series converges absolutely for all x in the open interval $(c - \rho, c + \rho)$, and diverges for all x outside the closed interval $[c - \rho, c + \rho]$. In addition, the series converges uniformly on any closed subinterval of $(c - \rho, c + \rho)$. To identify the interval of convergence, one must first check the endpoints $x = c - \rho$ and $x = c + \rho$, separately, to determine convergence or divergence. If any endpoint converges, then the interval of convergence is obtained by adding such an endpoint to the interval $(c - \rho, c + \rho)$. If no endpoint converges, then $(c - \rho, c + \rho)$ is the interval of convergence.

2. If $\rho = \infty$, then the series converges for all $x \in \mathbb{R}$; in this case, $(-\infty, +\infty)$ is the interval of convergence and ∞ is the radius of convergence. Furthermore, the series converges uniformly on any closed interval $[a, b]$.

Remark. When ρ is a radius of convergence, we will write $\rho > 0$ to mean that $\rho \in \mathbb{R}^+$ or $\rho = \infty$. If $\rho = \infty$ and $c \in \mathbb{R}$, we will interpret $(c - \rho, c + \rho)$ to be $(-\infty, \infty)$.

When applying Theorem 8.3.2, the value $\ell = \limsup_{k \to \infty} \sqrt[k]{|a_k|}$ may be difficult to evaluate. In this case, Corollaries 3.8.18 and 3.8.5 offer another way to find ℓ.

Theorem 8.3.3. Let $\sum_{k=0}^{\infty} a_k(x-c)^k$ be a power series where $a_k \neq 0$ for all $k \geq 0$. If $\lim_{k \to \infty} \left| \frac{a_{k+1}}{a_k} \right|$ is either a real number or infinity, then $\limsup_{k \to \infty} \sqrt[k]{|a_k|} = \lim_{k \to \infty} \left| \frac{a_{k+1}}{a_k} \right|$.

Example 8.3.4. Consider the power series $\sum_{k=0}^{\infty} \frac{k}{2^k}(x-3)^k$. By Corollary 3.2.10 and some algebra, we have that

$$\limsup_{k \to \infty} \sqrt[k]{\left| \frac{k}{2^k} \right|} = \frac{1}{2} \lim_{k \to \infty} \sqrt[k]{k} = \frac{1}{2},$$

and thus, by Theorem 8.3.2, the radius of convergence of the power series is 2. Alternatively, we can apply Theorem 8.3.3 to obtain the same result, namely,

$$\lim_{k \to \infty} \left| \frac{\frac{k+1}{2^{k+1}}}{\frac{k}{2^k}} \right| = \lim_{k \to \infty} \frac{k+1}{2^{k+1}} \frac{2^k}{k} = \frac{1}{2}.$$

In any case, we see that 2 is the radius of convergence and the power series converges for all x in the interval $(1, 5)$. For $x = 1$ and $x = 5$, we obtain the respective series

$$\sum_{k=0}^{\infty} (-1)^k k \quad \text{and} \quad \sum_{k=0}^{\infty} k$$

which both diverge. Therefore, $(1, 5)$ is the interval of convergence.

Let $\sum_{k=0}^{\infty} a_k(x-c)^k$ be a power series with interval of convergence I. One can now define a function $f\colon I \to \mathbb{R}$ by

$$f(x) = \sum_{k=0}^{\infty} a_k(x-c)^k = a_0 + a_1(x-c) + a_2(x-c)^2 + \cdots .$$

Can f be differentiated term by term, as a polynomial can be differentiated?

Theorem 8.3.5. Let $\sum_{k=0}^{\infty} a_k(x-c)^k$ be a power series with radius of convergence $\rho > 0$, let I be the interval $(c-\rho, c+\rho)$, and let $f\colon I \to \mathbb{R}$ be defined by $f(x) = \sum_{k=0}^{\infty} a_k(x-c)^k$. Then f is continuous and differentiable on I. Moreover, for all $x \in I$,

$$f'(x) = \sum_{k=1}^{\infty} k a_k(x-c)^{k-1} \qquad (8.16)$$

and the power series $\sum_{k=1}^{\infty} k a_k(x-c)^{k-1}$ also has radius of convergence equal to ρ.

Proof. As $\rho > 0$, (8.13) implies that $\ell = \limsup_{k\to\infty} \sqrt[k]{|a_k|}$ is a real number. Theorem 3.8.6 and Corollary 3.2.10 imply that

$$\limsup_{k\to\infty} \sqrt[k]{|k a_k|} = \lim_{k\to\infty} \sqrt[k]{k} \cdot \limsup_{k\to\infty} \sqrt[k]{|a_k|} = \ell.$$

Thus, $\sum_{k=1}^{\infty} k a_k(x-c)^{k-1}$ also has ρ as its radius of convergence (see Exercise 5). Let $I = (c-\rho, c+\rho)$, and let $f\colon I \to \mathbb{R}$ be defined by $f(x) = \sum_{k=0}^{\infty} a_k(x-c)^k$. To prove (8.16), let $x \in I$ and let $[a, b] \subseteq I$ be such that $x \in (a, b)$. Theorem 8.3.2 implies that $\sum_{k=0}^{\infty} a_k(x-c)^k$ and $\sum_{k=1}^{\infty} k a_k(x-c)^{k-1}$ converge uniformly on $[a, b]$. Corollary 8.2.7 implies that $f'(x) = \sum_{k=1}^{\infty} k a_k(x-c)^{k-1}$ for all $x \in [a, b]$. Thus, f is differentiable on I and hence, is continuous on I by Corollary 5.1.3. $\qquad \square$

Consider the power series in Example 8.3.4, with radius of convergence 2 and interval of convergence $(1, 5)$. Using this power series, we define $f\colon (1,5) \to \mathbb{R}$ by

$$f(x) = \sum_{k=0}^{\infty} \frac{k}{2^k}(x-3)^k = 0 + \frac{1}{2}(x-3) + \frac{2}{2^2}(x-3)^2 + \frac{3}{2^3}(x-3)^3 + \frac{4}{2^4}(x-3)^4 + \cdots$$

Theorem 8.3.5 implies that $f'\colon (1,5) \to \mathbb{R}$ exists and is equal to the power series, with radius of convergence 2, given by

$$f'(x) = \sum_{k=1}^{\infty} \frac{k^2}{2^k}(x-3)^{k-1} = \frac{1}{2} + \frac{2^2}{2^2}(x-3) + \frac{3^2}{2^3}(x-3)^2 + \frac{4^2}{2^4}(x-3)^3 + \cdots . \qquad (8.17)$$

Since the power series (8.17) has radius of convergence 2, Theorem 8.3.5 implies that $f'' \colon (1,5) \to \mathbb{R}$ exists and is equal to the following power series

$$f''(x) = \sum_{k=2}^{\infty} \frac{k^2(k-1)}{2^k}(x-3)^{k-2}$$

that also has radius of convergence 2. One can repeat Theorem 8.3.5 to obtain, for any $n \in \mathbb{N}$, a power series which is equal to $f^{(n)}$, on the interval $(1,5)$, where $f^{(n)}$ is the nth derivative of f. The following theorem also follows by repeatedly applying Theorem 8.3.5.

Theorem 8.3.6. Let $\sum\limits_{k=0}^{\infty} a_k(x-c)^k$ be a power series with radius of convergence $\rho > 0$. Let I be the open interval $(c-\rho, c+\rho)$, and define $f \colon I \to \mathbb{R}$ by $f(x) = \sum\limits_{k=0}^{\infty} a_k(x-c)^k$. Then, for all $n \in \mathbb{N}$, the nth derivative $f^{(n)} \colon I \to \mathbb{R}$ exists and satisfies

$$f^{(n)}(x) = \sum_{k=n}^{\infty} k(k-1)\cdots(k-n+1)a_k(x-c)^{k-n}$$

$$= n!a_n + \sum_{k=n+1}^{\infty} k(k-1)\cdots(k-n+1)a_k(x-c)^{k-n},$$

with radius of convergence ρ. So $f^{(n)}(c) = n!a_n$, for all $n = 0, 1, 2, \ldots$.

Remark. The term by term derivative of a power series $\sum\limits_{k=m}^{\infty} a_k(x-c)^k$ will have its starting value changed to $m+1$ only if the original power series starts with a constant term. For example, by differentiating $\sum\limits_{k=0}^{\infty} a_k(x-c)^k$, we obtain $\sum\limits_{k=1}^{\infty} a_k k(x-c)^{k-1}$; but, after differentiating $\sum\limits_{k=0}^{\infty} a_k(x-c)^{k+1}$, we obtain $\sum\limits_{k=0}^{\infty} a_k(k+1)(x-c)^k$.

We say that a function f is **infinitely differentiable** at a point x, if for all $n \in \mathbb{N}$, the nth derivative $f^{(n)}(x)$ exists. If f is infinitely differentiable at all points in an open interval I, then f is said to be infinitely differentiable on I. Theorem 8.3.6 shows that a function that is defined by a power series, with radius of convergence $\rho > 0$, is infinitely differentiable on an open interval.

Theorem 8.3.6 implies a uniqueness result concerning power series. Suppose that two power series have the same center and interval of convergence. Our next corollary shows that if these power series have equal values on this common interval, then they must be the same power series.

Corollary 8.3.7. Let $\sum\limits_{k=0}^{\infty} a_k(x-c)^k$ and $\sum\limits_{k=0}^{\infty} b_k(x-c)^k$ be power series with radius of convergence $\rho > 0$. Suppose that $\sum\limits_{k=0}^{\infty} a_k(x-c)^k = \sum\limits_{k=0}^{\infty} b_k(x-c)^k$ for all x in the interval $(c-\rho, c+\rho)$. Then $a_k = b_k$ for all $k \geq 0$.

Proof. Let $I = (c - \rho, c + \rho)$. Assume that $\sum_{k=0}^{\infty} a_k(x - c)^k = \sum_{k=0}^{\infty} b_k(x - c)^k$ for all x in I. Let $f: I \to \mathbb{R}$ be defined by $f(x) = \sum_{k=0}^{\infty} a_k(x - c)^k$. Thus, $f(x) = \sum_{k=0}^{\infty} b_k(x - c)^k$ for all $x \in I$, as well. So, $f(c) = a_0 = b_0$ (see (8.12)). Theorem 8.3.6 implies that $f^{(n)}(c) = n!a_n = n!b_n$ and so $a_n = b_n$, for all $n \geq 1$. Hence, $a_k = b_k$ for all $k \geq 0$. $\square$

Using Corollaries 8.2.2, 8.2.5, and Definition 6.2.12, the proof of Theorem 8.3.5 can be adapted to prove the next theorem.

Theorem 8.3.8. Let $\sum_{k=0}^{\infty} a_k(x-c)^k$ be power series with radius of convergence $\rho > 0$. Let I be the interval $(c-\rho, c+\rho)$, and let $f: I \to \mathbb{R}$ be defined by $f(x) = \sum_{k=0}^{\infty} a_k(x-c)^k$. Then f has an antiderivative on I. Moreover, for all $x \in I$,

$$\int_c^x f = \sum_{k=0}^{\infty} \frac{a_k}{k+1}(x - c)^{k+1}$$

and $\sum_{k=0}^{\infty} \frac{a_k}{k+1}(x - c)^{k+1}$ has ρ as its radius of convergence.

Exercises 8.3

1. Show that the power series $\sum_{k=0}^{\infty} \frac{x^k}{k!}$ converges for all x.

2. Find the radius of convergence of the power series: $\sum_{k=0}^{\infty} (3 + (-1)^k)^k x^k$.

3. Find the interval of convergence of the power series: (a) $\sum_{k=0}^{\infty} k! x^k$; (b) $\sum_{k=0}^{\infty} x^k$.

4. Suppose that for all $x \in \mathbb{R}$, $\sum_{k=0}^{\infty} a_k(x - c)^k$ converges if and only if $\sum_{k=0}^{\infty} b_k(x - c)^k$ converges. Show that $\limsup_{k\to\infty} \sqrt[k]{|a_k|} = \limsup_{k\to\infty} \sqrt[k]{|b_k|}$.

*5. Let $c \in \mathbb{R}$ and suppose that $\ell = \limsup_{k\to\infty} \sqrt[k]{|a_k|}$ is a real number.

   (a) For all $x \in \mathbb{R}$, show that $\limsup_{k\to\infty} \sqrt[k]{k |a_k|} |x - c|^{k-1} = |x - c| \ell$.
   [Note: If $x \neq c$, then $(x - c)^{k-1} = \frac{(x-c)^k}{x-c}$. See Corollary 3.2.9.]

   (b) Conclude that $\sum_{k=0}^{\infty} a_k(x - c)^k$ and $\sum_{k=1}^{\infty} ka_k(x - c)^{k-1}$ have the same radius of convergence. [See the proof of Theorem 8.3.2.]

6. Find the radius of convergence of the power series $\sum_{k=1}^{\infty} \frac{(k!)^2}{(2k)!}(x - 2)^k$.

7. Suppose that $\sum_{k=0}^{\infty} a_k(x-c)^k$ converges for all $x \in \mathbb{R}$. Prove that $\limsup_{k\to\infty} \sqrt[k]{|a_k|} = 0$.

8. Suppose that $\sum\limits_{k=1}^{\infty} a_k x^k$ has radius of convergence 2. Let $n \in \mathbb{N}$. Find the radius of convergence of (a) $\sum\limits_{k=1}^{\infty} a_k^n x^k$, (b) $\sum\limits_{k=1}^{\infty} a_k x^{nk}$, (c) $\sum\limits_{k=1}^{\infty} a_k x^{k^2}$.

9. Prove Theorem 8.3.8.

10. Suppose that a function f is defined by $f(x) = \sum\limits_{k=0}^{\infty} \frac{k+1}{4^k}(x-1)^k$. What is the domain of f? Find $f^{(90)}(1)$.

11. Find the radius of convergence of $\sum\limits_{k=0}^{\infty} \frac{(x-3)^k}{2^k}$ and find the sum of this series. Now find the sum of $\sum\limits_{k=1}^{\infty} k \frac{(x-3)^{k-1}}{2^k}$ and the sum of $\sum\limits_{k=0}^{\infty} \frac{(x-3)^{k+1}}{(k+1)2^k}$.

8.4   TAYLOR SERIES

In the previous section, we started with a power series centered at a point c with a radius of convergence $\rho > 0$. We then used this power series to define a function f and show that it is infinitely differentiable at all the points in an interval determined by c and ρ. In this section, we will attempt to reverse this process; that is, we will start with a function that is infinitely differentiable on an open interval and try to find a power series that is equal to the function on this interval.

Definition 8.4.1. Let f be a function that is defined on an open interval I that contains $c \in I$. Suppose that $\sum\limits_{k=0}^{\infty} a_k(x-c)^k$ is a power series that converges for all $x \in I$. If $f(x) = \sum\limits_{k=0}^{\infty} a_k(x-c)^k$ for all $x \in I$, then we say that f is **represented** by the power series $\sum\limits_{k=0}^{\infty} a_k(x-c)^k$. We also say that f has a **power series representation**.

Let f be a function that is infinitely differentiable at a point c. We would like to find a power series $\sum\limits_{k=0}^{\infty} a_k(x-c)^k$ such that $f(x) = \sum\limits_{k=0}^{\infty} a_k(x-c)^k$ for all x in some open interval I containing c. How can one find such a power series? If $f(x) = \sum\limits_{k=0}^{\infty} a_k(x-c)^k$ for all x in I, then Theorem 8.3.6 implies that $f^{(k)}(c) = k!a_k$ and thus, $a_k = \frac{f^{(k)}(c)}{k!}$ for all $k \geq 0$. Hence, we have found the only candidate; that is, if f has a power series representation, then this power series must be $\sum\limits_{k=0}^{\infty} \frac{f^{(k)}(c)}{k!}(x-c)^k$.

Definition 8.4.2. Let f be a function that is infinitely differentiable on an open interval I containing c. The **Taylor Series** for f centered at c is

$$\sum_{k=0}^{\infty} \frac{f^{(k)}(c)}{k!}(x-c)^k = f(c) + \frac{f'(c)}{1!}(x-c) + \frac{f''(c)}{2!}(x-c)^2 + \frac{f^{(3)}(c)}{3!}(x-c)^3 + \cdots,$$

where the coefficients are called the **Taylor coefficients**. When $c = 0$, the Taylor series is called the **Maclaurin Series** for f.

As noted above, if a function f has a power series representation centered at c, then that power series representation must be the Taylor series for f. On the other hand, there are infinitely differentiable functions f that do not have a power series representation and thus, the Taylor series for f will not be equal to the function f (see Example 8.4.11). In order to ensure that a function is equal to its Taylor series, the function must satisfy some additional conditions. Before identifying these conditions, we need to revisit a topic discussed in Section 5.3.

Let $f: I \to \mathbb{R}$ be infinitely differentiable on an open interval I containing c. For all $n \in \mathbb{N}$ and $x \in I$, let

$$P_n(x) = \sum_{k=0}^{n} \frac{f^{(k)}(c)}{k!}(x-c)^k \text{ and } R_n(x) = f(x) - P_n(x) \qquad (8.18)$$

where P_n is called the nth **Taylor polynomial** of f at c, and R_n is said to be the **remainder term**. The remainder term measures the difference between $f(x)$ and $P_n(x)$. We can now present a condition, which when satisfied, ensures that a function is equal to its Taylor series.

Theorem 8.4.3. Let $f: I \to \mathbb{R}$ be infinitely differentiable on an open interval I containing c. Let $x \in I$. Then $f(x) = \sum_{k=0}^{\infty} \frac{f^{(k)}(c)}{k!}(x-c)^k$ if and only if $\lim_{n\to\infty} R_n(x) = 0$.

Proof. For each $n \in \mathbb{N}$, note that $P_n(x)$ is the nth partial sum function for the Taylor series $\sum_{k=0}^{\infty} \frac{f^{(k)}(c)}{k!}(x-c)^k$ (see Definition 8.1.1). Thus, for all $x \in I$, we have the following equivalences:

$$f(x) = \sum_{k=0}^{\infty} \frac{f^{(k)}(c)}{k!}(x-c)^k \text{ iff } f(x) = \lim_{n\to\infty} P_n(x) \qquad \text{by Definition 7.1.1}$$

$$\text{iff } \lim_{n\to\infty}(f(x) - P_n(x)) = 0 \quad \text{by Theorem 3.1.10}$$

$$\text{iff } \lim_{n\to\infty} R_n(x) = 0 \qquad \text{by (8.18).} \qquad \square$$

Theorem 8.4.3 shows that a function f equals its the Taylor series on an interval I precisely when the sequence of remainder terms $\langle R_n(x) \rangle$ converges to 0, for all $x \in I$. However, when trying to prove that the sequence of remainder terms converges to 0, the definition of $R_n(x)$ in (8.18) is not very useful. Fortunately, there is another way to evaluate this remainder term. Taylor's Theorem 5.3.2 implies that for all $n \in \mathbb{N}$, there is a point d between c and x such that

$$f(x) = P_n(x) + \frac{f^{(n+1)}(d)}{(n+1)!}(x-c)^{n+1}. \qquad (8.19)$$

Since $R_n(x) = f(x) - P_n(x)$, Equation (8.19) implies that

$$R_n(x) = \frac{f^{(n+1)}(d)}{(n+1)!}(x-c)^{n+1}. \qquad (8.20)$$

We now present a result which can be used to show that the remainder terms will converge to 0.

Corollary 8.4.4. Let $f\colon I \to \mathbb{R}$ be infinitely differentiable on an open interval I containing c. If there is an $M > 0$ such that $\left|f^{(n)}(t)\right| \leq M$ for all $t \in I$ and all $n \in \mathbb{N}$, then the Taylor series for f, centered at c, is equal to f at all points $x \in I$.

Proof. Let M be such that $\left|f^{(n)}(t)\right| \leq M$ for all $t \in I$ and all $n \in \mathbb{N}$. Thus, for all $x \in I$ and all $n \in \mathbb{N}$, there is a $d \in I$ such that (see equation (8.20))

$$0 \leq |R_n(x)| = \frac{\left|f^{(n+1)}(d)\right|}{(n+1)!}|x-c|^{n+1} \leq \frac{M|x-c|^{n+1}}{(n+1)!}. \tag{8.21}$$

Exercise 16 on page 79 implies that $\lim\limits_{n\to\infty} \frac{M|x-c|^{n+1}}{(n+1)!} = 0$, for all $x \in I$. So $\lim\limits_{n\to\infty} R_n(x) = 0$ for every $x \in I$ by (8.21). Therefore, by Theorem 8.4.3, the Taylor series for f, centered at c, is equal to f at all points $x \in I$. $\square$

Example 8.4.5. Let $f\colon \mathbb{R} \to \mathbb{R}$ be defined by $f(x) = e^x$, the natural exponential function. We now apply Corollary 8.4.4 to show that f equals its Maclaurin series for all $x \in \mathbb{R}$. Since $f^{(n)}(0) = 1$ for all integers $n \geq 0$, we see that $\sum\limits_{k=0}^{\infty} \frac{x^k}{k!}$ is the Maclaurin series for f. By applying Exercise 12 on page 204, one can verify that the series converges for all x. Let $m \in \mathbb{N}$ and $M = e^m > 0$. Since $f^{(n)}(t) = e^t \leq M$ for all t in the interval $(-\infty, m)$. Corollary 8.4.4 implies that $e^x = \sum\limits_{k=0}^{\infty} \frac{x^k}{k!}$ for all $x \in (-\infty, m)$. Since this holds for any natural number m, it follows that $e^x = \sum\limits_{k=0}^{\infty} \frac{x^k}{k!}$ for all real numbers x. Substituting $x = 1$ into this Maclaurin series representation, we see that

$$e = \sum_{k=0}^{\infty} \frac{1}{k!} = 1 + \frac{1}{2!} + \frac{1}{3!} + \frac{1}{4!} + \frac{1}{5!} + \cdots.$$

In Examples 8.4.6–8.4.10 below, by applying differentiation, integration, substitution, or algebra on a given Taylor series representation, we obtain the Taylor series representation of a new function without having to compute its Taylor coefficients.

Example 8.4.6. Let $f\colon (-1,1) \to \mathbb{R}$ be defined by $f(x) = \frac{1}{1-x}$. Theorem 7.1.11 implies that $\frac{1}{1-x} = \sum\limits_{k=0}^{\infty} x^k$ for all $x \in (-1,1)$. Thus, $\frac{1}{1-x}$ equals its Maclaurin series for all $x \in (-1,1)$. Since $f'(x) = \frac{1}{(1-x)^2}$, Theorem 8.3.5 implies that $\frac{1}{(1-x)^2} = \sum\limits_{k=1}^{\infty} kx^{k-1}$ for all $x \in (-1,1)$. Thus, $\sum\limits_{k=1}^{\infty} kx^{k-1}$ is the Maclaurin series representation of the function $\frac{1}{(1-x)^2}$. As $\int_0^x f = \ln(1-x)$, we have that $\int_0^x f = \sum\limits_{k=0}^{\infty} \frac{x^{k+1}}{k+1}$ for $x \in (-1,1)$, by Theorem 8.3.8. Thus, $\sum\limits_{k=0}^{\infty} \frac{x^{k+1}}{k+1}$ is the Maclaurin series representation of the function $\ln(1-x)$ on the interval $(-1,1)$.

Example 8.4.7. In Example 8.4.5, it shown that $e^x = \sum_{k=0}^{\infty} \frac{x^k}{k!}$ for all real numbers x. Since $-x^2 \in \mathbb{R}$ for all $x \in \mathbb{R}$, we see that

$$e^{-x^2} = \sum_{k=0}^{\infty} \frac{(-x^2)^k}{k!} = \sum_{k=0}^{\infty} \frac{(-1)^k}{k!} x^{2k}, \text{ for all } x \in \mathbb{R}.$$

Thus, $xe^{-x^2} = \sum_{k=0}^{\infty} \frac{(-1)^k}{k!} x^{2k+1}$, for all $x \in \mathbb{R}$.

Example 8.4.8. In Example 8.4.6, it shown that $\frac{1}{1-x} = \sum_{k=0}^{\infty} x^k$ for all $x \in (-1,1)$. Observe that $1 - (-x) = 1 + x$. Since $-1 < -x < 1$ if and only if $-1 < x < 1$, we see that

$$\frac{1}{1+x} = \sum_{k=0}^{\infty}(-x)^k = \sum_{k=0}^{\infty}(-1)^k x^k, \text{ for all } x \in (-1,1).$$

Hence, $\frac{x}{1+x} = \sum_{k=0}^{\infty}(-1)^k x^{k+1}$, for all $x \in (-1,1)$.

Example 8.4.9. In Example 8.4.6, it shown that $\frac{1}{1-x} = \sum_{k=0}^{\infty} x^k$ for all $x \in (-1,1)$. Since $-1 < x^2 < 1$ if and only if $-1 < x < 1$, we see that

$$\frac{1}{1-x^2} = \sum_{k=0}^{\infty} x^{2k}, \text{ for all } x \in (-1,1).$$

Hence, $\frac{x}{1-x^2} = \sum_{k=0}^{\infty} x^{2k+1}$, for all $x \in (-1,1)$.

Example 8.4.10. In Example 8.4.8, we obtained the Maclaurin series for the function $f(x) = \frac{1}{1+x}$, that is, $f(x) = \sum_{k=0}^{\infty}(-1)^k x^k$ for all $x \in (-1,1)$. Since $\int_0^x f = \ln(1+x)$, we see that

$$\ln(1+x) = \sum_{k=0}^{\infty} \frac{(-1)^k}{k+1} x^{k+1}, \text{ for all } x \in (-1,1).$$

We end this section by identifying an infinitely differentiable function which is not equal to its Taylor Series. So not every infinitely differentiable function has a Taylor series representation.

Example 8.4.11. Let $f: \mathbb{R} \to \mathbb{R}$ be defined by

$$f(x) = \begin{cases} e^{-1/x^2}, & \text{if } x \neq 0; \\ 0, & \text{if } x = 0. \end{cases}$$

This function is clearly infinitely differentiable at x whenever $x \neq 0$. Moreover, it is infinitely differentiable at 0. For example, let us evaluate $f'(0)$:

$$f'(0) = \lim_{x \to 0} \frac{f(x) - f(0)}{x - 0} = \lim_{x \to 0} \frac{e^{-1/x^2}}{x} = 0$$

where the latter equality follows from Example 5.2.14 on page 142. In a similar manner, using Exercise 17 on page 147, one can show that $f^{(k)}(0) = 0$ for all $k \in \mathbb{N}$. Thus, $\sum_{k=0}^{\infty} f^{(k)}(0)x^k = 0$ for all x, which is not equal to $f(x)$ whenever $x \neq 0$.

Exercises 8.4

1. As shown in Example 8.4.5, $e^x = \sum_{k=0}^{\infty} \frac{x^k}{k!}$ for all real numbers x. Show that the term by term derivative of $\sum_{k=0}^{\infty} \frac{x^k}{k!}$ is equal to $\sum_{k=0}^{\infty} \frac{x^k}{k!}$.

2. Show that $\sin(x) = \sum_{k=0}^{\infty} \frac{(-1)^k}{(2k+1)!} x^{2k+1}$, for all $x \in \mathbb{R}$.

3. Using Exercise 2, find the Maclaurin series for $x \cos(x)$.

4. Express the integral $\int_0^1 e^{-x^2}\, dx$ as an infinite series.

5. Using the Maclaurin series for $\ln(1-x)$ in Example 8.4.6, evaluate $\sum_{k=1}^{\infty} \frac{(\frac{1}{3})^{k+1}}{k}$.

6. Find the sum of the series $\sum_{k=1}^{\infty} kx^k$ and $\sum_{k=1}^{\infty} k^2 x^k$, when x is in $(-1,1)$.

7. Find the sum of the series $\sum_{k=0}^{\infty} x^{2k+1}$, when x is in $(-1,1)$. Now find the sum of the series $2^{-1} + 2^{-3} + 2^{-5} + 2^{-7} + \cdots$.

8. Using the Maclaurin series for $\frac{1}{1+x}$ in Example 8.4.7, find the Maclaurin series for $\frac{1}{1+(2x-3)}$, and its interval of convergence.

9. Find the Taylor series for e^x centered at 1.

10. Let $\rho \in \mathbb{R}^+$ and $c \in \mathbb{R}$. Let $I = (c - \rho, c + \rho)$ and $f \colon I \to \mathbb{R}$. Suppose that $f(x) = \sum_{k=0}^{\infty} a_k(x-c)^k$ if and only if $x \in I$. Prove that $\limsup_{k \to \infty} \sqrt[k]{|a_k|} = \frac{1}{\rho}$.

11. Using the Maclaurin series for $\frac{1}{1+x}$ in Example 8.4.8, find the Maclaurin series for $\frac{1}{x}$ centered at 1 and over the interval $(0,2)$.

12. Let $f(x) = \sum_{k=0}^{\infty} a_k x^k$ for all x such that $|x| < \rho$, were $\rho \in \mathbb{R}^+$. Suppose that $f(x) = f(-x)$ whenever $|x| < \rho$. Prove that $a_k = 0$ for every odd integer k.

13. Suppose that the power series $\sum_{k=0}^{\infty} a_k(x-c)^k$ has radius of convergence $\rho > 0$. Let I be the interval $I = (c - \rho, c + \rho)$, and let $[a,b] \subseteq I$. Explain why the power series converges uniformly on $[a,b]$.

Proof of the Composition Theorem

Theorem 6.2.14 (Composition Theorem). If $f: [a, b] \to [c, d]$ is Riemann integrable and $g: [c, d] \to \mathbb{R}$ is continuous, then $(g \circ f): [a, b] \to \mathbb{R}$ is Riemann integrable.

Before proving this theorem, we present some technical notation and two technical lemmas that will be used in the proof.

Definition A.1. Let $f: [a, b] \to \mathbb{R}$ be bounded and $P = \{x_0, x_1, x_2, \ldots, x_n\}$ be a partition of $[a, b]$. Let $M_i = M_i(f)$ and $m_i = m_i(f)$ for $i = 1, \ldots, n$. Given $\delta > 0$, let

$$A_\delta = \{i : M_i - m_i < \delta\} \text{ and } B_\delta = \{i : M_i - m_i \geq \delta\}. \tag{A.1}$$

Let A_δ and B_δ be as defined in Definition A.1. Then

$$A_\delta \cup B_\delta = \{1, 2, 3, \ldots, n\} \text{ and } A_\delta \cap B_\delta = \varnothing. \tag{A.2}$$

Lemma A.2. Suppose that $f: [a, b] \to \mathbb{R}$ is bounded. For every $\delta > 0$, whenever $P = \{x_0, x_1, x_2, \ldots, x_n\}$ is a partition of $[a, b]$ such that

$$U(f, P) - L(f, P) < \delta^2, \tag{A.3}$$

then $\sum_{i \in B_\delta} \Delta x_i < \delta$ where B_δ is as defined in (A.1).

Proof. Let f, δ, and P be as stated. Assume that P satisfies (A.3). Let A_δ and B_δ be as in Definition A.1. From (A.2) and (A.3), we see that

$$U(f, P) - L(f, P) = \sum_{i=1}^n (M_i - m_i) \Delta x_i = \sum_{i \in A_\delta} (M_i - m_i) \Delta x_i + \sum_{i \in B_\delta} (M_i - m_i) \Delta x_i < \delta^2.$$

Therefore, $\sum_{i \in B_\delta} (M_i - m_i) \Delta x_i < \delta^2$. Since $\delta \leq M_i - m_i$ for all $i \in B_\delta$, we conclude that $\sum_{i \in B_\delta} \delta \Delta x_i \leq \sum_{i \in B_\delta} (M_i - m_i) \Delta x_i < \delta^2$. Thus, $\sum_{i \in B_\delta} \Delta x_i < \delta$. $\qquad \square$

Lemma A.3. Let $f\colon [a,b] \to [c,d]$, $g\colon [c,d] \to \mathbb{R}$, $\varepsilon' > 0$, and $\delta > 0$. Suppose that

$$\text{for all } y, y' \in [c,d], \text{ if } |y - y'| < \delta, \text{ then } |g(y) - g(y')| < \varepsilon'. \tag{A.4}$$

Then for any partition $P = \{x_0, x_1, x_2, \ldots, x_n\}$ of $[a,b]$, we have that

$$\sum_{i \in A_\delta} (M_i(g \circ f) - m_i(g \circ f))\, \Delta x_i \leq \varepsilon'(b - a),$$

where A_δ is as in Definition A.1.

Proof. Let $f\colon [a,b] \to [c,d]$, $g\colon [c,d] \to \mathbb{R}$, $\varepsilon' > 0$, and $\delta > 0$. Assume (A.4). Let $P = \{x_0, x_1, \ldots, x_n\}$ be a partition of $[a,b]$. Let $M_i^\circ = M_i(g \circ f)$, $m_i^\circ = m_i(g \circ f)$, $M_i = M_i(f)$, and $m_i = m_i(f)$ for each $i = 1, 2, \ldots, n$. Let $A_\delta = \{i : M_i - m_i < \delta\}$.

Claim. $M_i^\circ - m_i^\circ \leq \varepsilon'$ for all $i \in A_\delta$.

Proof of Claim. Let $i \in A_\delta$ and let $S = \{g(f(x)) : x \in [x_{i-1}, x_i]\}$. Observe that $M_i^\circ = \sup(S)$ and $m_i^\circ = \inf(S)$. Let $x, x' \in [x_{i-1}, x_i]$. Hence, $m_i \leq f(x) \leq M_i$ and $m_i \leq f(x') \leq M_i$. So $|f(x) - f(x')| \leq M_i - m_i$. Since $i \in A_\delta$, we see that $|f(x) - f(x')| < \delta$. Thus, (A.4) implies that $|g(f(x)) - g(f(x'))| < \varepsilon'$. Therefore, $M_i^\circ - m_i^\circ \leq \varepsilon'$ by Theorem 2.3.18. □

The Claim implies that

$$\sum_{i \in A_\delta} (M_i^\circ - m_i^\circ)\Delta x_i \leq \sum_{i \in A_\delta} \varepsilon' \Delta x_i = \varepsilon' \sum_{i \in A_\delta} \Delta x_i \leq \varepsilon'(b - a),$$

where the last inequality holds because $\sum\limits_{i \in A_\delta} \Delta x_i \leq \sum\limits_{i=1}^{n} \Delta x_i = b - a$. □

A Proof of the Composition Theorem

Proof of Theorem 6.2.14. Let $f\colon [a,b] \to [c,d]$ be integrable and $g\colon [c,d] \to \mathbb{R}$ be continuous. To prove that $(g \circ f)\colon [a,b] \to \mathbb{R}$ is integrable on $[a,b]$, we shall apply Theorem 6.1.18. Let $\varepsilon > 0$. Since g is continuous on $[c,d]$, we see that g is bounded. Let $K = \sup\{|g(x)| : x \in [c,d]\}$. Since $a < b$ and $K \geq 0$, there is an $\varepsilon' > 0$ such that

$$\varepsilon'(b - a + 2K) < \varepsilon. \tag{A.5}$$

By Theorem 4.5.5, g is uniformly continuous on $[c,d]$. So, there is a $\delta > 0$ such that $\delta \leq \varepsilon'$ and

$$\text{for all } y, y' \in [c,d], \text{ if } |y - y'| < \delta, \text{ then } |g(y) - g(y')| < \varepsilon'. \tag{A.6}$$

Since f is Riemann integrable on $[a,b]$, let $P = \{x_0, x_1, x_2, \ldots, x_n\}$ be a partition such that

$$U(f, P) - L(f, P) < \delta^2. \tag{A.7}$$

Let $A_\delta = \{i : M_i - m_i < \delta\}$ and $B_\delta = \{i : M_i - m_i \geq \delta\}$ where $M_i = M_i(f)$ and $m_i = m_i(f)$ for each $i = 1, \ldots, n$. Now, let $M_i^\circ = M_i(g \circ f)$ and $m_i^\circ = m_i(g \circ f)$, for all

$i = 1, \ldots, n$. Since $K = \sup\{|g(x)| : x \in [c, d]\}$, it follows that $-K \le m_i^{\circ} \le M_i^{\circ} \le K$ and thus,

$$M_i^{\circ} - m_i^{\circ} \le 2K \text{ for } i = 1, \ldots, n. \tag{A.8}$$

We now show that $U(g \circ f, P) - L(g \circ f, P) < \varepsilon$ as follows:

$$
\begin{aligned}
U(g \circ f,P) &- L(g \circ f, P) \\
&= \sum_{i=1}^{n} (M_i^{\circ} - m_i^{\circ}) \Delta x_i && \text{by Lemma 6.1.7} \\
&= \sum_{i \in A_\delta} (M_i^{\circ} - m_i^{\circ}) \Delta x_i + \sum_{i \in B_\delta} (M_i^{\circ} - m_i^{\circ}) \Delta x_i && \text{by (A.2)} \\
&\le \varepsilon'(b - a) + \sum_{i \in B_\delta} 2K \Delta x_i && \text{by (A.6), Lemma A.3, (A.8)} \\
&= \varepsilon'(b - a) + 2K \sum_{i \in B_\delta} \Delta x_i && \text{by algebra} \\
&\le \varepsilon'(b - a) + 2K\delta && \text{by (A.7) and Lemma A.2} \\
&\le \varepsilon'(b - a) + 2K\varepsilon' && \text{since } \delta \le \varepsilon' \\
&= \varepsilon'(b - a + 2K) < \varepsilon && \text{by algebra and (A.5).}
\end{aligned}
$$

Therefore, $g \circ f$ is Riemann integrable on $[a, b]$, by Theorem 6.1.18. $\qquad\square$

Topology on the Real Numbers

In this appendix, we will explore certain topological properties that concern certain subsets of $\mathbb{R}$. This involves extending the basic properties possessed by open intervals to a broader collection of sets of real numbers. We will do the same for closed intervals. These generalized properties are called open, closed, and compact.

B.1   OPEN AND CLOSED SETS

We begin by considering some relationships that hold between an individual real number and a set of real numbers. Recall that a neighborhood of a point $x \in \mathbb{R}$ is an open interval of the form $(x - \varepsilon, x + \varepsilon)$, centered at x, for some $\varepsilon > 0$. To simplify the notation, we will denote the neighborhood $(x - \varepsilon, x + \varepsilon)$ by U_ε^x. We will also write U^x to denote a neighborhood of x when it is not important to identify ε.

Now let $x \in \mathbb{R}$ and let $S \subseteq \mathbb{R}$. Think of x as being a point on a map and the set S as being a region on the map. The next definition gives meaning to the expressions "x is inside of S" and "x on the boundary of S."

Definition B.1.1. Let S be a set of real numbers.

1. The point $x \in \mathbb{R}$ is an **interior point** of S if there is a neighborhood U^x such that $U^x \subseteq S$. The set of all interior points of S is denoted by $\mathrm{int}(S)$.

2. The point x is a **boundary point** of S if *every* neighborhood U^x contains points in S and points not in S. The set of all boundary points of S is denoted by $\mathrm{bd}(S)$.

If x is an interior point of S, then $x \in S$. Whereas, a boundary point of S may be or may not be a member of S.

Example. We identify the sets $\mathrm{int}(S)$ and $\mathrm{bd}(S)$ for the following five sets:

1. $S = [0, 3)$: $\mathrm{int}(S) = (0, 3)$ and $\mathrm{bd}(S) = \{0\}$.
2. $S = \mathbb{N}$: $\mathrm{int}(S) = \varnothing$ and $\mathrm{bd}(S) = \mathbb{N}$.
3. $S = \{\frac{1}{n} : n \in \mathbb{N}\}$: $\mathrm{int}(S) = \varnothing$ and $\mathrm{bd}(S) = S \cup \{0\}$.

4. $S = \mathbb{Q}$: $\text{int}(S) = \varnothing$ and $\text{bd}(S) = \mathbb{R}$.
5. $S = \{q \in \mathbb{Q} : 0 < q < 3\}$: $\text{int}(S) = \varnothing$ and $\text{bd}(S) = [0, 3]$.

The concept of a topology on the set of real numbers involves certain subsets of $\mathbb{R}$ that are decreed to be "open sets." An open set of real numbers is an important concept in mathematics. The collection of all open sets is called a topology. Many topics that we previously studied (such as convergence and continuity) can be defined entirely in terms of open sets.

Definition B.1.2. Let S be a set of real numbers.

1. S is an **open set** if $\text{int}(S) = S$, that is, for every $x \in S$ there is a neighborhood U^x such that $U^x \subseteq S$.
2. S is a **closed set** if $\mathbb{R} \setminus S$ is an open set.

One can view a set S as being open, if every point in S is completely surrounded only by points in S.

Comment. An open set S contains **none** of the boundary points of S. A closed set S contains **all** of the boundary points of S.

Comment. The sets $\varnothing$ and $\mathbb{R}$ are both open and closed sets. These are the only subsets of $\mathbb{R}$ that have both of these properties.

We now show that an open interval is an open set.

Theorem B.1.3. Let $a, b \in \mathbb{R}$ be such that $a < b$. Then the open interval (a, b) is an open set.

Proof. Let $x \in (a, b)$, and so, $a < x < b$. Let $\varepsilon = \min\{x - a, b - x\}$. We will prove that $U_\varepsilon^x \subseteq (a, b)$. Let $y \in U_\varepsilon^x$ be arbitrary. Thus, $|y - x| < \varepsilon$. Since $\varepsilon = \min\{x - a, b - x\}$, we see that $|y - x| < x - a$ and $|y - x| < b - x$. Thus,

$$x - y < x - a \text{ and } y - x < b - x.$$

Hence, $a < y$ and $y < b$. Therefore, $a < y < b$. □

Thus, in particular, every neighborhood is an open set.

Theorem B.1.4. Let $a \in \mathbb{R}$. The open intervals $(-\infty, a)$ and $(a, +\infty)$ are open sets.

There is a simple relationship between open sets and closed sets.

Theorem B.1.5. Let S be a set of real numbers.

(a) The set S is open if and only if $\mathbb{R} \setminus S$ is closed.
(b) The set S is closed if and only if $\mathbb{R} \setminus S$ is open.

We will now explore how the set operations affect open and closed sets. For example, is the union or intersection of two open sets also open? What about the union or intersection of any family of open sets? We note that the operations $\bigcup \mathcal{F}$ and $\bigcap \mathcal{F}$ are defined in Section 1.2.5.

Theorem B.1.6. Let $\mathcal{F}$ be a family of open sets.

(a) Then $\bigcup \mathcal{F}$ is an open set.
(b) If $\mathcal{F}$ is finite and nonempty, then $\bigcap \mathcal{F}$ is an open set.

Proof. Let $\mathcal{F}$ be a family of open sets.

(a). Let $x \in \bigcup \mathcal{F}$. Thus, there is an $O \in \mathcal{F}$ such that $x \in O$. As O is an open set, there is a U^x such that $U^x \subseteq O$. Since $O \subseteq \bigcup \mathcal{F}$, it follows that $U^x \subseteq \bigcup \mathcal{F}$.

(b). Let $\mathcal{F} = \{O_1, \ldots, O_n\}$ where $n \in \mathbb{N}$. Let $x \in \bigcap \mathcal{F}$. Thus, $x \in O_i$ for all $O_i \in \mathcal{F}$. As every O_i is an open set, there is an $\varepsilon_i > 0$ such that $U_{\varepsilon_i}^{x_i} \subseteq O_i$. Let $\varepsilon = \min\{\varepsilon_1, \ldots, \varepsilon_n\} > 0$. As $U_\varepsilon^x \subseteq U_{\varepsilon_i}^{x_i} \subseteq O_i$ for all $O_i \in \mathcal{F}$, we have $U_\varepsilon^x \subseteq \bigcap \mathcal{F}$. $\square$

Corollary B.1.7. Let $\mathcal{G}$ be a nonempty family of closed sets.

(a) Then $\bigcap \mathcal{G}$ is a closed set.
(b) If $\mathcal{G}$ is finite, then $\bigcup \mathcal{G}$ is a closed set.

Proof. Let $\mathcal{G}$ be a nonempty family of closed sets.

(a). By Theorem ??(2), we have that (▲) $\mathbb{R} \setminus \bigcap \mathcal{G} = \bigcup\{\mathbb{R} \setminus B : B \in \mathcal{G}\}$. Since $\mathbb{R} \setminus B$ is an open set for each $B \in \mathcal{G}$, Theorem B.1.6(a) and (▲) imply that $\mathbb{R} \setminus \bigcap \mathcal{G}$ is an open set. Hence, $\bigcap \mathcal{G}$ is a closed set.

(b). See Exercise 6. Hint: Use Theorem B.1.6(b) and Theorem 1.2.11(1). $\square$

Theorem B.1.8. Let $a, b \in \mathbb{R}$. Then $(-\infty, a) \cup (b, +\infty)$ is an open set.

Proof. By Theorem B.1.4, $(a, +\infty)$ and $(-\infty, a)$ are open sets. Thus, Theorem B.1.6 implies that $(-\infty, a) \cup (b, +\infty)$ is an open set. $\square$

Theorem B.1.9. Let $a, b \in \mathbb{R}$ be such that $a \leq b$. Then $[a, b]$ is a closed set.

Proof. Since $\mathbb{R} \setminus [a, b] = (-\infty, a) \cup (b, +\infty)$, Theorem B.1.8 tells us that $\mathbb{R} \setminus [a, b]$ is an open set. Therefore, $[a, b]$ is a closed set. $\square$

B.1.1 Continuity Revisited

For $x \in \mathbb{R}$, recall that U^x denotes an open interval $(x - p, x + p)$ for some $p > 0$. As alluded to earlier, the continuity concept can be defined entirely in terms of open sets. This is established by the following two theorems.

Theorem B.1.10. Let $f \colon D \to \mathbb{R}$ and $c \in D$. The following are equivalent:

(1) f is continuous at c.
(2) For every $U^{f(c)}$ there exists a U^c such that $f[U^c \cap D] \subseteq U^{f(c)}$.
(3) For every $U^{f(c)}$ there exists a U^c such that $U^c \cap D \subseteq f^{-1}[U^{f(c)}]$.

Proof. Let $f \colon D \to \mathbb{R}$ be a function and let $c \in D$. Items (2) and (3) are equivalent by Exercise 13 on page 22. So we will show that (1) and (2) are equivalent.

(1) $\Rightarrow$ (2). Let f be continuous at c and let $U^{f(c)}$ be neighborhood of $f(c)$. Since f is continuous at c, there is a U^c such that $f[U^c \cap D] \subseteq U^{f(c)}$ (see Figure 4.1).

$(2) \Rightarrow (1)$. Assume that (▲) for each $U^{f(c)}$ there is a U^c such that $f[U^c \cap D] \subseteq U^{f(c)}$. Let $\varepsilon > 0$. So $U_\varepsilon^{f(c)}$ is a neighborhood. By (▲), there is a U_δ^c so that $f[U_\delta^c \cap D] \subseteq U_\varepsilon^{f(c)}$. Thus, for all $x \in D$, if $|x - c| < \delta$, then $|f(x) - f(c)| < \varepsilon$. So f is continuous at c. □

Theorem B.1.11. A every function $f \colon D \to \mathbb{R}$, the following are equivalent:

(1) f is continuous.

(2) For every open set V there exists an open set U such that $U \cap D = f^{-1}[V]$.

(3) For every closed set B there exists a closed set A such that $A \cap D = f^{-1}[B]$.

Proof. Let $f \colon D \to \mathbb{R}$ be a function.

$(1) \Rightarrow (2)$. Assume that f is continuous. Let V be an open set. Define the set $\mathcal{F}$ by

$$U^c \in \mathcal{F} \text{ iff } c \in f^{-1}[V] \text{ and } U^c \cap D \subseteq f^{-1}[U^{f(c)}] \text{ for some } U^{f(c)} \subseteq V. \qquad \text{(B.1)}$$

By Theorem B.1.6(a), $U = \bigcup \mathcal{F}$ is an open set. We now prove that $U \cap D = f^{-1}[V]$.

($\subseteq$). Let $x \in U \cap D$. So $x \in U^c \cap D$ for some $U^c \in \mathcal{F}$. Thus, by (B.1), we have that $x \in U^c \cap D \subseteq f^{-1}[U^{f(c)}]$ where $U^{f(c)} \subseteq V$. Since $f^{-1}[U^{f(c)}] \subseteq f^{-1}[V]$ (see Exercise 9 on page 22), we conclude that $x \in f^{-1}[V]$. Hence, $U \cap D \subseteq f^{-1}[V]$.

($\supseteq$). Let $c \in f^{-1}[V]$. So $c \in D$ and $f(c) \in V$. Since V is an open set, there is a neighborhood $U^{f(c)} \subseteq V$. Theorem B.1.10(3) implies there exists a U^c such that $U^c \cap D \subseteq f^{-1}[U^{f(c)}]$. Thus, by (B.1), $U^c \in \mathcal{F}$. So $c \in U \cap D$. Hence, $f^{-1}[V] \subseteq U \cap D$.

$(2) \Rightarrow (1)$. Assume (2). Let $c \in D$ and let $U^{f(c)}$ be a neighborhood of $f(c)$. By (2), there is an open set U such that $U \cap D = f^{-1}[U^{f(c)}]$. Since $f(c) \in U^{f(c)}$, we infer that $c \in U \cap D$. Since U is an open set, there is a U^c such that $U^c \subseteq U$. So

$$U^c \cap D \subseteq U \cap D = f^{-1}[U^{f(c)}].$$

Thus, by Theorem B.1.10(3), f is continuous at c. Therefore, f is continuous.

For the equivalence of items (2) and (3), see Exercise 5. □

B.1.2 Accumulation Points Revisited

Let $S \subseteq \mathbb{R}$. Recalling Definition 3.5.2, a point $x \in \mathbb{R}$ is an *accumulation point of S* if every neighborhood of x contains an infinite number of points from S. That is, for all neighborhoods U^x, the set $S \cap U^x$ is infinite. A point $x \in \mathbb{R}$ is called an *isolated point of S* if $x \in S$ and x is not an accumulation point of S. The set of all accumulation points of S is denoted by S'.

Comment. A point x is **not** an accumulation point of S if there is a neighborhood U^x such that the set $S \cap U^x$ is finite.

Definition B.1.12. Let $S \subseteq \mathbb{R}$. The **closure** of the set S is defined by $\mathrm{cl}(S) = S \cup S'$.

Lemma B.1.13. Let $S \subseteq \mathbb{R}$. Then $\mathrm{cl}(S)$ is a closed set.

Lemma B.1.14. Let $S \subseteq \mathbb{R}$, $x \in \mathbb{R}$ and $x \notin S$. Suppose that every neighborhood of x contains a point from S. Then every neighborhood of x must contain an infinite number of points from S.

Proof. Let $S \subseteq \mathbb{R}$, $x \in \mathbb{R}$ and $x \notin S$. Suppose that

$$\text{every neighborhood of } x \text{ contains a point from } S. \tag{A}$$

We will prove that every neighborhood of x must contain an infinite number of points from S. Suppose, for a contradiction, that some neighborhood of x, say U_ε^x (where $\varepsilon > 0$), contains only a finite number of points from S. So, $U_\varepsilon^x \cap S = \{x_1, x_2, \ldots, x_n\}$ where $n \in \mathbb{N}$. Let $\varepsilon^* = \min\{|x - x_1|, |x - x_2|, \ldots, |x - x_n|\} > 0$. Because $x \notin S$, it follows that $\varepsilon^* > 0$. Now, since $\varepsilon^* < \varepsilon$ it follows that $U_{\varepsilon^*}^x \subseteq U_\varepsilon^x$; and hence, $U_{\varepsilon^*}^x \cap S \subseteq U_\varepsilon^x \cap S$. Thus, $U_{\varepsilon^*}^x \cap S \subseteq \{x_1, x_2, \ldots, x_n\}$. In addition, since $\varepsilon^* \leq |x - x_i|$ for each $i \leq n$, it follows that $x_i \notin U_{\varepsilon^*}^x$ for all $i \leq n$. So, $U_{\varepsilon^*}^x \cap S = \varnothing$. Therefore, $U_{\varepsilon^*}^x$ is a neighborhood of x which contains no points from S. This contradicts (A). □

Theorem B.1.15. Let $S \subseteq \mathbb{R}$. Then S is closed if and only if S contains all of its accumulation points.

Proof. Let S be a set of real numbers.

($\Rightarrow$). Assume that S is closed; thus, $\mathbb{R} \setminus S$ is open. Let x be an accumulation point of S. Suppose, for a contradiction, that $x \notin S$. Since $\mathbb{R} \setminus S$ is open, there is a neighborhood U^x such that U^x contains no points from S. However, since x is an accumulation point of S, U^x contains an infinite number of points from S.

($\Leftarrow$) Assume that S contains all of its accumulation points. Let $x \in \mathbb{R} \setminus S$. We must show that there exists a neighborhood of x which contains no points in S. Suppose, for a contradiction, that *every neighborhood of x contains some points from S*. Thus, by Lemma B.1.14, every neighborhood of x contains an infinite number of points from S. So x is an accumulation point of S. Hence, by assumption, $x \in S$. □

Theorem B.1.15 and Theorem 3.5.4 imply the following useful equivalence.

Theorem B.1.16. Let $S \subseteq \mathbb{R}$. Then S closed if and only if for every sequence $\langle s_n \rangle$ such that $s_n \in S$ for all $n \in \mathbb{N}$, if $\lim_{n \to \infty} s_n = c$, then $c \in S$.

Exercises B.1

1. Let A be open and B be closed. Prover that $A \setminus B$ is open and $B \setminus A$ is closed.

2. Let $S \subseteq \mathbb{R}$. Prove that $\mathrm{bd}(S)$ is a closed set.

3. Let $S \subseteq \mathbb{R}$ and let $\mathcal{F} = \{A : S \subseteq A \text{ and } A \text{ is closed}\}$. Prove that $\mathrm{cl}(S) = \bigcap \mathcal{F}$.

4. Let $f : D \to \mathbb{R}$ be a function where $D \subseteq \mathbb{R}$. Let $A \subseteq \mathbb{R}$ and $B \subseteq \mathbb{R}$. Show that $(\mathbb{R} \setminus A) \cap D = f^{-1}[B]$ if and only if $A \cap D = f^{-1}[\mathbb{R} \setminus B]$.

*5. Using Exercise 4, complete the proof of Theorem B.1.11 by showing that items (2) and (3) are equivalent.

*6. Prove Corollary B.1.7(b).

7. Prove Lemma B.1.13.

8. Prove Theorem B.1.16.

9. Let $f: D \to \mathbb{R}$ be continuous and let $a \in \mathbb{R}$.

   (a) Show that there is an open set U such that $U \cap D = \{x \in D : f(x) > a\}$. Observe that $f^{-1}[V] = \{x \in D : f(x) > a\}$, where $V = \{y \in \mathbb{R} : y > a\}$.

   (b) Show that there is an open set U such that $U \cap D = \{x \in D : f(x) < a+1\}$.

   (c) Show that there is a closed set A such that $A \cap D = \{x \in D : f(x) \geq a\}$.

   (d) Show that there is a open set U such that $A \cap D = \{x \in D : f(x) \neq a\}$.

   (e) Show that there is a closed set A such that $A \cap D = \{x \in D : f(x) = a\}$.

10. Let $f: D \to \mathbb{R}$ and $g: D \to \mathbb{R}$ be continuous. Show that there is an open set U such that $U \cap D = \{x \in D : f(x) \neq g(x)\}$.

B.2 COMPACT SETS

Closed intervals are very important in real analysis. Theorem 4.4.8 shows that the continuous image of a closed interval is a closed interval. Moreover, Theorem 4.5.5 states that a continuous function on a closed interval is uniformly continuous. These theorems may not hold if one replaces "closed interval" with "open interval." What is so special about closed intervals? *Closed intervals are compact.* As we will see, the previously cited theorems hold if we replace "closed interval" with "compact set."

Before we can define the concept of a compact set, we need to discuss the notion of an open cover.

Definition B.2.1. Let $S \subseteq \mathbb{R}$. An **open cover** of S is a family $\mathcal{F}$ of open sets such that $S \subseteq \bigcup \mathcal{F}$, that is, $\mathcal{F}$ **covers** S. If $\mathcal{F}'$ is a finite subset of $\mathcal{F}$ such that $S \subseteq \bigcup \mathcal{F}'$, then $\mathcal{F}'$ is called a **finite subcover** of S.

Example. Consider the subset $S = [2,7]$ of $\mathbb{R}$. For each $n \in \mathbb{N}$, let $O_n = (1, 7 - \frac{1}{n})$. In addition, let $O = (6,8)$. Let $\mathcal{F} = \{O, O_1, O_2, O_3, \dots\}$. Thus, $\mathcal{F}$ is an open cover of S because $S \subseteq \bigcup \mathcal{F}$. Let $\mathcal{F}' = \{O, O_2\}$. Note that $O \cup O_2 = (6,8) \cup (1, 6 + \frac{1}{2}) = (1,8)$. Thus, $S \subseteq \bigcup \mathcal{F}'$ and $\mathcal{F}'$ is a finite subcover of S.

Example. Consider the subset $S = [2,7)$ of $\mathbb{R}$. For each $n \in \mathbb{N}$, let O_n be the open interval $O_n = (1, 7 - \frac{1}{n})$. Let $\mathcal{F} = \{O_n : n \in \mathbb{N}\}$. Then $S \subseteq \bigcup_{n \in \mathbb{N}} O_n$ and so $\mathcal{F}$ is an open cover of S. But any finite subset of $\mathcal{F}$, say, $\{O_{n_1}, O_{n_2}, \dots, O_{n_k}\}$ will not cover S, because for $m = \max\{n_1, n_2, \dots, n_k\}$

$$O_{n_1} \cup O_{n_2} \cup \cdots \cup O_{n_k} = (1, 7 - \frac{1}{m})$$

and $S = [2,7) \not\subseteq (1, 7 - \frac{1}{m})$. Therefore, $\mathcal{F}$ has no finite subcover of S.

Definition B.2.2 (Compactness). A set $S \subseteq \mathbb{R}$ is said to be **compact** if *every* open cover of S has a finite subcover.

It follows, vacuously, that the empty set is compact.

Lemma B.2.3. Suppose that $C \subseteq \mathbb{R}$ is a nonempty, closed, and bounded set. Then $\alpha = \inf(C)$ and $\beta = \sup(C)$ exist. In addition, $\alpha \in C$ and $\beta \in C$. Hence, $\alpha = \min(C)$ and $\beta = \max(C)$.

Proof. Let $C \subseteq \mathbb{R}$ be nonempty, closed, and bounded. By the completeness axiom, $\beta = \sup(C)$ exists. We will now show that $\beta \in C$. Suppose, for a contradiction, that $\beta \notin C$. Since $\mathbb{R} \setminus C$ is open, there is a neighborhood U^β such that $U^\beta \cap C = \varnothing$. Thus, there is an $x \in U^\beta$ such that $x < \beta$ and for all y, if $x < y \le \beta$, then $y \notin C$. So, $x < \beta$ and $s \le x$ for all $s \in C$. But this contradicts that fact that β is the least upper bound for C. Thus, $\beta \in C$ and so $\beta = \max(C)$. The argument proving that $\alpha = \inf(C)$ exists and $\alpha = \min(C)$, is similar. □

Theorem B.2.4. Let $C \subseteq \mathbb{R}$ be closed and bounded. Then C is compact.

Proof. As $\varnothing$ is compact, let $C \subseteq \mathbb{R}$ be nonempty, closed, and bounded. Let $\mathcal{F}$ be an open covering of C. For each $x \in \mathbb{R}$, define $C_x = \{z \in C : z \le x\}$. Consider the set

$$S = \{x \in \mathbb{R} : \mathcal{F} \text{ has a finite subcover of } C_x\}. \tag{B.2}$$

Claim 1. S is nonempty.

Proof of Claim 1. By Lemma B.2.3, $\min(C) = m \in C$. Thus, $C_m = \{m\}$. Since $\mathcal{F}$ is an open cover of C, there is an $O \in \mathcal{F}$ such that $m \in F$. Therefore, $\{O\}$ is a finite subcover of C_m. Hence, $m \in S$ and $S \ne \varnothing$. This completes the proof of Claim 1. □

Claim 2. S is not bounded above.

Proof of Claim 2. Suppose, for a contradiction, that S is bounded above. Thus, by the completeness axiom, $\beta = \sup(S)$ exists. Therefore,

$$(\forall y > \beta) \,(\mathcal{F} \text{ does } \mathbf{not} \text{ have a finite subcover of } C_y) \tag{B.3}$$
$$(\forall x < \beta) \,(\mathcal{F} \text{ does have a finite subcover of } C_x). \tag{B.4}$$

(To see that (B.4) holds, note that if $x < y$, then $C_x \subseteq C_y$.) Since either $\beta \notin C$ or $\beta \in C$, we consider these two cases.

CASE 1: $\beta \notin C$. Since $\mathbb{R} \setminus C$ is open, there is a U^β such that $U^\beta \cap C = \varnothing$. So, there exists $x, y \in U^\beta$ such that $x < \beta < y$ and for all z, if $x \le z \le y$, then $z \in U^\beta$ and hence, $z \notin C$. Thus, $C_x = C_y$. Now since $x < \beta$, (B.4) implies that $\mathcal{F}$ has a finite subcover of C_x. So $\mathcal{F}$ has a finite subcover of C_y. As $y > \beta$, this contradicts (B.3).

CASE 2: $\beta \in C$. Since $\mathcal{F}$ is an open cover of C, there is an $O \in \mathcal{F}$ such that $\beta \in O$. Since O is an open set, let U^β be such that $U^\beta \subseteq O$. So, there exists $x, y \in U^\beta$ such that $x < \beta < y$ and for all z, if $x \le z \le y$, then $z \in O$. As $x < \beta$, (B.4) implies that $\mathcal{F}$ has a finite subcover $\mathcal{F}'$ of C_x. Hence, $\mathcal{F}' \cup \{O\}$ is a finite subcover of C_y. Since $y > \beta$, this contradicts (B.3). This completes the proof of Claim 2. □

Therefore, S is not bounded above. Since C is bounded, let $x \in S$ be such that $z \le x$ for all $z \in C$. Therefore $C = C_x$ and thus, $\mathcal{F}$ has a finite subcover of C. □

Theorem B.2.4 implies that a closed interval is a compact set, as is the union of finitely many closed intervals. Thus, we can now present and prove generalizations of Theorems 4.4.8 and 4.5.5.

Theorem B.2.5. If $f: D \to \mathbb{R}$ is continuous and D is compact, then $f[D]$ is compact.

Proof. Let $f: D \to \mathbb{R}$ be continuous and D be compact. Let $\mathcal{G}$ be an open covering of $f[D]$. Let $\mathcal{F}$ be defined by

$$U \in \mathcal{F} \text{ iff } U \text{ is open and } U \cap D = f^{-1}[V] \text{ for some } V \in \mathcal{G}. \tag{B.5}$$

Claim. $\mathcal{F}$ is an open covering of D.

Proof of Claim. Let $c \in D$. Thus, $f(c) \in f[D]$. Since $\mathcal{G}$ covers $f[D]$, there is a $V \in \mathcal{G}$ such that $f(c) \in V$. Theorem B.1.11 implies that there is an open set U such that $U \cap D = f^{-1}[V]$. Thus, $U \in \mathcal{F}$ and, as $c \in f^{-1}[V]$, we see that $c \in U$. Hence, $\mathcal{F}$ is an open covering of D. This completes the proof of the claim. $\square$

Since D is compact, let $\mathcal{F}' = \{U_1, U_2, \dots, U_n\} \subseteq \mathcal{F}$ be a finite subcover of D. For each $U_i \in \mathcal{F}'$, let $V_i \in \mathcal{G}$ confirm that $U_i \in \mathcal{F}$ according to (B.5), the definition of $\mathcal{F}$. Now let $\mathcal{G}' = \{V_1, V_2, \dots, V_n\} \subseteq \mathcal{G}$. We now prove that $\mathcal{G}'$ covers $f[D]$. Let $y \in f[D]$ and let $c \in D$ be such that $y = f(c)$. Since $\mathcal{F}'$ covers D, let $U_i \in \mathcal{F}'$ be such that $c \in U_i$. Since $U_i \cap D = f^{-1}[V_i]$, it follows that $c \in f^{-1}[V_i]$. Thus, $y = f(c) \in V_i$. Therefore, $\mathcal{G}'$ covers $f[D]$ and $f[D]$ is compact. $\square$

Theorem B.2.6. If $f: D \to \mathbb{R}$ is continuous and D is compact, then f is uniformly continuous on D.

Proof. Let $f: D \to \mathbb{R}$ be continuous, D be compact, $\varepsilon > 0$, and $\varepsilon^* = \frac{\varepsilon}{2}$. Define $\mathcal{F}$ by

$$U_\delta^c \in \mathcal{F} \text{ iff } c \in D \text{ and } f[U_{2\delta}^c \cap D] \subseteq U_{\varepsilon^*}^{f(c)}. \tag{B.6}$$

Let $c \in D$. Since f is continuous, Theorem B.1.10(2) implies that there is a U_d^c such that $f[U_d^c \cap D] \subseteq U_{\varepsilon^*}^{f(c)}$. Let $\delta = \frac{d}{2}$. So $d = 2\delta$ and $c \in U_\delta^c \in \mathcal{F}$. Thus, $\mathcal{F}$ is an open covering of D. As D is compact, let $\mathcal{F}' = \{U_{\delta_1}^{c_1}, U_{\delta_2}^{c_2}, \dots, U_{\delta_n}^{c_n}\} \subseteq \mathcal{F}$ be a finite subcover of D. Let $\delta = \min\{\delta_1, \delta_2, \dots, \delta_n\} > 0$. Let $x, y \in D$ be such that $|x - y| < \delta$. Since $\mathcal{F}'$ covers D, we have that $x \in U_{\delta_i}^{c_i}$ for some $U_{\delta_i}^{c_i} \in \mathcal{F}'$. Thus, $|x - c_i| < \delta_i$. Since

$$|c_i - y| \le |c_i - x| + |x - y| < \delta_i + \delta \le \delta_i + \delta_i = 2\delta_i,$$

we see that $y \in U_{2\delta_i}^{c_i}$. We also have that $x \in U_{2\delta_i}^{c_i}$. Thus, by (B.6), we have that $|f(x) - f(c_i)| < \varepsilon^*$ and $|f(y) - f(c_i)| < \varepsilon^*$. Hence,

$$|f(x) - f(y)| \le |f(x) - f(c_i)| + |f(y) - f(c_i)| < \varepsilon^* + \varepsilon^* = \varepsilon.$$

Therefore, f is uniformly continuous on D. $\square$

Exercises B.2

1. Show that each of the following subsets S of $\mathbb{R}$ are **not** compact by finding an open cover of S which has no finite subcover.

1. $S = [0, 3)$.

2. $S = \mathbb{N}$.

3. $S = \{\frac{1}{n} : n \in \mathbb{N}\}$.

4. $S = \{q \in \mathbb{Q} : 0 < q < \sqrt{2}\}$.

2. Let $\langle s_n \rangle$ be a sequence that converges to c. Show that $\{c, s_1, s_2, \dots\}$ is compact.

3. Let K be a compact set and let $\langle s_n \rangle$ be such that $s_n \in K$ for all $n \in \mathbb{N}$. Show that $\langle s_n \rangle$ has a subsequence that converges to a point in K.

4. Let K be a compact set and let $S \subseteq K$ be a closed set. Prove that S is compact.

5. Let K be a compact set and let S be a closed set. Prove that $K \cap S$ is compact.

6. Let K be a compact set and let S be an open set. Prove that $K \setminus S$ is compact.

*7. Let $\mathcal{F}$ be a nonempty set of compact sets. Prove that $\bigcap \mathcal{F}$ is compact.

8. Let $\mathcal{F}$ be a finite set of compact sets. Prove that $\bigcup \mathcal{F}$ is compact.

B.3   THE HEINE–BOREL THEOREM

In this section we present a theorem that provides an alternative characterization of compactness. This theorem was discovered during the time when mathematicians were developing a logically sound foundation for real analysis. Much of this work was influenced by Cantor's work in set theory and topology.

Theorem B.3.1 (Heine–Borel). Let $S \subseteq \mathbb{R}$. Then S is compact if and only if S is closed and bounded.

Proof. Let $S \subseteq \mathbb{R}$. Theorem B.2.4 shows that if S is closed and bounded, then S is compact. So we just need to prove the converse. To do this, assume that S is compact. We will prove that S is closed and bounded.

Claim 1. S is a closed set.

Proof of Claim 1. To show that S is closed, we will show that $\mathbb{R} \setminus S$ is an open set. So let $x \in \mathbb{R} \setminus S$. For each $n \in \mathbb{N}$, the set

$$U^x_{1/n} = \left\{ y \in \mathbb{R} : |x - y| < \frac{1}{n} \right\}$$

is a neighborhood of x. We will prove that there is an $n \in \mathbb{N}$ such that $U^x_{1/n} \subseteq \mathbb{R} \setminus S$. For each $n \in \mathbb{N}$, let O_n be the open set defined by

$$O_n = \left\{ y \in \mathbb{R} : |x - y| > \frac{1}{n} \right\}.$$

Note that for all $m, n \in \mathbb{N}$,

$$m \leq n \text{ implies } O_m \subseteq O_n. \tag{B.7}$$

Let $\mathcal{F} = \{O_n : n \in \mathbb{N}\}$. Since $x \notin S$, we see that $S \subseteq \bigcup \mathcal{F}$. So $\mathcal{F}$ is an open cover of S. As S is compact, there is a finite $\mathcal{F}' \subseteq \mathcal{F}$ that covers S. Let $\mathcal{F}' = \{O_{n_1}, \dots O_{n_k}\}$

be such that $S \subseteq \bigcup \mathcal{F}'$. Let $n \in \mathbb{N}$ be the largest of the natural numbers $n_1, \ldots, n_k$. Thus, by (B.7), $\bigcup \mathcal{F}' = O_n$. Hence, $S \subseteq O_n$. As $O_n \cap U^x_{1/n} = \varnothing$, it follows that $U^x_{1/n} \subseteq \mathbb{R} \setminus S$. Hence, $\mathbb{R} \setminus S$ is an open set. This completes the proof of Claim 1. □

Claim 2. S is a bounded set.

Proof. We will show that there is an $n \in \mathbb{N}$ such that $S \subseteq (-n, n)$. For each $n \in \mathbb{N}$, let O_n be the open interval $O_n = (-n, n)$. Note that for all $m, n \in \mathbb{N}$,

$$m \leq n \text{ implies } O_m \subseteq O_n. \tag{B.8}$$

Let $\mathcal{F} = \{O_n : n \in \mathbb{N}\}$. Since $S \subseteq \mathbb{R}$, it follows that $S \subseteq \bigcup \mathcal{F}$. Since S is compact, there is a finite $\mathcal{F}' \subseteq \mathcal{F}$ that covers S. Let $\mathcal{F}' = \{O_{n_1}, \ldots O_{n_k}\}$. Let $n \in \mathbb{N}$ be the largest of the natural numbers $n_1, \ldots, n_k$. Thus, $\bigcup \mathcal{F}' = O_n$ by (B.8). Hence, $S \subseteq O_n$; that is, $S \subseteq (-n, n)$. Thus, S is bounded. This completes the proof of Claim 2. □

Claim 1 and Claim 2 complete our proof of the Heine–Borel Theorem. □

The Heine–Borel Theorem is a topological tool that can be used to produce different proofs of several of the theorems previously established in this text. To confirm this assertion, we now give a topological proof of Theorem 3.5.3.

Theorem 3.5.3 (Bolzano–Weierstrass for Sets). Let $S \subseteq \mathbb{R}$ be infinite. If S is bounded, then there is a point $x \in \mathbb{R}$ such that x is an accumulation of S.

Proof. Let S be an infinite bounded subset of $\mathbb{R}$. So, there exists $a, b \in \mathbb{R}$ such that $S \subseteq [a, b]$. We shall now show that there is an $x \in [a, b]$ which is an accumulation point of S. Suppose, for a contradiction, that for all $x \in [a, b]$, x is not an accumulation point of S. So for all $x \in [a, b]$, there exists a neighborhood of x, say U^x, such that $S \cap U^x$ is finite. Now consider the following set of open sets

$$\mathcal{F} = \{U^x : x \in [a, b] \text{ and } S \cap U^x \text{ is finite.}\}$$

Hence, $\mathcal{F}$ is an open cover of $[a, b]$. By Theorem B.1.9, $[a, b]$ is a closed and bounded subset of $\mathbb{R}$. Thus, the Heine–Borel Theorem implies that $[a, b]$ is compact. As $\mathcal{F}$ is an open cover of $[a, b]$, let $\mathcal{F}' = \{U^{x_1}, U^{x_2}, \ldots, U^{x_n}\} \subseteq \mathcal{F}$ be a finite subcover of $[a, b]$. So

$$[a, b] \subseteq \bigcup \mathcal{F}' = U^{x_1} \cup U^{x_2} \cup \cdots \cup U^{x_n}$$

and, because $S \subseteq [a, b]$, we have that

$$S \subseteq U^{x_1} \cup U^{x_2} \cup \cdots \cup U^{x_n}.$$

Therefore,

$$S = (S \cap U^{x_1}) \cup (S \cap U^{x_2}) \cup \cdots \cup (S \cap U^{x_n}).$$

Since each $S \cap U^{x_i}$ is finite, we conclude that S is a finite union of finite sets. Thus, S is a finite set. Contradiction. □

The Heine–Borel Theorem also offers a much easier way to show that a set is compact.

B.3.1 The Finite Intersection Property

Definition B.3.2. A family of sets $\mathcal{K}$ has the **finite intersection property** if the intersection of any nonempty finite subset of $\mathcal{K}$ is nonempty.

We now present a generalization of Theorem 2.5.2, the nested intervals theorem.

Theorem B.3.3. Let $\mathcal{K}$ be a nonempty family of compact sets. If $\mathcal{K}$ has the finite intersection property, then $\bigcap \mathcal{K} \neq \varnothing$.

Proof. Let $\mathcal{K}$ be a nonempty family of compact sets and with the finite intersection property. As each $K \in \mathcal{K}$ is compact, we see that K is closed and bounded by the Heine–Borel Theorem B.3.1. So for all $K \in \mathcal{K}$, $\widetilde{K} = \mathbb{R} \setminus K$ is an open set. Define the set $\mathcal{F} = \{\widetilde{K} : K \in \mathcal{K}\}$, which consists of open sets. We see that (▲) $\mathbb{R} \setminus \bigcap \mathcal{K} = \bigcup \mathcal{F}$, by Theorem 1.2.11(2). Let $K_0 \in \mathcal{K}$ be a fixed element of $\mathcal{K}$. So K_0 is a compact set.

We will now prove that $\bigcap \mathcal{K} \neq \varnothing$. Suppose, for a contradiction, that $\bigcap \mathcal{K} = \varnothing$. From (▲) we see that $\mathbb{R} = \bigcup \mathcal{F}$. So $K_0 \subseteq \bigcup \mathcal{F}$. Since K_0 is compact, there is a finite $\mathcal{F}' \subseteq \mathcal{F}$ such that (▶) $K_0 \subseteq \bigcup \mathcal{F}'$. Let

$$\mathcal{G} = \{K_0\} \cup \{K : \widetilde{K} \in \mathcal{F}'\}.$$

Since $\mathcal{G}$ is finite, $\mathcal{G} \subseteq \mathcal{K}$, and $\mathcal{K}$ has the finite intersection property, there is a y such that (▼) $y \in \bigcap \mathcal{G}$. So, in particular, $y \in K_0$. Thus, by (▶), $y \in \widetilde{K}$ for some $\widetilde{K} \in \mathcal{F}'$. Hence, $K \in \mathcal{G}$ and $y \notin K$. This contradicts (▼). Therefore, $\bigcap \mathcal{K} \neq \varnothing$. ☐

B.3.2 The Cantor Set

We end this appendix, by establishing the existence of a remarkable set called the Cantor set. This set was introduced by Georg Cantor in 1884 and has many curious properties, some of which will be discussed here.

In order to identify the Cantor set, we must first construct a nested set of compact sets $\{K_n : n \in \mathbb{N}\}$ where $K_n \supseteq K_{n+1}$ for all $n \in \mathbb{N}$. We being this construction by letting $K_1 = [0,1]$. Divide $[0,1]$ into three subintervals of equal length and remove the middle open interval $(\frac{1}{3}, \frac{2}{3})$, and let $K_2 = [1, \frac{1}{3}] \cup [\frac{2}{3}, 1]$. Now divide each of the intervals that comprise K_2 into three subintervals of equal length and remove all of the middle open intervals, to obtain

$$K_3 = \left[0, \frac{1}{3^2}\right] \cup \left[\frac{2}{3^2}, \frac{3}{3^2}\right] \cup \left[\frac{6}{3^2}, \frac{7}{3^2}\right] \cup \left[\frac{8}{3^2}, 1\right].$$

Continuing in this manner, we obtain $K_1, K_2, K_3, \ldots, K_n, \ldots$, where $K_n \supseteq K_{n+1}$ for all $n \in \mathbb{N}$ (see Figure B.1). We make an initial observation: Every K_n is the union of 2^{n-1} many disjoint closed intervals each of length $\frac{1}{3^{n-1}}$. Now let $\mathcal{K} = \{K_n : n \in \mathbb{N}\}$. The **Cantor set**, which we denote by $\mathcal{C}$, is the set $\mathcal{C} = \bigcap \mathcal{K}$.

Since K_n is a finite union of closed intervals, we see that each K_n is compact. Thus, Exercise 7 on page 247 implies that $\mathcal{C}$ is compact. Since $\mathcal{K}$ is nested, it follows that $\mathcal{K}$ has the finite intersection property. Thus, the Cantor set $\mathcal{C} = \bigcap \mathcal{K}$ is nonempty

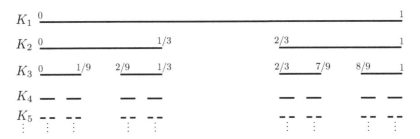

Figure B.1: Constructing the Cantor Set.

by Theorem B.3.3. However, the fact that $\mathcal{C}$ is nonempty is not surprising, since the endpoints that appear in each K_n are all in $\mathcal{C}$; that is,

$$0, 1, \frac{1}{3}, \frac{2}{3}, \frac{1}{9}, \frac{2}{9}, \frac{7}{9}, \frac{8}{9}, \cdots \in \mathcal{C}.$$

One might conjecture the following: *Every point in $\mathcal{C}$ appears in the above countable list of endpoints.* As we will see, this conjecture is actually false.

We can now pose a question: Does the Cantor set $\mathcal{C}$ contain any intervals? This seems unlikely, but to verify that $\mathcal{C}$ contains no intervals, we must provide a proof. Let $x \in \mathcal{C}$ and $y \in \mathcal{C}$ be such that $x < y$. If $(x, y) \subseteq \mathcal{C}$, then $(x, y) \subseteq K_n$ for every $n \in \mathbb{N}$. Let $n \in \mathbb{N}$ be such that $\frac{1}{3^{n-1}} < y - x$. Since K_n is the union of disjoint closed intervals of length $\frac{1}{3^{n-1}}$, we cannot have that $(x, y) \subseteq K_n$. So $\mathcal{C}$ contains no intervals. For this reason, $\mathcal{C}$ is said to have no interior points.

We can now pose another question: What is the total length of all the intervals removed in the process of constructing $\mathcal{C}$? Note that

- K_2 was obtained by removing one interval of length $\frac{1}{3}$,
- K_3 was obtained by removing two intervals of length $\frac{1}{3^2}$,
- K_4 was obtained by removing four intervals of length $\frac{1}{3^3}$,
- K_5 was obtained by removing eight intervals of length $\frac{1}{3^4}$.

Continuing in this way, we see that the total length of the removed intervals is

$$\frac{1}{3} + 2\frac{1}{3^2} + 4\frac{1}{3^3} + 8\frac{1}{3^4} + \cdots = \sum_{k=0}^{\infty} \frac{1}{3}\left(\frac{2}{3}\right)^k = \frac{\frac{1}{3}}{1 - \frac{2}{3}} = 1.$$

Thus, in the construction of the Cantor set, the total length of all of the "middle" open intervals removed is 1. Since the Cantor set consists of the points in $[0, 1]$ that remain after removing all of these intervals, we conclude that the "length" of the Cantor set must be 0. We know that a point has zero length. Thus, the Cantor set is "small," because it has the same length as a point. On the other hand, one can show that $\mathcal{C}$ is uncountable. So the Cantor set is large in the sense of cardinality, but small in the sense of length.

A *infinite binary sequence* is a function $\alpha \colon \mathbb{N} \to \{0, 1\}$. Thus, we can write α as

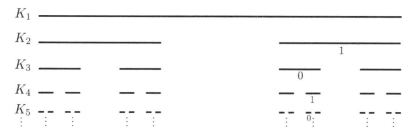

Figure B.2: Intervals in $\mathcal{I}^\alpha$ where $\alpha = \langle 1, 0, 1, 0, \ldots \rangle$.

$\langle \alpha(1), \alpha(2), \alpha(3), \ldots \rangle$ where $\alpha(n) = 0$ or $\alpha(n) = 1$, for all $n \in \mathbb{N}$. Such sequences can be used to identify the elements in the Cantor set.

Each point of the Cantor set lies in the intersection of an infinite nested family of closed intervals. To illustrate this assertion, let $\alpha = \langle 1, 0, 1, 0, \ldots \rangle$ be a specific infinite binary sequence. Think of the number 1 as representing "the right subinterval" and 0 as "the left subinterval." Using the sequence α, we can define a nested set of closed intervals in stages. This can be viewed in Figure B.2. We begin by letting $I_1^\alpha = [0, 1]$. Since $\alpha(1) = 1$, we let $I_2^\alpha = [\frac{2}{3}, 1]$ which is the right subinterval of I_1^α. As $\alpha(2) = 0$, we next select $I_3^\alpha = [\frac{2}{3}, \frac{7}{9}]$, the left subinterval of I_2^α. Since $\alpha(3) = 1$, let $I_4^\alpha = [\frac{20}{27}, \frac{7}{9}]$, the right subinterval of I_3^α. Continuing in this way, we obtain the nested set $\mathcal{I}^\alpha = \{I_n^\alpha : n \in \mathbb{N}\}$ of closed intervals where each interval I_n^α has length $\frac{1}{3^{n-1}}$. Since $I_n^\alpha \subseteq K_n$ for all $n \in \mathbb{N}$, we see that $\bigcap \mathcal{I}^\alpha \subseteq \bigcap \mathcal{K} = \mathcal{C}$. We know by Theorem 2.5.2 that $\bigcap \mathcal{I}^\alpha \neq \varnothing$. In fact, $\bigcap \mathcal{I}^\alpha$ has exactly one element which we will denote by c_α. Thus, α acts as a sequence of instructions which allows one to acquire a point c_α in $\mathcal{C}$, the Cantor set.

So any infinite infinite binary sequence α can be viewed as an address to an element $c_\alpha \in \mathcal{C}$, and the associated set of closed intervals $\mathcal{I}^\alpha = \{I_n^\alpha : n \in \mathbb{N}\}$ can also be viewed as a road map by which to locate c_α. Moreover, if α and β are two distinct infinite binary sequences, then $c_\alpha \neq c_\beta$. In addition, for each point $c \in \mathcal{C}$ there is an infinite binary sequence α such that $c = c_\alpha$. These observations imply that $\mathcal{C}$ is uncountable and that $\mathcal{C}$ has the same cardinality as $\mathbb{R}$ [2, Chapter 5]. So the Cantor set $\mathcal{C}$ contains no intervals, has zero length and yet, it has as many points as the set of all real numbers. One can also show the every $c \in \mathcal{C}$ is an accumulation point of $\mathcal{C}$ (see Exercise 9). So $\mathcal{C}$ has no isolated points. These are a few of the paradoxical properties that are possessed by the Cantor set.

Exercises B.3

1. Let $\mathcal{K} = \{A_n : n \in \mathbb{N}\}$ be a nested family of nonempty sets; that is, $A_n \neq \varnothing$ and $A_{n+1} \subseteq A_n$ for all $n \in \mathbb{N}$.

   (a) Show that $\mathcal{K}$ has the finite intersection property.

   (b) For all $n \in \mathbb{N}$, let $A_n = (0, \frac{1}{n}]$. Show that $\mathcal{K} = \{A_n : n \in \mathbb{N}\}$ is a nested family of nonempty sets. By (a), $\mathcal{K}$ has the finite intersection property. Now evaluate $\bigcap \mathcal{K}$. Does your answer contradict Theorem B.3.3?

2. Let $f \colon D \to \mathbb{R}$ be continuous where D is compact. Prove that f is bounded.

3. Let $f \colon D \to \mathbb{R}$ be continuous where $D \neq \varnothing$ is compact. Prove that there is a $c \in D$ such that $f(x) \leq f(c)$ for all $x \in D$.

4. Let $f \colon D \to \mathbb{R}$ be continuous and D is compact. Let $\mathcal{G}$ be a nonempty set of closed subsets of $f[D]$. Suppose that $\mathcal{G}$ has the finite intersection property. Prove that there exists a $c \in D$ such that $f(c) \in \bigcap \mathcal{G}$.

5. Let α be an infinite binary sequence and let $\mathcal{I}^\alpha = \{I_n^\alpha : n \in \mathbb{N}\}$ be the associated nested set of closed intervals (see Section B.3.2). Let $\langle s_n \rangle$ be a sequence of real numbers such that $s_n \in I_n^\alpha$ for all $n \in \mathbb{N}$. Prove that $\langle s_n \rangle$ is a Cauchy sequence.

6. Let α be an infinite binary sequence and let $\mathcal{I}^\alpha = \{I_n^\alpha : n \in \mathbb{N}\}$ be the associated nested set of closed intervals (see Section B.3.2). Show that there cannot be two distinct points in $\bigcap \mathcal{I}^\alpha$.

7. Let α and β are two distinct infinite binary sequences. Show that $c_\alpha \neq c_\beta$.

8. Let $c \in \mathcal{C}$, the Cantor set. Inductively define an infinite binary sequence α such that $c \in I_n^\alpha$ for all $n \in \mathbb{N}$. Conclude that $c = c_\alpha$.

*9. Let $c \in \mathcal{C}$, the Cantor set. Let α be such that $c = c_\alpha$. Let $\varepsilon > 0$ and $n \in \mathbb{N}$ be such that $\frac{1}{3^{n-1}} < \varepsilon$. Prove that $U_\varepsilon^c \cap \mathcal{C}$ is infinite.

Review of Proof and Logic

LOGIC: THE BASIS FOR PROOFS

Some Logical Connectives

Given a list of "propositions" or statements $P, Q, R, \ldots$ we can form sentences using the logical connectives $\wedge, \vee, \neg$. For example,

1. $P \wedge Q$ (means "P and Q" and is called a *conjunction*).
2. $P \vee Q$ (means "P or Q" and is called a *disjunction*).
3. $\neg P$ (means "not P" and is called a *negation*).

Using these connectives as building blocks, one can construct more complex sentences, for example, $(P \wedge \neg Q) \vee (\neg S \wedge R)$.

Truth Tables

The truth value of a sentence of the propositional logic can be evaluated from the truth value of its components. We shall explain what this "means" by using truth tables. The above logical connectives have the natural truth values given by the following tables, where T means "true" and F means "false."

P	Q	$P \wedge Q$
T	T	T
T	F	F
F	T	F
F	F	F

P	Q	$P \vee Q$
T	T	T
T	F	T
F	T	T
F	F	F

P	$\neg P$
T	F
F	T

Logical Equivalence

Definition. Let ψ and φ be sentences of propositional logic. Then ψ and φ are **logically equivalent**, denoted by $\psi \Leftrightarrow \varphi$, if they are both true at the same time and both false at the same time.

Example. Let ψ be the sentence $\neg(P \vee Q)$ and let φ be the sentence $\neg P \wedge \neg Q$. Show that ψ and φ are logically equivalent, that is, show $\neg(P \vee Q) \Leftrightarrow \neg P \wedge \neg Q$.

Solution. It is sufficient to show that truth values of $\neg(P \vee Q)$ and $\neg P \wedge \neg Q$ are always the same. This is done by constructing the following parallel truth tables:

Truth table for $\neg(P \vee Q)$:

	P	Q	$P \vee Q$	$\neg(P \vee Q)$
	T	T	T	F
	T	F	T	F
	F	T	T	F
	F	F	F	T
Step #	1	1	2	3

Truth table for $\neg P \wedge \neg Q$:

	P	Q	$\neg P$	$\neg Q$	$\neg P \wedge \neg Q$
	T	T	F	F	F
	T	F	F	T	F
	F	T	T	F	F
	F	F	T	T	T
Step #	1	1	2	2	3

Since the final columns of the truth tables for $\neg(P \vee Q)$ and $\neg P \wedge \neg Q$ are the same, we conclude that $\neg(P \vee Q) \Leftrightarrow \neg P \wedge \neg Q$.

The Conditional and Biconditional Connectives

The Conditional Connective. Given two propositions P and Q, the conditional connective $\to$ means "implies," and can be used to form the sentence $P \to Q$. This sentence can be read as "if P, then Q." Given two propositions P and Q, the sentence $P \to Q$ has the following truth table:

P	Q	$P \to Q$
T	T	T
T	F	F
F	T	T
F	F	T

Definition. Let P and Q be statements. The formula $\neg Q \to \neg P$ is called the **contrapositive** of $P \to Q$.

Definition. Let P and Q be statements. The formula $Q \to P$ is called the **converse** of $P \to Q$.

Note: the two statements $P \to Q$ and $Q \to P$ are **not equivalent**. This can be shown by comparing their truth tables:

P	Q	$P \to Q$
T	T	T
T	F	F
F	T	T
F	F	T

P	Q	$Q \to P$
T	T	T
T	F	T
F	T	F
F	F	T

The Biconditional Connective. Given two propositions P and Q, the biconditional connective $\leftrightarrow$ means "if and only if," and can be used to form the sentence $P \leftrightarrow Q$. This new connective has the following truth table:

P	Q	$P \leftrightarrow Q$
T	T	T
T	F	F
F	T	F
F	F	T

Some Propositional Logic Laws

DeMorgan's Laws

1. $\neg(P \vee Q) \Leftrightarrow \neg P \wedge \neg Q$
2. $\neg(P \wedge Q) \Leftrightarrow \neg P \vee \neg Q$

Conditional Laws

1. $(P \to Q) \Leftrightarrow (\neg P \vee Q)$
2. $(P \to Q) \Leftrightarrow \neg(P \wedge \neg Q)$

Contrapositive Law

1. $(P \to Q) \Leftrightarrow (\neg Q \to \neg P)$.

Quantifiers

The quantifier $\forall$ means "for all" and is called the *universal quantifier*. The quantifier $\exists$ means "there exists" and is called the *existential quantifier*. For example, we can form the sentences

1. $\forall x P(x)$ [means "for all x, $P(x)$"].
2. $\exists x P(x)$ [means "there exists an x such that $P(x)$"].

Quantifier Negation Laws

1. $\neg \exists x P(x) \Leftrightarrow \forall x \neg P(x)$.
2. $\neg \forall x P(x) \Leftrightarrow \exists x \neg P(x)$.

Definition (Bounded Quantifiers). We write $(\forall x \in A)\, P(x)$ to mean that *for every x in A, $P(x)$ is true*. Also $(\exists x \in A)\, P(x)$ means that *for some x in A, $P(x)$ is true*.

Bounded Quantifier Negation Laws

1. $\neg(\exists x \in A)\, P(x) \Leftrightarrow (\forall x \in A)\, \neg P(x)$.
2. $\neg(\forall x \in A)\, P(x) \Leftrightarrow (\exists x \in A)\, \neg P(x)$.

Definition (Bounded Number Quantifiers). We write $(\forall x < a)P(x)$ to mean that *for every number $x < a$, $P(x)$ is true*. We also write $(\exists x < a)P(x)$ to assert that *for some number $x < a$, $P(x)$ is true*.

Negation Laws for Bounded Number Quantifiers

1. $\neg(\forall x > a)P(x) \Leftrightarrow (\exists x > a)\neg P(x)$.   3. $\neg(\forall x < a)P(x) \Leftrightarrow (\exists x < a)\neg P(x)$.
2. $\neg(\exists x > a)P(x) \Leftrightarrow (\forall x > a)\neg P(x)$.   4. $\neg(\exists x < a)P(x) \Leftrightarrow (\forall x < a)\neg P(x)$.

These laws also hold for $\leq, >, \geq$.

PROOF STRATEGIES

1. To prove an algebraic equation, try one of the following:

    (a) Transform one side of the equation into the other side of the equation.

    (b) Derive the equation from any previously given, or assumed, equations.

2. To prove $P \rightarrow Q$, try one of the following:

    (a)   Assume P
                                Prove Q.

    (b)   Assume $\neg Q$
                                Prove $\neg P$.

3. To prove $P \wedge Q$, try the following:

                                Prove P
                                Prove Q.

4. To prove $P \vee Q$, try one of the following:

    (a)   Assume $\neg P$
                                Prove Q.

    (b)   Assume $\neg Q$
                                Prove P.

    (c) Try using a division by cases. In each case, prove P or prove Q.

5. To prove $P \leftrightarrow Q$, try the following:

                              Prove $P \rightarrow Q$
                                Prove $Q \rightarrow P$.

6. To prove P by contradiction:

                              Assume $\neg P$
                         Derive "a contradiction."

7. To prove $\forall x P(x)$, or $(\forall x \in A)P(x)$:

        Let x, or respectively $x \in A$, be arbitrary. Now prove $P(x)$.

8. To prove $\exists x P(x)$, or $(\exists x \in A)P(x)$, try one of the following:

    (a)   Let $x =$ (the value you found)
                                  Prove $P(x)$.

    (b) Prove $P(x)$ for some x, or respectively for some $x \in A$.

ASSUMPTION STRATEGIES

1. When assuming $P \to Q$:

   (a) If you are assuming or can prove P, then you can conclude Q.

   (b) If you are assuming or can prove $\neg Q$, then you can conclude $\neg P$.

2. When assuming $P \vee Q$:

   (a) If required to prove R, try the following division by cases,

   > Case 1: Assume P.
   > Prove R.
   > Case 2: Assume Q.
   > Prove R.

   (b) If you are assuming or can prove $\neg P$, then you can conclude Q.

   (c) If you are assuming or can prove $\neg Q$, then you can conclude P.

3. When assuming $P \wedge Q$:

   > You can assume P and assume Q.

4. When assuming $P \leftrightarrow Q$:

   (a) If you are assuming or can prove P, then you can conclude Q.

   (b) If you are assuming or can prove Q, then you can conclude P.

5. When assuming $\neg P$:

   (a) In a proof by contradiction try to prove P and thereby derive a contradiction.

   (b) Reexpress $\neg P$ as a positive statement, and try to *use* the positive statement.

6. When assuming $\forall x P(x)$, or $(\forall x \in A)P(x)$:

   > Take *any useful* value for x, or respectively $x \in A$; say a, and assume $P(a)$.

7. When assuming $\exists x P(x)$, or $(\exists x \in A)P(x)$:

   > Introduce a **new** term x_0, or respectively $x_0 \in A$, and assume $P(x_0)$.

8. Any established theorem can be assumed and used in a proof.

Well-Ordering and Induction Proof Strategies

Let b be a fixed integer and let n be an integer variable.

1. To prove $(\forall n \geq b)\, P(n)$ by the well-ordering principle, use:

   > Assume that $\neg P(n)$ holds for some integer $n \geq b$.
   > Let $N \geq b$ be the smallest such integer satisfying $\neg P(N)$.
   > Derive "a contradiction."

2. To prove $(\forall n \geq b)\, P(n)$ by mathematical induction, use:

Base step:	Prove $P(b)$.
Inductive step:	Let $n \geq b$ be an integer.
	Assume $P(n)$.
	Prove $P(n+1)$.

3. To prove $(\forall n \geq b)\, P(n)$ by strong induction with one base step, use:

Base step:	Prove $P(b)$.
Inductive step:	Let $n > b$ be an integer.
	Assume $P(k)$ whenever $b \leq k < n$.
	Prove $P(n)$.

4. To prove $(\forall n \geq b)\, P(n)$ by strong induction with multiple base steps, identify the integer $c > b$ and use:

Base step:	Prove $P(b)$.
Base step:	Prove $P(b+1)$.
	$\vdots$
Base step:	Prove $P(c)$.
Inductive step:	Let $n > c$ be an integer.
	Assume $P(k)$ whenever $b \leq k < n$.
	Prove $P(n)$.

Proof and Assumption Strategies for Set Theory

1. To prove $A \subseteq B$, use the form:

Let $x \in A$.
Prove $x \in B$.

2. To prove $A = B$, try one of the following:

(a)
Prove $A \subseteq B$
Prove $B \subseteq A$.

(b)
Let x be arbitrary.
Prove $x \in A \leftrightarrow x \in B$.

3. When *assuming* $A \subseteq B$, if you know or can prove $x \in A$, then you can conclude $x \in B$. If you know or can prove $x \notin B$, then you can conclude $x \notin A$.

4. When *assuming* $A = B$, if you know or can prove $x \in A$, then you can conclude $x \in B$. If you know or can prove $x \notin A$, then you can conclude $x \notin B$.

Proof and Assumption Strategies for Functions

1. To prove $f = g$ where $f\colon A \to B$ and $g\colon A \to B$, use:

> Let $x \in A$.
> Prove $f(x) = g(x)$.

2. To prove that a function $f\colon A \to B$ is one-to-one, use:

> Let $x \in A$ and $y \in A$.
> Assume $f(x) = f(y)$.
> Prove $x = y$.

3. To prove that a function $f\colon A \to B$ is onto, use:

> Let $y \in B$.
> Let $x = $ (the element in A you found).
> Prove $f(x) = y$.

4. When *assuming* $f\colon A \to B$ is one-to-one. If you are also assuming or can prove that $f(x) = f(y)$, then you can conclude that $x = y$ whenever $x, y \in A$.

5. When *assuming* $f\colon A \to B$ is onto, then for any $y \in B$ you can conclude that $f(x) = y$ for some $x \in A$.

Bibliography

[1] D.W. Cunningham. *A Logical Introduction to Proof*. SpringerLink: Bücher. Springer New York, 2012.

[2] D.W. Cunningham. *Set Theory: A First Course*. Cambridge Mathematical Textbooks. Cambridge University Press, 2016.

[3] D.W. Cunningham. Why does trigonometric substitution work? *International Journal of Mathematical Education in Science and Technology*, 49(4):588–593, 2018.

[4] R.A. Gordon. *Real Analysis: A First Course*. Addison-Wesley Higher Mathematics. Addison-Wesley, 2001.

[5] I. Grattan-Guinness and H.J.M. Bos. *From the Calculus to Set Theory, 1630-1910: An Introductory History*. Princeton paperbacks. Princeton University Press, 2000.

[6] S.R. Lay. *Analysis with an Introduction to Proof*. Pearson Education, 2015.

[7] P.R. Mercer. *More Calculus of a Single Variable*. Undergraduate Texts in Mathematics. Springer New York, 2014.

[8] M. Rosenlicht. *Introduction to Analysis*. Dover books on mathematics. Dover Publications, 1986.

[9] H.L. Royden and P. Fitzpatrick. *Real Analysis*. Prentice Hall, 2010.

[10] W. Rudin. *Principles of Mathematical Analysis*. International Series in Pure and Applied Mathematics. McGraw-Hill, 1976.

List of Symbols

$\displaystyle\sum_{k=1}^{\infty} a_k$, 189

$\displaystyle\sum_{k=1}^{n} a_k$, 189

$\displaystyle\limsup_{n\to\infty} x_k = \infty$, 224

$\rho > 0$, 226

$\mathrm{bd}(S)$, 239

$\mathrm{int}(S)$, 239

S', 242

$(\forall x < a), (\exists x < a)$, 255

Index

Printed in the United States
By Bookmasters